BIOS INSTANT NOTES

Genetics

FOURTH EDITION

BIOS INSTANT NOTES

Genetics

FOURTH EDITION

Hugh Fletcher
School of Biological Sciences,
The Queens University of Belfast, UK

Ivor Hickey
St Mary's University College,
Belfast, UK

GS Garland Science
Taylor & Francis Group

LONDON AND NEW YORK

Garland Science
Vice President: Denise Schanck
Editor: Elizabeth Owen
Editorial Assistant: Louise Dawnay
Production Editor: Georgina Lucas
Copyeditor: Suzanne Dalgleish
Typesetting and illustrations: Cenveo Publisher Services
Proofreader: Dawn Booth
Printed by: TJ International

ISBN 978-0-4156-9314-1

Library of Congress Cataloging-in-Publication Data

Fletcher, H. L. (Hugh L.)
  Genetics / Hugh Fletcher, Ivon Hickey. — 4th ed.
     p. ; cm. — (BIOS instant notes)
  Rev. ed. of: Genetics / G.I. Hickey, H.L. Fletcher, and P. Winter. 3rd ed. 2007.
  Includes bibliographical references and index.
  ISBN 978-0-415-69314-1 (alk. paper)
  I. Hickey, G. I. (G. Ivor) II. Hickey, G. I. (G. Ivor). Genetics. III. Title. IV. Series:
BIOS instant notes.
  [DNLM: 1. Genetics—Outlines. QU 18.2]

  576.5—dc23   2012013036

Published by Garland Science, Taylor & Francis Group, LLC, an informa business, 711 Third Avenue, 8th Floor, New York NY 10017, USA, and 3 Park Square, Milton Park, Abingdon, OX14 4RN, UK.

15 14 13 12 11 10 9 8 7 6 5 4 3 2 1

Garland Science
Taylor & Francis Group
LONDON AND NEW YORK

Visit our web site at http://www.garlandscience.com

Contents

Preface to Edition 4

This edition of *Instant Notes in Genetics* sees some large rearrangements and additions in content, particularly a reduction in molecular detail. All sections on molecular genetics are thoroughly revised. The molecular detail of transcription and translation has been considerably reduced and consolidated while DNA mutation and repair have moved into this section. Section E, DNA Technology, has also been revised to emphasize the various ways that the ability of one strand of DNA to find a matching strand and pair with it have been exploited to study DNA. We have concentrated on the concepts and principles and left the intricacies to *Instant Notes in Molecular Biology*. A new section on DNA sequencing has been introduced to highlight how the technology has developed over three decades. Initially it took years to find a gene and obtain its sequence, now it is possible to sequence a whole human genome in a matter of weeks. There is now so much sequence data available that information storage and analysis are more critical than the sequencing. There are also risks of producing data for its own sake, and problems with quality that accompany mass production. It is apparent that 'the human reference sequence' does not exist. Every individual haploid genome contains multiple recessive mutations, and the 1000 Genomes Project is intended to collect human genome data from 1000 people around the world so that meaningful comparisons can be made. Currently, when researchers want to investigate links between variant sequences, that is mutations, with a particular multifactorial disease such as coronary heart disease they have to sequence large areas of genome to look for the variants. The 1000 Genomes Project already documents about 95% of the total variation present in humans in the more interesting parts of the genome, so anyone starting a new project can determine the variation and know what to look for from the start. We confidently predict that some genes will be found to be nonfunctional in the large majority of people.

The sections on biotechnology and transgenics have been updated and a section on cloning animals and stem cells has been added. These sections reflect a cutting edge of applied genetics. Genetically engineered potatoes are now grown in Europe to provide chemicals for biodegradable plastics, among other things, but the fear of protests still limits production for food. Another new topic is 'Genes in Development' which demonstrates broadly how regulation of genes is responsible for producing bodies of a particular shape. It also highlights the common ancestry of all animals, because the same genes work on the same parts of the bodies of flies, mammals, and fishes. Mendelian genetics hasn't changed since Mendel, just as gravity hasn't changed since Newton, but we are still finding out more and more about the mechanisms by which genetics operates. It is still a rapidly advancing science, attracting a cluster of recent Nobel prizes, and it is a privilege to be able to communicate some of this to the next generation of geneticists.

A1 DNA structure

Key Notes

Nucleotides

DNA is a polymer containing chains of nucleotide monomers. Each nucleotide contains a sugar, a base, and a phosphate group. The sugar is 2'-deoxyribose which has five carbons named 1' (one-prime) 2', etc. There are four types of base: adenine and guanine have two carbon–nitrogen rings and are purines; thymine and cytosine have a single ring and are pyrimidines. The bases are attached to the 1' carbon of the deoxyribose. A sugar plus a base is termed a nucleoside. A nucleotide has one, two or three phosphate groups attached to the 5' carbon of the sugar. Nucleotides occur as individual molecules or polymerized as DNA or RNA.

DNA polynucleotides

Nucleotide triphosphates of the four bases are joined to form DNA polynucleotide chains. Two phosphates are lost during polymerization and the nucleotides are joined by the remaining phosphate. A phosphodiester bond forms between the 5' phosphate of one nucleotide and the 3' hydroxyl of the next nucleotide. The polynucleotide has a free 5' phosphate at one end (5' end) and a free 3' OH (3' end) at the other end. The sequence of bases encodes the genetic information. It is read 5'→3'. Polynucleotides are extremely long. It is possible to have 4^n different sequences.

The double helix

DNA molecules are composed of two polynucleotide strands wrapped around each other to form a double helix. The sugar–phosphate part of the molecule forms a backbone. The bases face inwards and are stacked on top of each other. The two polynucleotide chains run in opposite directions (antiparallel). The double helix is right-handed and executes a turn every 10 bases. The helix has a major groove which interacts with proteins and a minor grove. Variant DNA structures have been identified including Z DNA which has a left-handed helix.

Complementary base pairing

Hydrogen bonds between bases on the two DNA strands stabilize the double helix. The available space between the strands restricts the bases that can interact such that a purine always interacts with a pyrimidine. Thus, A interacts only with T, and G only with C. This is called complementary base pairing. The restriction on base pairing means that the sequences of bases on the two strands are related to each other, such that the sequence of one determines and predicts the sequence of the other. This allows genetic information to be preserved during replication of the DNA and expression of the genes. Disruption of the hydrogen bonds between

	the bases by heat or chemicals or by the action of enzymes causes the strands of the double helix to separate.
RNA structure	In RNA thymine is replaced by uracil and 2-deoxyribose by ribose. RNA normally exists as a single polynucleotide strand; however, short stretches of base pairing may occur between complementary sequences.
DNA replication	In this process a cell copies its DNA prior to dividing. The DNA is copied 5'→3' by DNA polymerases using single-stranded DNA as a template. Replication is semi-conservative. A helicase separates the double helix. DNA is synthesized continuously on the leading strand and discontinuously as segments (Okazaki fragments) on the lagging strand. DNA synthesis by DNA polymerase is initiated at a short RNA primer synthesized by primase RNA polymerase. Primer sequence is replaced with DNA later. DNA ligase joins the Okazaki fragments by a phosphodiester bond.
Related topics	(A5) DNA mutations (A4) DNA to protein

Nucleotides

The ability of DNA to carry the genetic information required by a cell to reproduce itself is closely related to the structure of DNA molecules. DNA is a polymer and consists of a long chain of monomers called **nucleotides**. The DNA molecule is said to be a polynucleotide. Each nucleotide has three parts: a sugar, a nitrogen-containing ring-structure called a **base**, and a phosphate group. The sugar present in DNA is a five carbon pentose called 2'-deoxyribose in which the –OH group on carbon 2 of ribose is replaced by hydrogen (Figure 1).The carbon atoms in the sugar are numbered 1 to 5. The numbers are given a dash ('), referred to as **prime**, to distinguish them from the numbers of the atoms in the base. The numbering is important because it indicates where other components of the nucleotide are attached to the sugar.

Nucleotides contain one of four bases: **adenine**, **guanine**, **thymine**, or **cytosine** (Figure 2). These are complex molecules containing carbon and nitrogen ring structures. Adenine and guanine contain two carbon–nitrogen rings and are known as **purines**. Cytosine and thymine contain a single ring and are called **pyrimidines**. The bases are attached to the sugar by a bond between the 1' carbon of the sugar and a nitrogen at position 9 of the purines or position 1 of the pyrimidines. A sugar plus a base is called a **nucleoside** (Figure 3a).

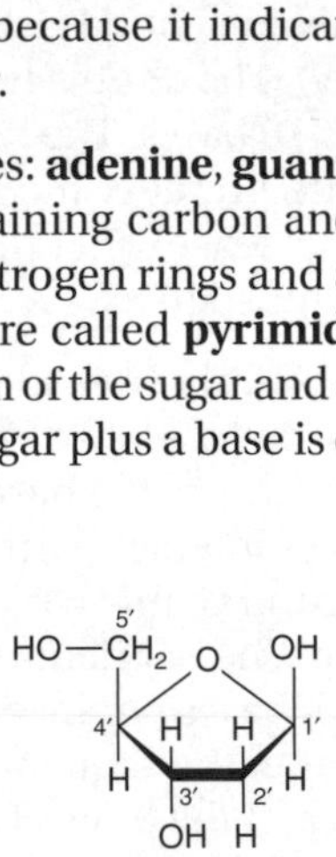

2'-Deoxyribose

Figure 1. Structure of 2'-deoxyribose.

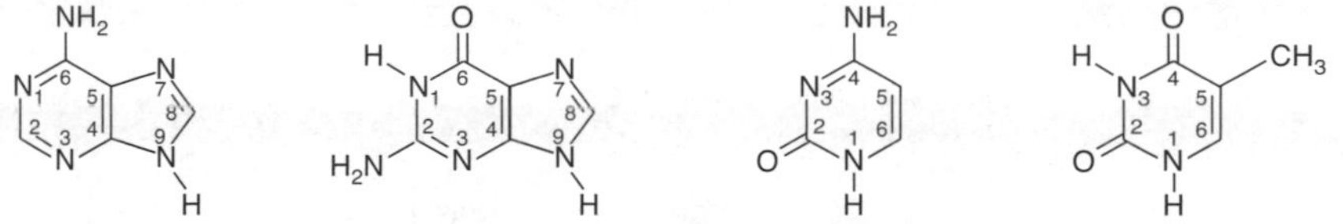

Figure 2. Bases in DNA.

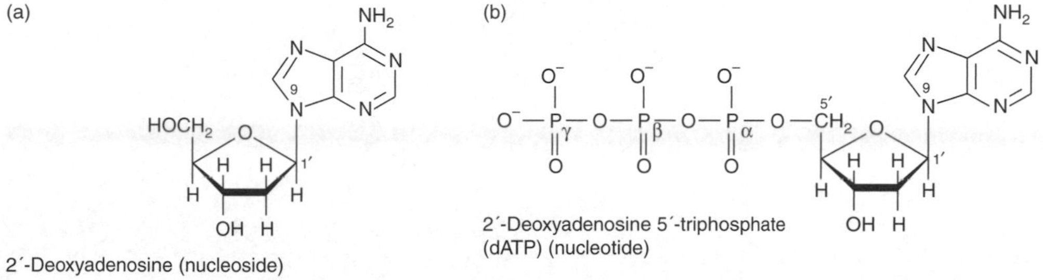

Figure 3. Structure of (a) nucleosides, (b) nucleotides.

Nucleotides contain phosphate groups (PO_4) attached to the 5′ carbon of the sugar (Figure 3b). A nucleoside is called a nucleotide when a phosphate group is attached. The attachment can consist of one, two, or three phosphate groups joined together. The phosphate groups are called α, β, and γ, with α directly attached to the sugar. Nucleotides may exist in cells as individual molecules (nucleotide triphosphates play an important role in cells as the carriers of energy used to power enzymatic reactions) or polymerized as nucleic acids (DNA or RNA).

DNA polynucleotides

Nucleotide triphosphates are joined together to give polynucleotides. There are four used to synthesize DNA polynucleotides, 2′-deoxyadenosine 5′-triphosphate (dATP or A), 2′-deoxythymidine 5′-triphosphate (dTTP or T), 2′-deoxycytosine 5′-triphosphate (dCTP or C) and 2′-deoxyguanosine 5′-triphosphate (dGTP or G). The β and γ phosphates are lost during polymerization and the nucleotide units are joined together by the remaining phosphate. The 5′ phosphate of one nucleotide forms a bond with the 3′ carbon of the next nucleotide, eliminating the –OH group on the 3′ carbon during the reaction. The bond is called a **3′–5′ phosphodiester bond** (C–O–P) (Figure 4). The polynucleotide chain has a free 5′ triphosphate at one end known as the 5′ end and a free 3′ hydroxyl group at the other end called the 3′ end. This distinction gives the DNA polynucleotide polarity so that a DNA molecule can be described as running 5′→3′ or 3′→5′.

It is the sequence of the bases in the DNA polynucleotide that encodes the genetic information. This sequence is always written in the 5′→3′ direction (polymerase enzymes copy DNA molecules in this direction). Polynucleotides can be extremely long with no apparent limit to the number of nucleotides and no restrictions on the sequence of the nucleotides. The maximum number of possible base sequences for a polynucleotide is 4^n, where n is the number of nucleotides. This is an enormous number. For example, a polynucleotide containing just six bases could be arranged as $4^6 = 4096$ different sequences.

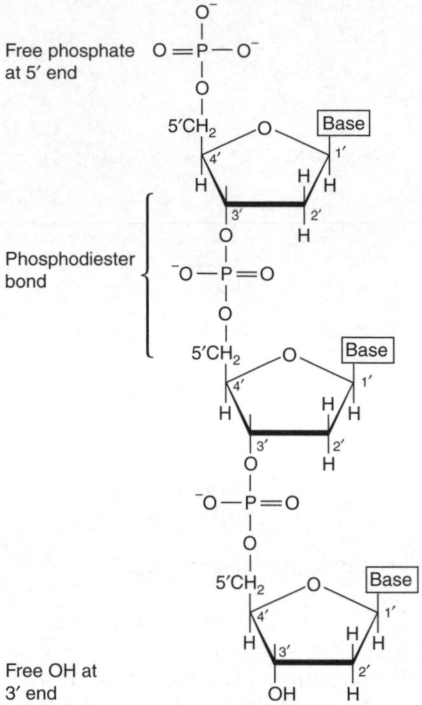

Free phosphate at 5′ end

Phosphodiester bond

Free OH at 3′ end

Figure 4. Phosphodiester bonds join nucleotides in a DNA polynucleotide.

The double helix

DNA molecules have a very distinct and characteristic three-dimensional structure known as the double helix (Figure 5). The structure of DNA was discovered in 1953 by Watson and Crick working in Cambridge using X-ray diffraction pictures taken by Franklin and Wilkins. DNA exists as two polynucleotide chains wrapped around each other to form the double helix. The sugar–phosphate part of the molecule forms a spine

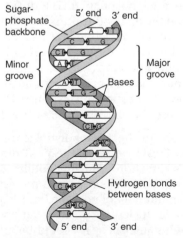

Figure 5. The double helix.

or backbone which is on the outside of the helix. The bases, which are flat molecules, face inwards towards each other, slightly off-center, and are stacked on top of each other like a pile of plates.

X-ray diffraction pictures of the double helix show repeated patterns of bands that reflect the regularity of the structure of the DNA. The double helix executes a turn every 10 base pairs. The pitch of the helix is 34 Å so the spacing between bases is 3.4 Å. The diameter of the helix is 20 Å. The double helix is said to be **antiparallel**. One of the strands runs in the 5′→3′ direction and the other in the 3′→5′ direction. Only antiparallel polynucle-otides form a stable helix. The double helix is not absolutely regular and when viewed from the outside a **major groove** and a **minor groove** can be seen. These are important for interaction with proteins, for replication of the DNA and for expression of the genetic information. The double helix is right-handed. This means that if the double helix were a spiral staircase and you were climbing up, the sugar–phosphate backbone would be on your right.

A number of variant forms of DNA occur when crystals of the molecule are formed under different conditions. The form present in cells is called the **B form**. Another form, called the A form, has a slightly more compact structure. Other forms that exist are C, D, E, and Z, which is striking because it exists as a left-handed helix. Regions in chromosomes containing nonstandard structures such as Z-DNA have been identified.

Complementary base pairing

The bases of the two polynucleotide chains interact with each other. The space between the polynucleotides is such that a two-ring purine interacts with a single-ring pyrimidine. Thus, thymine always interacts with adenine and guanine with cytosine. Hydrogen bonds form between the bases and help to stabilize the interaction. Two bonds form between A and T, and three between G and C. Thus, G–C bonds are stronger than A–T bonds. The way in which the bases form pairs between the two DNA strands is known as **comple-mentary base pairing** and is of fundamental importance (Figure 6). Combinations other than G–C and A–T do not work because they are too large or too small to fit inside the helix or they do not align correctly to allow hydrogen-bond formation. Because G must always bond to C and A to T, the sequences of the two strands are related to each other and are said to be complementary with the sequence of one strand predicting and determining the sequence of the other. This means that one strand can be used to rep-licate the other. This is a vital mechanism for retaining genetic information and passing it on to other cells following cell division. Complementary base pairing is also essential

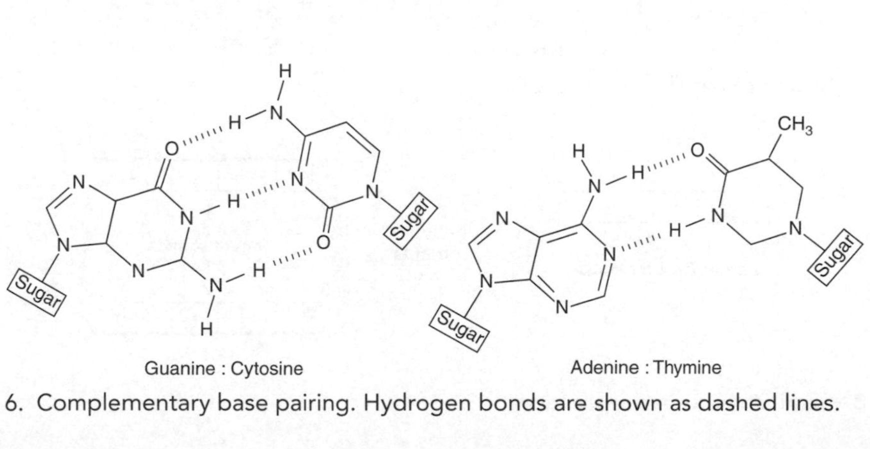

Guanine : Cytosine   Adenine : Thymine

Figure 6. Complementary base pairing. Hydrogen bonds are shown as dashed lines.

for the expression of genetic information and is central to the way DNA sequences are transcribed into mRNA and translated into protein.

The double helix is stabilized by hydrogen bonds between the base pairs. These can be disrupted by heat and some chemicals. This results in separation of the double helix into two strands and the molecule is said to be **denatured**. In cells, enzymes can separate the strands of the double helix for the purposes of copying the DNA and for expression of the genetic information.

RNA structure

The structure of RNA is similar to that of DNA but a number of important differences exist. In RNA, ribose replaces 2′-deoxyribose and the base thymine is replaced by another base, uracil, which can also base pair with adenine (Figure 7). In addition, RNA molecules normally exist as a single polynucleotide strand and do not form a double helix. However, it is possible for base pairing to occur between complementary parts of the same RNA strand, resulting in short double-stranded regions.

DNA replication

DNA is copied by enzymes called DNA polymerases. These act on single-stranded DNA synthesizing a new strand complementary to the original strand. DNA synthesis always occurs in the 5′→3′ direction. Replication is said to be **semi-conservative** (Figure 8). This means that each copied DNA molecule contains one strand derived from the parent molecule and one newly synthesized daughter strand.

The mechanism of DNA replication is very similar in most organisms. Replication needs to be very accurate because even a small error rate would result in the loss of important genetic information. Accuracy is ensured by the ability of the DNA polymerases to check that the correct bases have been inserted in the newly synthesized strand. This is referred to as **proofreading**. It is estimated that just one base in five billion is inserted incorrectly.

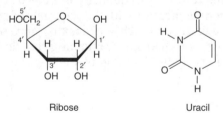

Ribose   Uracil

Figure 7. Structures of ribose and uracil.

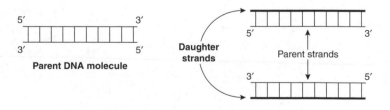

Figure 8. Semi-conservative replication.

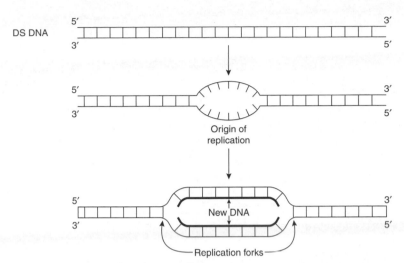

DS DNA

Origin of
replication

New DNA

Replication forks

Figure 9. Replication origin and replication forks.

During DNA replication the double helix of a cell's entire DNA is progressively unwound producing segments of single-stranded DNA which can be copied by DNA polymerases. Unwinding of the double helix begins at a distinct position called the **replication origin** and gradually progresses along the molecule, usually in both directions. Replication origins usually contain sequences rich in weaker A–T base pairs. The region where the helix unwinds and new DNA is synthesized is called the **replication fork** (Figure 9). At the replication fork a number of distinct events occur:

- **Separation of the double helix**. This is achieved by the action of a **helicase** enzyme.
- **Synthesis of leading and lagging strands**. Synthesis of DNA by DNA polymerases occurs only in the 5′→3′ direction. As the two strands of the double helix run in opposite directions (one strand runs 5′→3′ and the other 3′→5′) slightly different mechanisms are required to replicate each. One strand, called the **leading strand**, is copied in the same direction as the unwinding helix and so can be synthesized continuously (Figure 10a). The other strand, known as the lagging strand, is synthesized in the opposite direction and must be copied discontinuously The lagging strand is synthesized as a series of segments known as **Okazaki fragments** (Figure 10b).
- **Priming**. DNA polymerases require a short double-stranded region to initiate or prime DNA synthesis. This is produced by an RNA polymerase, called **primase**, which is able to initiate synthesis on single-stranded DNA. The primase synthesizes a short RNA primer sequence on the DNA template creating a short double-stranded region. DNA polymerase then synthesizes DNA beginning at the RNA primer. On the lagging strand, synthesis ends when the next RNA primer is encountered. At this point a different DNA polymerase takes over and removes the RNA primer replacing it with DNA (Figure 11).
- **Ligation**. The final step required to complete synthesis of the lagging strand is for the Okazaki fragments to be joined together by phosphodiester bonds. This is carried out by a **DNA ligase** enzyme (Figure 11).

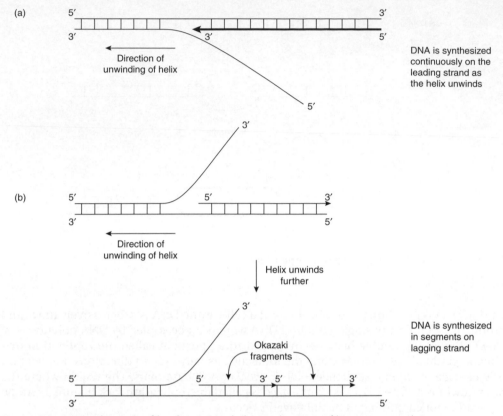

Figure 10. (a) Replication of leading strand. (b) Replication of lagging strand.

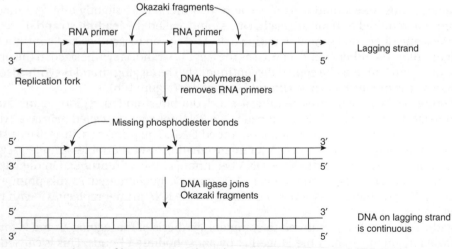

Figure 11. Completion of lagging strand synthesis.

A2 Genes

Key Notes

Structure of genes	A gene is a unit of information and corresponds to a discrete segment of DNA that encodes the amino acid sequence of a polypeptide. Human cells contain about 21 000 genes arranged on 23 chromosomes. The genes are dispersed and are separated by noncoding intergenic DNA. Information is encoded on the template strand which directs the synthesis of an RNA molecule. Both DNA strands can act as the template strand. DNA molecules have an enormous capacity to store genetic information.
Gene families	Some genes are arranged as clusters known as operons and multigene families. Operons occur in bacteria and contain coregulated genes with a related function. Multigene families occur in higher organisms and contain genes that are identical or similar that are not regulated coordinately. Simple multigene families contain identical genes whose product is required in large amounts. Complex multigene families contain genes that are very similar and encode proteins with a related function.
Gene expression	The biological information encoded in genes is made available by gene expression. In this process, an RNA copy of a gene is synthesized which then directs the synthesis of a protein. The central dogma states that information is always transferred from DNA to RNA to protein. The functioning of cells is dependent on the coordinated activity of many proteins. Gene expression ensures that proteins are synthesized in the correct place at the correct time.
Gene promoters	Gene expression is highly regulated. Not all of the genes present in a cell are active and different types of cell express different genes. The expression of a gene is regulated by a segment of DNA upstream of the coding sequence called the promoter. This binds RNA polymerase and associated transcription factor proteins and initiates synthesis of an RNA molecule.
Introns and exons	The coding sequence of a gene is split into a series of segments called exons, which are separated by noncoding sequences called introns which usually account for most of the gene sequence. The number and sizes of the introns vary between genes. Introns are removed from RNA transcripts by a process called splicing, prior to protein synthesis. Introns are not usually present in bacterial genes.
Pseudogenes	Copies of some genes exist which contain sequence errors acquired during evolution that prevent them from producing proteins. These are called pseudogenes and they represent evolutionary relics of original genes. Examples include the globin pseudogenes.

Related topics	(A7) Regulation of gene expression	(B2) Prokaryote genomes
	(A8) Epigenetics and chromatin modification	(B3) Eukaryote genomes

Structure of genes

The biological information needed by an organism to reproduce itself is contained in its DNA. The information is encoded in the base sequence of the DNA and is organized as a large number of genes, each of which contains the instructions for the synthesis of a polypeptide. In physical terms, a gene is a discrete segment of DNA with a base sequence that encodes the amino acid sequence of a polypeptide. Genes vary greatly in size from less than 100 base pairs to several million base pairs. In higher organisms the genes are present on a series of extremely long DNA molecules called **chromosomes**. In humans there are currently 20 563 protein coding genes identified on 23 chromosomes. The genes are very dispersed and are separated from each other by sequences that do not appear to contain useful information; this is called **intergenic DNA**. The intergenic DNA is very long, such that in humans gene sequences account for less than about 30% of the total DNA. Only one of the two strands of the DNA double helix carries the biological information: this is called the **template strand** and it is used to produce an RNA molecule of complementary sequence which directs the synthesis of a polypeptide. The other strand is called the **nontemplate strand**. Both strands of the double helix have the potential to act as the template strand: individual genes may be encoded on different strands. Other terms are used to describe the strands of the double helix as alternatives to template and nontemplate. These include **sense/antisense** and **coding/noncoding**: the terms antisense and noncoding are equivalent to the template strand.

The capacity of DNA molecules to store information is enormous. For a DNA molecule n bases long, the number of different combinations of the four bases is 4^n. Even for very short DNA molecules the number of different sequences possible is very large. In practice, there are limitations to the sequences that can contain useful information. However the capacity to encode information remains vast.

Gene families

Most genes are spread out randomly along the chromosomes, however some are organized into groups or clusters. Two types of cluster occur: these are **operons** (Figure 1) and **multigene families** (Figure 2).

Operons are gene clusters found in bacteria. They contain genes that are regulated in a coordinated way and encode proteins with closely related functions. An example is the *lac* operon in *E. coli*, which contains three genes encoding enzymes required by the bacterium to break down lactose. When lactose is available as an energy source, the enzymes encoded by the *lac* operon are required together. The clustering of the genes within the operon allows them to be switched on or off at the same time, allowing the organism to use its resources efficiently (Figure 1).

In higher organisms, operons are absent and clustered genes exist as multigene families. Unlike operons, the genes in a multigene family are identical or are very similar and are not regulated coordinately. The clustering of genes in multigene families probably reflects a requirement for multiple copies of that gene which was fulfilled by duplication during evolution. Some multigene families exist as separate clusters on different chromosomes;

Figure 1. The *lac* operon. Three genes (*lac Z, Y,* and *A*) are arranged and regulated together.

this probably arose by rearrangements of the DNA during evolution which resulted in the breaking up of clusters. Multigene families may be simple or complex. In **simple multigene families** (Figure 2a) the genes are identical. An example is the gene for the 5S ribosomal RNA. In humans, there are about 2000 clustered copies of this gene reflecting the high demand of cells for the gene product. **Complex multigene families** (Figure 2b) contain genes that are very similar but not identical. An example is the globin gene family that encodes a series of polypeptides (α, β, γ, ε, ζ globins) that differ from each other by just a few amino acids. Globin polypeptides form complexes with each other and with a cofactor molecule called heme to give the adult and embryonic forms of the oxygen-carrying blood protein, hemoglobin.

Gene expression

The biological information in a DNA molecule is contained in its base sequence. Gene expression is the process by which this information is made available to the cell. The use of the information is described by the **central dogma**, originally proposed by Crick, which states that information is transferred from DNA to RNA to protein (Figure 3). During gene expression, DNA molecules copy their information by directing the synthesis of an RNA molecule of complementary sequence. This process is known as **transcription**. The RNA then directs the synthesis of a polypeptide whose amino acid sequence is determined by the base sequence of the RNA. This process is known as **translation**. The amino acid sequence of the protein determines its three-dimensional structure which in turn dictates its function. The central dogma states that the transfer of information can

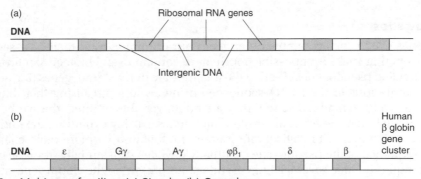

Figure 2. Multigene families. (a) Simple. (b) Complex.

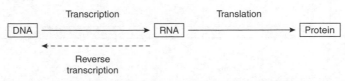

Figure 3. The central dogma above and reverse transcription below.

only occur in one direction – from DNA to RNA to protein – and cannot occur in reverse. An exception to this rule is found in retroviruses, which have an enzyme called **reverse transcriptase** that can copy RNA into DNA. The functioning of cells, and in turn of living organisms, is dependent on the coordinated activity of many different proteins. The biological information contained within the genes acts as a set of instructions for synthesizing proteins at the correct time and in the correct place.

Gene promoters

The expression of the biological information present in genes is highly regulated. Not all the genes present in a cell's DNA are expressed and different genes are active in different cell types. The overall complement of genes that are active determines the characteristics of a cell and its function within the organism. Thus, for example, many of the genes that are active in muscle cells are different from those that are active in blood cells. Expression of genes is regulated by a segment of DNA sequence present upstream of the coding sequence known as the **promoter**. Conserved DNA sequences in the promoter are recognized and bound by the RNA polymerase and other associated proteins called **transcription factors** that bring about the synthesis of an RNA transcript of the gene. The expression of a gene in a cell is determined by the promoter sequence and its ability to bind RNA polymerase and transcription factors.

Introns and exons

One of the more surprising features of genes is that in higher organisms the coding information is usually split into a series of segments of DNA sequence called **exons**. These are separated by sequences that do not contain useful information called **introns** (Figure 4). The number of introns varies greatly, from zero to more than 50 in some genes. The length of the exons and introns also varies but the introns are usually much longer and account for the majority of the sequence of the gene. Before the biological information in a gene can be used to synthesize a protein, the introns must be removed from RNA molecules by a process called **splicing** which leaves the exons and the coding information continuous. Introns are a feature of higher organisms only and are not usually found in bacteria.

Pseudogenes

Some genes exist which resemble other genes but examination of their base sequence shows errors that make it impossible for them to contain useful biological information. These are called pseudogenes (Section B3) and they are derived from genes that acquired errors or mutations in their DNA sequence during evolution, causing their biological information to be scrambled so that they are no longer able to direct the synthesis of a protein. As such, pseudogenes are evolutionary relics. During evolution, the initial base changes causing loss of biological information are followed by more rapid changes so that the sequence of the pseudogene eventually deviates substantially from the original gene. Examples include several globin pseudogenes present in the globin gene clusters.

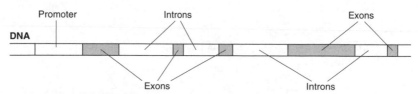

Figure 4.   Introns and exons.

A3 The genetic code

Key Notes

Genetic expression

Genetic information is encoded in the base sequence of DNA molecules as a series of genes. Gene expression is the term used to describe how cells decode the information to synthesize proteins required for cellular function. The expression of a gene involves the synthesis of a complementary RNA molecule whose sequence specifies the amino acid sequence of a protein. The DNA sequence of the gene is colinear with the amino acid sequence of the polypeptide.

Genetic code

Amino acids are encoded by 64 base triplets called codons which encode the 20 amino acids. Most amino acids have more than one codon. This is known as the degeneracy of the genetic code and it helps to minimize the effect of mutations. Codons that specify the same amino acid are called synonyms and differ at their third base, known as the 'wobble' position. AUG is the initiation codon and encodes methionine. There are three stop codons: UAG, UGA, and UAA.

Reading frames

Three possible sets of codons can be read from any sequence depending on which base is chosen as the start. Each set of codons is known as a reading frame. The initiation codon determines the reading frame of a protein coding sequence. Other reading frames tend to contain stop codons and are not used for protein synthesis. An open reading frame is a sequence of codons bounded by start and stop codons.

Universality of the code

The genetic code applies universally with all organisms using the same codons for each amino acid. However, some exceptions to the standard codon usage occur in mitochondrial genomes and in some unicellular organisms.

Related topics

(A4) DNA to protein

Gene expression

The information required by an organism to reproduce itself is carried by its DNA, encoded in the base sequence and organized as a series of genes. Gene expression is the term used to describe the process by which cells decode and make use of this information to synthesize the proteins that are responsible for cellular function. During gene expression, information is copied from DNA to RNA by the synthesis of an RNA molecule whose base sequence is complementary to that of the DNA template. The RNA then directs the synthesis of a protein whose amino acid sequence is specified by the base sequence of the RNA. For every gene the DNA sequence is colinear with the amino acid sequence of the polypeptide it encodes such that the 5'→3' base sequence of the coding strand specifies the amino acid sequence of the encoded polypeptide from the amino to the carboxy terminus.

Genetic code

The genetic code describes how base sequences are interpreted into amino acid sequences during protein synthesis. The DNA sequence of a gene is divided into a series of units of three bases. Each set of three bases is called a codon and specifies a particular amino acid. The four bases in DNA and RNA can combine as a total of $4^3 = 64$ codons which specify the 20 amino acids found in proteins (Table 1). Because the number of codons is greater, all of the amino acids, with the exceptions of methionine and tryptophan, are encoded by more than one codon. This feature is referred to as the **degeneracy** or the **redundancy** of the genetic code. Codons which specify the same amino acid are called synonyms and tend to be similar. For example, ACU, ACC, ACA, and ACG all specify the amino acid threonine. Variations between synonyms tend to occur at the third position of the codon, which is known as the **wobble position**. The degeneracy of the genetic code minimizes the effects of mutations so that alterations to the base sequence are less likely to change the amino acid encoded and possible deleterious effects on protein function are avoided. Of the 64 possible codons, 61 encode amino acids. The remaining three, UAG, UGA, and UAA, do not encode amino acids but instead act as signals for protein synthesis to stop and as such are known as **termination codons** or **stop codons**. The codon for methionine, AUG, is the signal for protein synthesis to start and is known as the **initiation codon**. Thus all polypeptides start with methionine, although this is sometimes removed later.

Table 1. The genetic code

First position (5′ end)	Second position				Third position (3′ end)
	U	C	A	G	
U	Phe UUU	Ser UCU	Tyr UAU	Cys UGU	U
	Phe UUC	Ser UCC	Tyr UAC	Cys UGC	C
	Leu UUA	Ser UCA	Stop UAA	Stop UGA	A
	Leu UUG	Ser UCG	Stop UAG	Trp UGG	G
C	Leu CUU	Pro CCU	His CAU	Arg CGU	U
	Leu CUC	Pro CCC	His CAC	Arg CGC	C
	Leu CUA	Pro CCA	Gln CAA	Arg CGA	A
	Leu CUG	Pro CCG	Gln CAG	Arg CGG	G
A	Ile AUU	Thr ACU	Asn AAU	Ser AGU	U
	Ile AUC	Thr ACC	Asn AAC	Ser AGC	C
	Ile AUA	Thr ACA	Lys AAA	Arg AGA	A
	Met AUG	Thr ACG	Lys AAG	Arg AGG	G
G	Val GUU	Ala GCU	Asp GAU	Gly GGU	U
	Val GUC	Ala GCC	Asp GAC	Gly GGC	C
	Val GUA	Ala GCA	Glu GAA	Gly GGA	A
	Val GUG	Ala GCG	Glu GAG	Gly GGG	G

Reading frames

In addition to identifying the start of protein synthesis, the initiation codon determines the reading frame of the RNA sequence. Depending on which base is chosen as the start of a codon, three possible sets of codons may be read from any base sequence. In practice, during protein synthesis, normally only one reading frame contains useful information; the other two reading frames usually contain stop codons which prevent them from being used to direct protein synthesis (Figure 1). A set of codons that runs continuously and is bounded at the start by an initiation codon and at the end by a termination codon is known as an **open reading frame** (ORF). This characteristic is used to identify protein-coding DNA sequences in genome-sequencing projects.

Universality of the code

Initially the genetic code was believed to apply universally, that is all organisms would recognize individual codons as the same amino acids. However, it has now been shown that some variation in the code exists, although this is rare. For example, animal mitochondria have a small DNA genome containing about 20 genes in which deviations from the genetic code occur. Changes are mostly associated with start and stop codons. For example, UGA, which is normally a termination codon, codes for tryptophan, whereas AGA and AGG, which normally encode arginine, are termination codons, and AUA, normally isoleucine, specifies methionine. It is thought that these changes tend to be viable because the mitochondrion is a closed system. A few examples of nonstandard codon usage have now been found outside mitochondrial genomes in unicellular organisms. For example UAA and UAG, which are normally stop codons, encode glutamic acid in some protozoa.

Reading frame 1. 5′ –AUG ACU AAG AGA UCC GG –3′
                        Met   Thr   Lys   Arg   Ser

Reading frame 2. 5′ –A UGA CUA AGA GAU CCG G –3′
                         Stop Leu   Arg   Asp   Pro

Reading frame 3. 5′ –AU GAC UAA GAG AUC CGG –3′
                          Asp **Stop** Glu   Ile   Arg

Figure 1.   Every DNA sequence can be read as three separate reading frames depending on which base is chosen as the start of the codon.

A4 DNA to protein

Key Notes

Introduction   Expression of genes takes place in three steps: transcription, RNA processing, and translation. In translation the genetic information encoded in nucleic acids is converted to amino acid sequences at ribosomes.

RNA transcription   RNA polymerases transcribe RNA copies from the template strand of DNA sequences. The RNA product is referred to as a transcript. The process is similar to the replication of DNA by DNA polymerases. Transcription commences at a promoter sequence in the DNA and continues until termination sequences are reached. Different polymerase enzymes transcribe specific sets of genes in eukaryotes.

RNA processing   The initial pre-mRNA transcript is large and contains introns. These are removed from the transcript in spliceosomes. Differential splicing allows the production of variants of a protein from the same gene. Prokaryote mRNA does not require to be processed. Modifications to 5′ and 3′ regions of the mRNA also take place during RNA processing.

Translation   This takes place on ribosomes and commences at an AUG initiation codon. The nucleotide sequence is read in triplets by tRNA molecules each charged with their respective amino acid. The anticodon of the tRNA molecule binds to the complementary codon in the mRNA. Peptidyl transferase forms peptide bonds between the aligned amino acids to create nascent polypeptides. Proteins may be chemically modified after translation.

Related topics   (A2) Genes   (F3) Genes and cancer
(A7) Regulation of gene expression

Introduction

The procedure of expressing genes, that is going from the information encoded in the nucleotide sequence to the synthesis of protein, has three steps. The sequence of bases in DNA is copied into RNA by the process of **transcription**. The information remains in nucleic acid code. The RNA molecule is processed and eventually directs the synthesis of proteins at the ribosomes where the genetic information stored in nucleic acid sequence is **translated** to produce the appropriate sequence of amino acids.

RNA transcription

RNA is synthesized by enzymes called **RNA polymerases** using DNA as a template. The two strands of the double helix are called the template and the nontemplate strands.

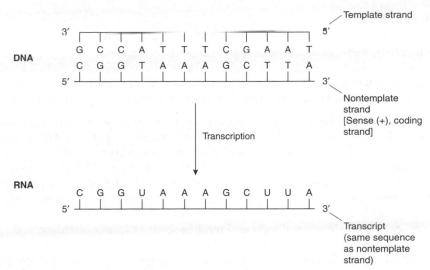

Figure 1. Transcription of the DNA template strand produces an RNA transcript with the same sequence as the nontemplate strand.

RNA is copied from the **template strand** and the RNA molecule synthesized is similar in sequence to the **nontemplate strand** (Figure 1). Gene sequences usually refer to the nontemplate strand. Other names for the nontemplate strand are the **sense (+) strand** or the **coding strand.** The RNA molecule synthesized is called a **transcript**.

During transcription RNA is synthesized by the polymerization of ribonucleoside triphosphate subunits (ATP, UTP, GTP, CTP). The 3′–OH of one ribonucleotide reacts with the 5′ phosphate of another to form a phosphodiester bond. The order in which the ribonucleotides are added to the growing RNA chain is determined by the order of the bases in the template DNA. As RNA polymerase synthesizes RNA in the 5′→3′ direction, new ribonucleotides are added to the growing chain at the free 3′ end in a process very similar to DNA synthesis (Section A1).

Transcription **initiates** specifically at the start of a gene. Signals for the initiation of transcription occur in the **promoter** sequence which lies directly upstream of the transcribed sequence of a gene. The promoter contains specific DNA sequences where the double helix partially dissociates to give an **open promoter complex** that acts as a region of attachment for the RNA polymerase.

Transcription continues in the **elongation phase.** During transcription, only a small portion of the double helix is unwound at any one time. The unwound area contains the newly synthesized RNA base-paired with the template DNA strand and extends over 12–17 bases. The final transcript complements not only the coding region of the DNA but also the regions that lie 5′ and 3′ of it. In eukaryotes the initial transcript is complex as it contains both coding **exons** and noncoding **introns** (Section A2) Specific sequences in the DNA signal termination of transcription.

In eukaryotes three separate RNA polymerases transcribe different sets of genes. **RNA polymerase II** transcribes genes that encode proteins and also the regulatory microRNA genes. Most of the genes that code for the nucleic acid structure of the ribosome, **ribosomal RNA**, are transcribed by **RNA polymerase III**, and the genes that encode the tRNAs that are involved in translation are transcribed by **RNA polymerase I**.

RNA processing

In eukaryotes the initial transcript (pre-mRNA) produced from protein coding genes is large and mostly composed of sequences transcribed from noncoding introns. These regions are removed by a process known as **splicing**. In prokaryotes, mRNA is not processed and translation of the message begins even before transcription is complete. Prokaryotic genes do not normally contain introns and so splicing is unnecessary.

Splicing takes place in the nucleus. Type 1 introns contain an active site in the RNA and splice themselves out. For type 2 introns, specific nucleotide sequences at the 5′ and the 3′ ends of each intron are recognized by **small nuclear ribonucleoproteins** (snRNPs), complexes of protein and specific RNA molecules. The combination of the pre-mRNA and the snRNPs is called the **spliceosome** and this is responsible for folding the pre-mRNA into the correct conformation for splicing when the intronic RNA sequence is excised and the neighboring exonic sequences brought together (Figure 2). The snRNPs can vary within or between cells and allow regulation and modification of splicing. Differential splicing patterns can be observed from the same mRNA in different tissues and this allows multiple variants of a protein to be produced by a single 'gene'. The patterns are often altered in tumors (Section F3).

In addition to being spliced, eukaryote pre-mRNAs are also altered at their 3′ and 5′ termini. At their 5′ end a modification known as capping takes place which involves addition of the modified nucleotide, **7-methylguanosine**. A long tail of adenines is added to the 3′ end in almost all mRNAs. The mature mRNA molecule can now be translated at ribosomes in the cytoplasm.

Translation

Translation takes place on ribosomes. Transfer RNA molecules play a key role in this process by delivering amino acids to the ribosome in an order specified by the mRNA sequence; this ensures that the amino acids are joined together in the correct order (Figure 3). Cells usually contain between 31 and 40 individual species of tRNA, each of which binds specifically to one of the 20 amino acids. Consequently, there may be more than one tRNA for each amino acid. Transfer RNAs that bind the same amino acid are called **isoacceptors**. Before translation begins, amino acids become covalently linked to their tRNAs which then recognize codons in the mRNA specifying that amino acid. The attachment of an amino acid to its tRNA is called **aminoacylation** or **charging**. The amino acid is covalently attached to the end of the acceptor arm of the tRNA by the enzyme

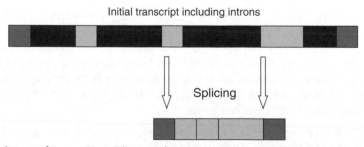

Figure 2.   Splicing of pre-mRNA. The initial transcript contains introns (black), exons (light gray) and 3′ and 5′ untranslated regions (dark gray). After splicing the mRNA molecule the introns have been removed and the molecule is ready to undergo capping at the 5′ end and addition of poly-A at the 3′ end before being exported to the cytoplasm for translation.

aminoacyl tRNA synthetase. A separate enzyme exists for each amino acid and each enzyme can charge all the isoacceptor tRNAs for that amino acid.

When the appropriate amino acid has been attached to the tRNA it recognizes the codon for that amino acid in the mRNA allowing it to place the amino acid in the correct position, as specified by the sequence of the mRNA. This ensures that the amino acid sequence encoded by the mRNA is translated faithfully. Codon recognition takes place via the anti-codon loop of the tRNA and specifically by three nucleotides in the loop known as the **anticodon** which binds to the codon by complementary base-pairing. When two amino acids are closely aligned the enzyme **peptidyl transferase** forms a peptide bond between them and the nascent polypeptide chain continues to grow in this fashion. The four bases present in DNA can combine as 64 codons. Three codons act as signals for translation to stop and the remaining 61 encode the 20 amino acids present in proteins (Section A3).

Ribosomes exist as separate large and small subunits. The first step in translation involves the binding of the small ribosomal subunit to the mRNA. Translation usually begins at the sequence **AUG** (bacteria sometimes use GUG or UUG) which encodes methionine and is known as the translation initiation codon.

Translation proceeds along the mRNA away from the initiation site, and amino acids are linked by peptide bonds. The initial methionine is removed from the nascent polypeptide. Messenger RNAs can be translated by several ribosomes at once, forming structures known as polysomes. Translation is stopped at termination codons (Section A3).

The polypeptide is further processed by enzymes that may modify some amino acids. This may involve adding small chemical groups; for example, methylation, hydroxylation, or formylation. Chains of sugars (polypeptides) are often added, a process called glycosylation. Generally, all proteins projecting outside the cell are glycosylated. This occurs in the Golgi apparatus. A specific example occurs on human red blood cells where blood type A has N-acetylgalactosamine on a polysaccharide chain called the H substance. Blood type B has galactose instead, AB cells have both on separate chains, and blood group O has unmodified H substance.

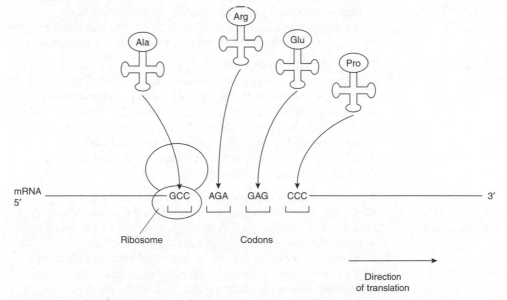

Figure 3.   Role of tRNAs in translation.

A5 DNA mutation

Key Notes

Mutations

Mutations are alterations to the usual DNA sequence of an organism that result from the action of chemical and physical agents or errors of DNA replication. Mutations are perpetuated by cell division. The nature of a mutation and its effect on an organism are described by the genotype and phenotype. Organisms may have wild-type or mutant phenotypes. Point and gross mutations exist. Only mutations in the coding regions of genes are likely to affect protein function.

Point mutations

These involve the alteration of a single base. Missense mutations change a single encoded amino acid. Nonsense mutations create stop codons and produce shortened polypeptides. Frameshift mutations involve insertion or deletion of a base producing an altered reading frame. Silent mutations occur at the third base of a codon and do not change the encoded amino acid. Silent mutations accumulate in the DNA as polymorphisms. Point mutations have varying consequences for the biological activity of the encoded protein.

Gross mutations

These involve alteration of longer DNA sequences. Deletion mutations involve the removal of bases. The deleted sequence may vary from a single base to an entire gene sequence. Insertion mutations involve the addition of extra bases. Small insertions and deletions are collectively called 'indels.' Rearrangements involve segments of DNA sequence within or outside a gene exchanging position with each other. Gross mutations usually result in a complete loss of the biological activity of the encoded protein.

Mutation and disease

DNA mutation is the underlying cause of genetic diseases and of cancer. Genetic diseases are caused by inherited mutations. Usually a single gene is involved. The mutation arises in a germ cell and results in an individual who carries the mutation in all of his/her cells and may pass it on to subsequent generations. Mutations associated with the development of cancer occur in somatic cells and often in genes that control cell division.

Mutations at the level of the organism

Mutations can be defined in terms of the phenotype they produce. This usually refers to bacterial mutants. Auxotrophic mutants fail to synthesize essential metabolites. Temperature-sensitive mutants fail to grow when the temperature is raised. Antibiotic-resistant mutants survive in the presence of antibiotics. Regulatory mutants loose the ability to regulate the expression of genes or operons.

Related topics   (A3) The genetic code   (F3) Genes and cancer
(F1) Genetic diseases

Mutations

The DNA sequence of a gene determines the amino acid sequence of its encoded protein. It is very important that the DNA sequence is preserved because alterations to the amino acid sequence may affect the ability of the protein to function, which in turn may have a deleterious effect on the organism. Alterations to the DNA sequence do occur as a result of the action of a number of chemical and physical agents on DNA and also due to rare errors in DNA replication. These changes are known as **mutations**. Once introduced, the DNA sequence changes are made permanent by DNA replication and are passed on to daughter cells following cell division.

Two important terms that describe an organism carrying a mutation are **genotype** and **phenotype**. Genotype is used to describe the mutation and the gene it occurs in. Phenotype describes the effect on the organism of the mutation. An organism that displays the usual phenotype for that species is called the **wild type**. An organism whose usual phenotype has changed as the result of a mutation is called a **mutant**. Mutations occur in two forms: **point mutations** which involve a change in the base present at any position in a gene, and **gross mutations** which involve alterations of longer stretches of DNA sequence. The location of the mutation within a gene is important. Only mutations that occur within the coding region are likely to affect the protein. Mutations in noncoding or intergenic regions do not usually have an effect.

Point mutations

Point mutations fall into a number of categories, each with different consequences for the protein encoded by the gene.

Missense mutations

These point mutations involve the alteration of a single base which changes a codon such that the encoded amino acid is altered (Figure 1a). Such mutations usually occur in one of the first two bases of a codon. The redundancy (degeneracy) of the genetic code means that mutation of the third base is less likely to cause a change in the amino acid. The effect of a missense mutation on the organism varies. Most proteins will tolerate some change in their amino acid sequence. However, alterations of amino acids in parts of the protein that are important for structure or function are more likely to have a deleterious effect and to produce a mutant phenotype.

Nonsense mutations

These are point mutations that change a codon for an amino acid into a termination codon (Figure 1b). The mutation causes translation of the messenger RNA to end prematurely, resulting in a shortened protein which lacks part of its carboxyl-terminal region. Nonsense mutations usually have a serious effect on the activity of the encoded protein and often produce a mutant phenotype.

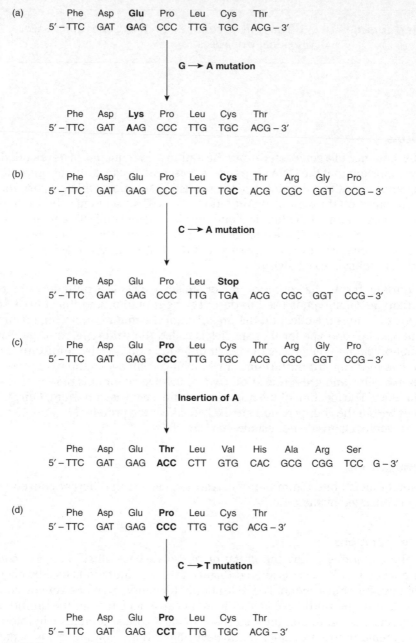

Figure 1. (a) Missense point mutation. (b) Nonsense point mutation. (c) Frameshift point mutation. (d) Silent point mutation.

Frameshift mutations

These result from the insertion of extra bases or the deletion of existing bases from the DNA sequence of a gene. If the number of bases inserted or deleted is not a multiple of three the reading frame will be altered and the ribosome will read a different set of codons downstream of the mutation, substantially altering the amino acid sequence of the encoded protein (Figure 1c). Frameshift mutations usually have a serious effect on the encoded protein and are associated with mutant phenotypes.

Silent mutations

Mutations may occur at the third base of a codon and, due to the degeneracy of the genetic code, the amino acid will not be altered (Figure 1d). Silent mutations have no effect on the encoded protein and do not result in a mutant phenotype. They tend to accumulate in the DNA of organisms where they are known as **polymorphisms**. They contribute to variability in the DNA sequence of individuals of a species.

Point mutations that involve replacement of a purine with a pyrimidine or vice-versa are known as **transversions**. Replacements involving two purines or two pyrimidines are **transitions**.

Gross mutations

Gross mutations cause substantial alterations to the DNA often involving long stretches of sequence. A number of gross mutations can occur.

Deletions

These involve the loss of a portion of the DNA sequence (Figure 2a). The amount lost varies greatly. Deletions can be as small as a single base or much larger – in some cases corresponding to the entire gene sequence.

Insertions

In this case the mutation occurs as a result of insertion of extra bases, usually from another part of a chromosome (Figure 2b). As for deletions, the amount inserted may be one or two bases or may be much larger.

The term 'indel' is used to indicate small deletions and insertions as a group.

Rearrangements

These mutations involve segments of DNA sequence within or outside a gene exchanging position with each other (Figure 2c). A simple example is the inversion mutation in which a portion of the DNA sequence is excised then re-inserted at the same position but in the opposite orientation.

Gross mutations, because they involve major alterations to gene sequences, invariably have a serious effect on the encoded protein and are frequently associated with a mutant phenotype.

Mutation and disease

In humans and other higher organisms DNA mutation plays a significant role in the development of disease. Specifically, mutation is the underlying cause in genetic diseases such as hemophilia and cystic fibrosis and also contributes to development of cancers.

(a)

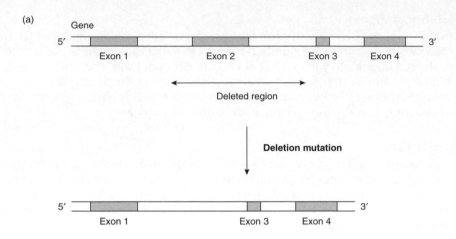

(b)

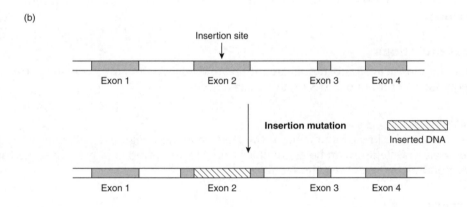

(c)

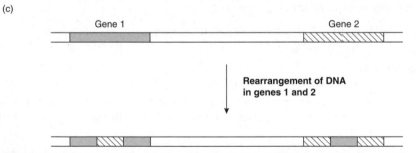

Figure 2.   (a) Deletion mutation. (b) Insertion mutation. (c) Rearrangement mutation.

Genetic diseases are caused by mutations that are inherited and are passed on from parents to offspring. Mutations can occur at random in any cell but normally a mutation in a single cell has no effect on the organism because cells are continually being replaced. However, a mutation in a germ cell (sperm or ovum) can be passed on after conception and will be present in every cell of the resulting offspring which then becomes a carrier of the mutation and can in turn pass the mutation on to subsequent generations. Genetic diseases usually involve mutation of a single gene. The mutated genes vary widely; this accounts for the diverse nature of genetic diseases.

Mutations associated with the development of cancer occur in the normal cells of the body (somatic cells). Cancers often develop in association with mutations in genes that regulate cell division. Such mutations often confer the ability for cells to divide in an uncontrolled way leading to the development of a tumor.

Mutations at the level of organism

In addition to defining mutations in terms of alterations to the DNA sequence, it is also possible to use definitions that describe the phenotype of the mutant organism. This approach applies to both prokaryotic and eukaryotic cells but is most commonly associated with bacterial mutants. Several mutant phenotypes have been defined.

Auxotrophic mutants

These mutants lack a gene product involved in the synthesis of an essential metabolite, such as an amino acid. They can only be isolated by being grown in a medium supplemented with the metabolite. For example, the tryptophan auxotroph of *E. coli* cannot synthesize tryptophan for itself and must obtain it from the growth medium. The opposite of an auxotroph is a prototroph which has no special nutritional requirements.

Temperature-sensitive mutants

These mutants are limited by the conditions used to grow the cells and can only survive when cultured at a given temperature. If the temperature is raised, the cells are unable to grow and they die. The mutation carried by a temperature-sensitive mutant usually affects an amino acid that is important for maintaining the structure of the protein. When the temperature is raised the mutant protein becomes denatured and loses its activity resulting in the mutant phenotype.

Antibiotic-resistant mutants

Antibiotics kill wild-type bacteria but have no effect on resistant mutants. Antibiotic resistance can arise in several ways but often involves mutation of a gene encoding a protein that is a target for the antibiotic.

Regulatory mutants

In this case mutants lose the ability to regulate expression of a gene or operon. For example, regulatory mutants of *E. coli* exist which express genes required for the metabolism of lactose even when lactose is absent from the medium. These are called constitutive mutants and they usually arise by mutation of the lac repressor protein.

A6 Mutagens and DNA repair

Key Notes

Mutagens

Mutations arise due to rare errors in DNA replication or by the action of chemical and physical agents called mutagens on DNA. Changes in the structure of nucleotides cause altered base-pairing which becomes permanent after DNA replication. Other mutagens produce physical distortions in the DNA that block replication or transcription.

Chemical mutagens

Many different chemicals act as mutagens. Base analogs substitute for normal bases during DNA replication and cause mutation by having altered base-pairing patterns. Intercalating agents slip between the bases in the double helix. They cause the insertion of an extra base during replication producing a frameshift mutation. Many chemical mutagens modify bases, often by the addition of alkyl or aryl groups or by deamination. DNA also undergoes spontaneous mutation by reaction with normal chemical species in cells. Reactive oxygen species (ROS) present in aerobic cells also damage bases.

Physical mutagens

Ionizing radiation in the form of X-rays and γ-rays damage DNA molecules extensively. Ultraviolet radiation is absorbed by bases and leads to the formation of cyclobutyl dimers between adjacent pyrimidine bases. Heat is a significant mutagen.

DNA repair

The presence of numerous agents that mutate DNA has led organisms to develop extensive DNA repair mechanisms. The repair mechanisms are complex but essentially three main types occur: excision repair, direct repair, and mismatch repair.

Excision repair

Single-strand nicks are created in the DNA adjacent to a damaged nucleotide by a repair enzyme. A nuclease removes the damaged base and adjacent bases. The gap is then filled with new DNA by a DNA polymerase and closed by DNA ligase.

Direct repair

This mechanism involves the reversal of structural alterations that occur in nucleotides. Photoreactivation is an important example which involves repair of thymine dimers by enzymes called DNA photolyases which break the links formed on dimerization.

Mismatch repair

This system corrects errors of DNA replication by identifying mismatched nucleotides and replacing the incorrectly inserted base. The system determines which of the

	mismatched bases is correct by identifying the parental DNA strand which is methylated.
Genotoxicity	Genotoxicity refers to the detection of agents that damage DNA and hence cause mutations. Commercial tests using animals have now been largely replaced by systems using bacteria or animal cells in culture.
	The best known test is the Ames test which assays the capacity of potential mutagens to revert mutations in the histidine operon of *Salmonella typhimurium*. Mutagens induce increased numbers of revertants. The test can be modified to detect procarcinogens by adding rat liver extract. Animal cell lines can be used in similar tests. Clastogens are mutagens that cause chromosome damage. They can be detected by assessing their ability to induce sister chromatid exchanges (SCEs). These occur spontaneously but the rate of induction is very sensitive to mutagens. SCEs are only detectable when two chromatids of the same chromosome stain differently. This is achieved by labeling cells for two divisions with bromodeoxyuridine prior to staining.
Related topics	(A1) DNA structure (A5) DNA mutation

Mutagens

Mutations can arise due to errors introduced during DNA replication. These are normally very rare, occurring in bacteria about once in every 10^{10} bases incorporated and probably more frequently in higher organisms. This is called the **spontaneous mutation rate**. The rate of mutation increases when cells are exposed to chemical or physical agents known as **mutagens** that interact directly with the DNA and alter the structure of individual nucleotides. This may lead to an alteration or a failure of base-pairing such that when DNA replication occurs an incorrect base is inserted opposite the modified base changing the DNA sequence. Subsequent rounds of replication make the change permanent leaving the DNA sequence mutated (Figure 1). Some mutagens work in a different way and cause serious physical distortions in the DNA which block DNA replication or transcription.

A wide variety of natural and synthetic, organic and inorganic chemicals can react with DNA altering its structure and causing mutation. Most chemical mutagens are **carcinogenic** and cause cancer. In addition, a range of physical agents can cause mutation. These include ionizing radiation in the form of X-rays and γ-rays, nonionizing radiation, particularly ultraviolet light, and also heat.

Chemical mutagens

A wide variety of chemicals can cause mutation of DNA. Some chemical mutagens are **base analogs**. These are structurally similar to the normal bases found in DNA and can be incorporated by DNA polymerases into the DNA during replication. They cause mutation

Figure 1. Mechanisms of mutation. (a) Mutations arise due to errors introduced during DNA replication. (b) Mutations arise due to structural changes in nucleotides leading to altered base-pairing.

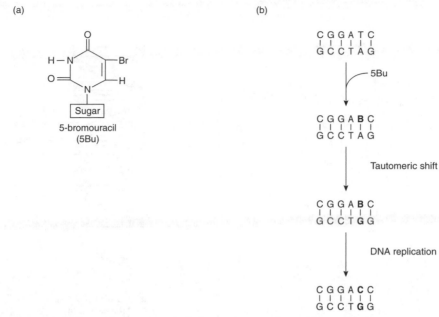

Figure 2.   (a) Structure of 5-bromouracil. (b) Mutation of DNA by 5-bromouracil.

by producing altered base-pairing. An example is **5-bromouracil (5BU)**, a base analog derived from thymine. 5BU normally base-pairs with adenine; however, it can undergo a slight change in its structure called a tautomeric shift which causes it to base-pair with guanine. After DNA replication, the original TA base pair is replaced by GC on one of the daughter strands leading to a point mutation (Figure 2).

Other chemical mutagens are **intercalating agents**. These are flat molecules that disrupt DNA replication by slipping between adjacent base pairs of the double helix. An example is **ethidium bromide**, a molecule containing four rings that has dimensions similar to a purine–pyrimidine base-pair. Ethidium bromide is said to intercalate into the double helix causing adjacent base-pairs to move apart slightly (Figure 3). It is not clear exactly how this disrupts replication but the overall effect is to cause the insertion of a single nucleotide at the intercalation position causing a frameshift mutation in the gene.

Many mutagens act by chemically modifying bases. Some cause the addition of alkyl or aryl groups. Examples include **methylmethane sulfonate** and **ethylnitrosourea** which add methyl groups to bases at a variety of positions (Figure 4). Other base modifications include **deamination**. For example, **nitrous acid** deaminates cytosine to produce uracil (Figure 5) which base-pairs with adenine, resulting in an alteration of the base-pair from GC to AT after replication. Deamination of adenine to a guanine analog called hypoxanthine causes an AT base-pair to be replaced by GC.

Mutations can also occur spontaneously due to the chemical reactivity of DNA with some of the normal chemical species in cells. For example, cytosine undergoes spontaneous deamination to uracil. In addition, cytosine, which is sometimes present as 5-methylcytosine, may be deaminated to produce thymine. A very common modification which occurs spontaneously is **depurination** which results from breakage of the link between a deoxyribose sugar and its purine base (Figure 6). Damage to bases can also occur due

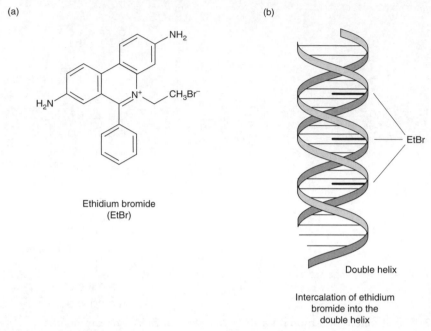

(a)

Ethidium bromide
(EtBr)

(b)

EtBr

Double helix

Intercalation of ethidium
bromide into the
double helix

Figure 3.   (a) Structure of ethidium bromide. (b) Mutagenic effect of ethidium bromide.

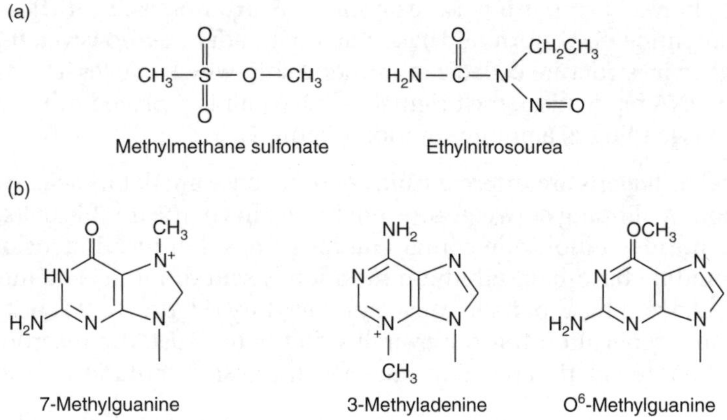

(a)

Methylmethane sulfonate   Ethylnitrosourea

(b)

7-Methylguanine   3-Methyladenine O^6-Methylguanine

Figure 4.   Examples of (a) alkylating agents and (b) alkylated bases.

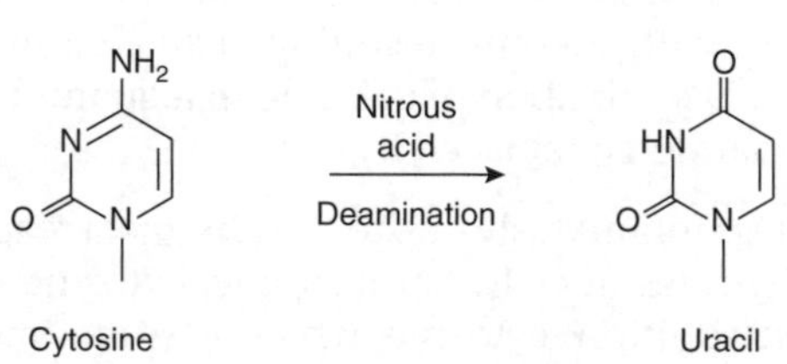

Nitrous
acid
———————→
Deamination

Cytosine   Uracil

Figure 5.   Deamination of cytosine to uracil by nitrous acid.

Figure 6.   Depurination of DNA.

to the presence of reactive oxygen species (ROS) which occur in all aerobic cells. These include the superoxide, hydrogen peroxide, and hydroxyl radicals.

Physical mutagens

High energy ionizing radiation, such as X-rays and γ-rays, cause extensive damage to DNA molecules producing strand breaks and the destruction of sugars and bases. Non-ionizing radiation in the form of ultraviolet (UV) light is absorbed by bases and can induce structural changes. In particular, UV light can cause the formation of structures called **cyclobutyl dimers** between adjacent pyrimidines, especially thymines (Figure 7). Dimerization causes the bases to stack closer together and can result in deletion mutations following DNA replication. Heat is also a significant environmental mutagen. Its effect on DNA molecules is to produce apurinic sites in the DNA polynucleotide which can cause point or deletion mutations when the DNA is replicated.

DNA repair

The requirement that the DNA sequence of a gene is preserved, despite the presence of numerous agents that can mutate DNA, has caused organisms to develop methods which prevent mutation by repairing damage to DNA. The repair mechanisms are complex but essentially three main types occur.

Excision repair

This is a complex system which is probably the most common form of DNA repair. Many different types of damage are repaired including pyrimidine dimers. Initially, one of a number of enzymes recognizes nucleotides that are damaged and marks them for repair. The mark can take the form of a nick in one of the strands of the double helix adjacent to the area of damage or a damaged base may be removed leaving a gap. In the next stage, a nuclease removes the marked nucleotide as well as a number of its neighbors. A DNA polymerase (DNA polymerase I in *E. coli*) then synthesizes new DNA to replace the

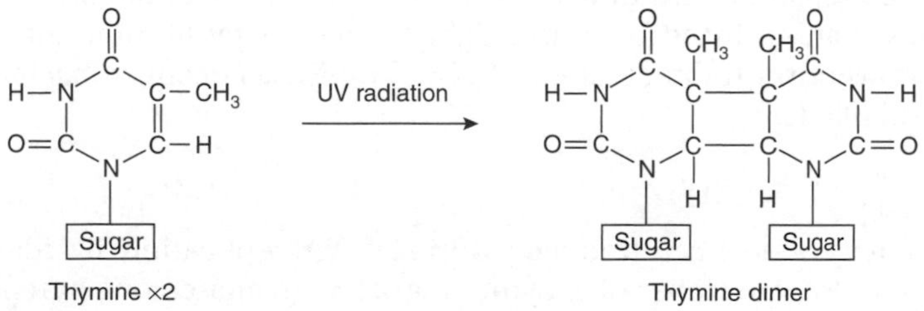

Figure 7.   Formation of thymine dimers by UV radiation.

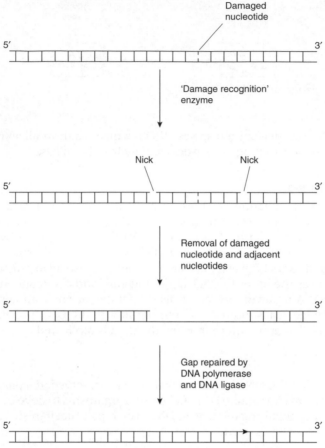

Figure 8.   Excision DNA repair.

missing bases and DNA ligase joins the new DNA to the existing molecule, restoring the DNA to its original structure (Figure 8).

Direct repair

This is a much less common form of DNA repair which involves reversal of structural alterations that occur in nucleotides. An important example of direct repair is **photoreactivation** which repairs pyrimidine dimers produced by UV radiation. Enzymes called DNA photolyases are induced by visible light and repair pyrimidine dimers by breaking the links that form on dimerization. DNA photolyases occur in bacteria, microbial eukaryotes, and plants.

Mismatch repair

This system corrects errors introduced during DNA replication by identifying mismatched nucleotides. A number of enzymes mark the mismatch or can repair it directly. It is important that the system can determine which of the mismatched bases is correct and it does this by distinguishing between the parental DNA strand which has the correct sequence and the daughter strand containing the mutated sequence. In *E. coli* the parental strand is easily recognized because it is tagged with methyl groups attached to adenine bases.

Genotoxicity

This is a recently developed branch of toxicology which identifies mutagens in the environment. Most agents that cause cancer, **carcinogens**, are also mutagens. Identification of mutagens is recognized as important in many areas including pharmaceuticals, food additives, agriculture, pollution analysis, and in many industrial processes. For this reason it has been necessary to develop tests to identify dangerous compounds. Early testing systems relied on small mammals such as rats or mice, but these tests were time consuming, very expensive and attracted considerable ethical criticism. A number of *in vitro* tests have been developed which use bacteria or animal cells grown in tissue culture. In many countries there are legal requirements for new chemicals to be tested by a series of different tests before being licensed by government agencies.

The best known test is the Ames test. This is a very rapid test for mutagens. It utilizes bacteria of the species *Salmonella typhimurium* that have a mutation in the histidine operon (Section A7), and hence cannot synthesize histidine. They are referred to as histidine auxotrophs (*his⁻*). Compounds are tested to determine whether they can induce reversion of this mutation. The test can be carried out, very simply, by plating out *his⁻* mutants on an agar plate that contains only a trace of histidine. A crystal, or a filter disc containing a solution of the compound to be tested, is placed on the surface of the agar. The bacteria grow for a short time until the histidine is depleted. After that point, the only bacteria capable of continuing growth to form colonies are those which have undergone reversion and are capable of synthesizing their own histidine, histidine prototrophs (*his⁺*). If the test compound is not mutagenic a few revertant colonies will be found randomly scattered across the agar plate. If it is mutagenic then the number of colonies will be increased and they will be clustered around the point where the compound was placed on the plate. Obviously the test can be constructed in a more quantitative manner to give a dose response curve for any compound under test.

In order to identify the maximum number of mutagens two types of *his⁻* mutants are used, one is a single-base substitution and the other a frameshift mutation (Section A5). This allows detection of mutagens which have different effects on DNA. It is also possible to genetically alter both the permeability of the bacteria to test compounds, and to decrease their ability to repair damaged DNA. This again increases the likelihood of detecting mutagenic activity.

Some compounds which are known to cause cancer are only capable of doing so after they have been converted to mutagens by the action of enzymes within the body. These are known as **procarcinogens**. Enzyme action converts them to **ultimate carcinogens**. If procarcinogens are used in the Ames test they will give a negative result, however the test can be adapted to take account of this. The liver is a rich source of activating enzymes. Liver extracts, containing these enzymes, can be added along with the test compound. Activation of a procarcinogen will then occur resulting in increased numbers of revertants. The Ames test is rapid; it can be carried out in 48 hours. It is cheap and easily quantifiable. It has identified many compounds as mutagens including certain hair dyes, flame-retardants and food colorings.

One deficiency of the Ames test is that the target organism is a bacterium rather than a mammal. For this reason a number of tests have been developed using cultured animal cell lines. These are very similar to the Ames test but use different selection systems and test genes.

Agents that damage chromosomes are known as **clastogens**. These are also identified by test systems. These tests can be carried out on cell lines, in laboratory animals or even in plants. The tests consist of scoring chromosome aberrations such as breaks, exchanges, ring chromosomes, dicentrics, and translocations. The frequency of aberration is often low. An alternative to enumeration of such gross chromosomal aberrations is to count **sister chromatid exchanges** (SCEs). SCE involves exchange of material between two chromatids of the same chromosome, and is a process that takes place spontaneously at low frequencies in all cell types. It can occur both in mitosis and meiosis. Its usefulness is that the frequency of SCE increases much more rapidly than gross chromosomal aberrations, as a response to clastogen treatment. SCE can be detected by a number of procedures, all of which are dependent on the semi-conservative nature of DNA replication (Section A1). The essence of the technique is to make sister chromatids stain differently so that exchanges can easily be observed. This is achieved by a complex staining method after culturing cells for two rounds of division in the presence of the thymidine analog bromodeoxyuridine (BrdU). This is incorporated into the newly synthesized DNA in place of thymidine and alters the staining properties of the chromatid so that one stains dark and the other light.

A7 Regulation of gene expression

Key Notes	
Regulation of gene expression	Some proteins are present in all cells of an organism while others are only present in some cells or at certain times. Which genes are 'on' or 'off' is determined by a number of regulatory pathways. These operate at different levels but control of transcription is the most important. This requires some genes to control the transcription of others.
Regulation of transcription in prokaryotes	Many bacterial genes are arranged as coordinately regulated operons that encode proteins with related functions. Inducible operons such as the *lac* operon encode enzymes involved in metabolic pathways and are induced by the substrate for the pathway. Repressible operons such as the *trp* operon encode enzymes involved in biosynthetic pathways. Gene expression is regulated by the pathway end product, or by attenuation.
The *lac* operon	This operon contains three genes (*lac Z, Y, A*) encoding enzymes required by *E. coli* to metabolize lactose. The genes are transcribed from a single promoter and their expression is induced by lactose. In the presence of lactose, allolactose binds the *lac* repressor preventing it from binding the *lac* operator and allowing the operon to be transcribed. When lactose has been used up the *lac* repressor regains its ability to bind the *lac* operator and transcription is blocked.
Catabolite repression	This regulatory mechanism allows *E. coli* to repress the *lac* operon in the presence of glucose. Catabolite activator protein (CAP) binds cAMP and stimulates transcription of the *lac* operon by binding upstream of the *lac* promoter. Levels of cAMP are regulated by glucose which inhibits adenylate cyclase. When glucose is available cAMP levels are low, CAP fails to bind the *lac* promoter and the operon is transcribed at a low level. When glucose levels are low, cAMP levels rise, CAP binds the *lac* promoter and stimulates transcription of the operon. Catabolite repression ensures that when glucose and lactose are both available, glucose is used first.
The *trp* operon	This operon contains five genes transcribed from a single promoter encoding enzymes required for the biosynthesis of tryptophan. The *trp* repressor binds the *trp* operator in the presence of tryptophan and blocks transcription of the operon. In the absence of tryptophan the *trp* repressor fails to bind and transcription of the operon proceeds.

Attenuation	This regulatory mechanism allows fine adjustment of expression of the *trp* operon and other operons. mRNA sequences between the *trp* promoter and the first *trp* operon gene are capable of forming either a large stem-loop structure that does not influence transcription or a smaller terminator loop. A short coding region upstream contains tryptophan codons. When tryptophan levels are adequate RNA polymerase transcribes the region closely followed by a ribosome which prevents formation of the larger stem-loop, allowing the terminator loop to form ending transcription. If tryptophan is lacking, the ribosome is stalled, the RNA polymerase moves ahead and the large stem-loop forms. Formation of the terminator loop is blocked and transcription of the operon proceeds.
Regulation by alternative sigma factors	Alternative σ factors alter the specificity of bacterial RNA polymerase allowing it to recognize different gene promoters.
Regulation of transcription in eukaryotes	Eukaryotic cells regulate gene expression mostly by varying the rate of gene transcription. Interactions between RNA polymerase II and basal transcription factors lead to the formation of the transcription initiation complex (TIC) at the TATA box. Other transcription factors change the rate of transcription initiation by binding to promoter sequences and influencing the stability of the TIC. Distant sequences called enhancers and silencers also influence the rate of transcription.
Transcription factors	Gene promoters have multiple binding sites for transcription factors each of which can influence transcription. The overall effect on transcription depends on the complement of transcription factors bound. Transcription factors have a modular structure containing DNA binding, dimerization, and transactivation domains with characteristic structural motifs. Transcription factors operate in a combinatorial manner. They form hetero- and homodimers. Transcription factors can also repress transcription by direct or indirect mechanisms.
Regulation of gene expression by hormones and cytokines	Hormones and cytokines influence target cells by altering the patterns of gene transcription. Steroid hormones enter cells and bind steroid hormone receptor protein releasing it from an inhibitory protein. The receptor dimerizes and is translocated to the nucleus where it binds target gene promoters activating transcription. Polypeptide hormones and cytokines bind receptor proteins on the surface of target cells. Gene activation is triggered by signal transduction in which a network of proteins is sequentially activated by protein phosphorylation.

Post-transcriptional regulation by RNA interference	RNA interference (RNAi) occurs when double-stranded RNA is processed by the enzymes Drosha and Dicer into microRNA (miRNA) or short interfering RNA (siRNA) fragments, usually 22 bp long. These complex with protein, including an argonaute protein, to form a RNA-induced silencing complex (RISC). The RNA becomes single stranded and guides the complex to complementary mRNA where it binds and can catalyze cutting of the mRNA. In cases where base-matching is imperfect, it may not cut but blocks translation. Genome sequence data can be used to generate antisense RNA and is proving very effective as a procedure for reducing expression of genes experimentally to observe phenotypic effects.
Differentiation and development	Cell and tissue differentiation and development in multicellular organisms occurs by long-term epigenetic regulation of genes. This involves molecular modifications to DNA and histone rendering DNA inaccessible to RNA polymerase.
Related topics	(A2) Genes (B2) Prokaryote genomes (A4) DNA to protein (B3) Eukaryote genomes (A8) Epigenetics and (C11) Genes in development chromatin modification

Regulation of gene expression

With only a small number of exceptions all genes code for proteins and gene expression refers to which proteins are present in a cell or tissue. Some proteins are present in all cells but others are only present at certain times or in certain cell types. Hence, the pattern of gene expression must be regulated. This is true for single-celled prokaryotes and multicellular eukaryotes alike. Genes may be 'on' or 'off' depending on whether the proteins they code for are present or not and there are also quantitative changes in patterns of expression.

Gene expression can be regulated at several levels, including protein activity and stability and translation of mRNA on ribosomes, but the most important mechanisms concern transcription (Section A4). In both prokaryotes and eukaryotes this involves genes that code for proteins whose function is to regulate the transcription of others, but the mechanisms differ greatly between the two kingdoms.

Regulation of transcription in prokaryotes

An important feature which determines how gene transcription is regulated in bacteria is the organization of the genes as **operons**. These are transcriptional units in which several genes, usually encoding proteins with related functions, are regulated together. Other genes also occur which encode regulatory proteins that control gene expression in operons. Many different operons have been identified in *E. coli*. Most contain genes that encode proteins involved in the biosynthesis of amino acids or the metabolism of nutrients. Operons are classified as **inducible** or **repressible**. Inducible operons contain genes that encode enzymes involved in metabolic pathways. Expression of the genes is controlled by a substrate of the pathway. An example of an inducible operon is the *lac* **operon** which encodes enzymes required for the metabolism of lactose. Repressible operons contain genes that encode enzymes involved in biosynthetic pathways and gene expression is controlled by the end product of the pathway which may repress expression of the operon or control it by an alternative mechanism called **attenuation.** An example of a repressible operon is the *trp* **operon** which encodes enzymes involved in the biosynthesis of tryptophan.

The *lac* operon

This operon contains three genes encoding enzymes required by the *E. coli* bacterium for the utilization of the disaccharide sugar lactose. These are lactose permease, which transports lactose into the cell, β-galactosidase, which hydrolyzes lactose into its component sugars (glucose and galactose), and β-galactoside transacetylase, which is also involved in the hydrolysis of lactose. These enzymes are normally present in *E. coli* at very low levels but in the presence of lactose their levels rise rapidly. The three genes in the *lac* operon are known as *lac Z, Y,* and *A* and encode β-galactosidase, lactose permease, and β-galactoside transacetylase, respectively. The genes are sequential and are transcribed as a single mRNA from a single promoter. Another regulatory gene, *lac I,* which is expressed separately, lies upstream of the operon and encodes a protein called the *lac* **repressor** which regulates the expression of the *lac Z, Y* and *A* genes. In the absence of lactose the *lac* repressor binds to a DNA sequence, called the **operator**, positioned between the *lac* promoter and the beginning of the *lac Z* gene. When bound to the operator, the *lac* repressor blocks the path of the RNA polymerase bound to the *lac* promoter upstream of it and prevents transcription of the *lac* genes. When the cell encounters lactose a few molecules of the *lac* enzymes present in the cell allow lactose to be taken up and metabolized. **Allolactose**, an isomer of lactose produced as an intermediate during the metabolism of lactose, acts as an inducer. It binds to the lactose repressor and changes its conformation such that it can no longer bind to the operator. The path of the RNA polymerase is no longer blocked and the operon is transcribed. Large numbers of enzyme molecules are produced which take up lactose and metabolize it. The presence of lactose thus induces the expression of the enzymes needed to metabolize it. When the lactose is used up the *lac* repressor returns to its original conformation and again binds the *lac* operator, preventing transcription and switching off the operon (Figure 1).

Catabolite repression

This term describes an additional regulatory mechanism which allows the *lac* operon to sense the presence of glucose, an alternative and preferred energy source to lactose. If glucose and lactose are both present, cells will use up the glucose first and will not expend energy splitting lactose into its component sugars. The presence of glucose in the cell switches off the *lac* operon by a mechanism called **catabolite repression** which involves a regulatory protein called the **catabolite activator protein** (CAP) (Figure 2). CAP binds to a DNA sequence upstream of the *lac* promoter and enhances binding of the RNA polymerase leading to increased transcription of the operon. However, CAP only binds in the presence of a derivative of ATP called cyclic adenosine monophosphate (cAMP) whose levels are influenced by glucose. The enzyme **adenylate cyclase** catalyzes

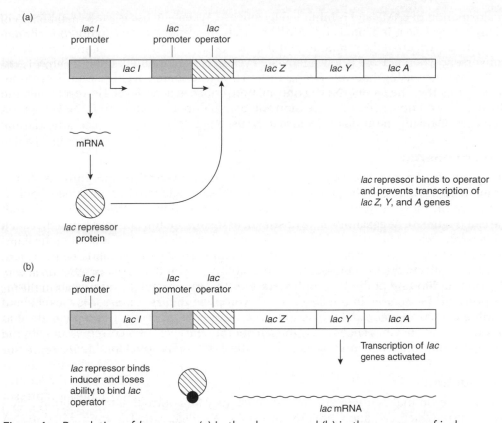

Figure 1.   Regulation of *lac* operon (a) in the absence and (b) in the presence of inducer.

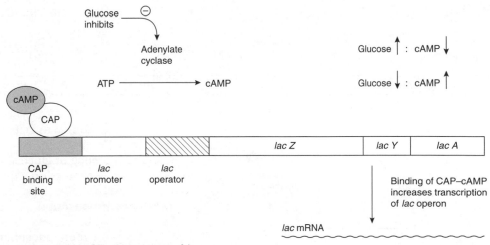

Figure 2.   Catabolite repression of *lac* operon.

the formation of cAMP and is inhibited by glucose. When glucose is available to the cell, adenylate cyclase is inhibited and cAMP levels are low. Under these conditions CAP does not bind upstream of the promoter and the *lac* operon is transcribed at a very low level. Conversely, when glucose is low adenylate cyclase is not inhibited, cAMP is higher and CAP binds, increasing the level of transcription from the operon. If glucose and lactose are present together the *lac* operon will only be transcribed at a low level. However when the glucose is used up catabolite repression will end and transcription from the *lac* operon increases, allowing the available lactose to be used up.

The *trp* operon

This operon contains five genes encoding enzymes involved in biosynthesis of the amino acid tryptophan. The genes are expressed as a single mRNA transcribed from an upstream promoter. Expression of the operon is regulated by the level of tryptophan in the cell (Figure 3). A regulatory gene upstream of the *trp* operon encodes a protein called the **trp repressor**. This protein binds a DNA sequence called the **trp operator** which lies just downstream of the *trp* promoter partly overlapping it. When tryptophan is present in the cell it binds to the *trp* repressor protein enabling it to bind the *trp* operator sequence, obstructing binding of the RNA polymerase to the *trp* promoter and preventing transcription of the operon. In the absence of tryptophan the *trp* repressor is incapable of binding the *trp* operator, and transcription of the operon proceeds. Tryptophan, the end product of the enzymes encoded by the *trp* operon, thus acts as a co-repressor with the *trp* repressor protein and inhibits its own synthesis by end product inhibition.

Attenuation

The *trp* operon makes use of an alternative strategy for controlling transcription called attenuation which can finely tune expression levels (Figure 4). The transcribed mRNA sequence between the *trp* promoter and the first *trp* gene is capable of forming two stem-loop structures. The relative positions of the sequences mean that both stem-loops cannot form at once: just one or the other may be present at any time. The larger, more stable structure does not influence transcription and occurs upstream of the smaller stem-loop which acts as a transcription terminator. If this structure forms it will terminate

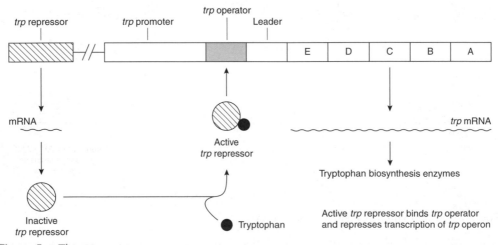

Figure 3.   The *trp* operon.

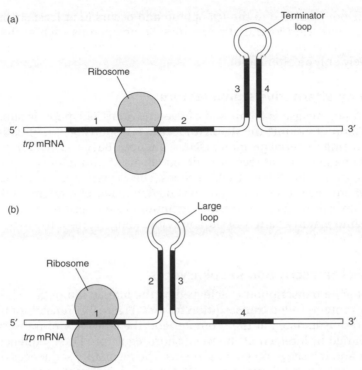

Figure 4.   Attenuation in the *trp* operon. (a) When tryptophan is abundant, the ribosome obstructs formation of the large loop, allowing the terminator loop to form; (b) when tryptophan is scarce, the ribosome is stalled, allowing the large loop to form, preventing formation of the terminator loop.

transcription before the first gene is reached eliminating gene expression. Attenuation depends on the fact that transcription and translation are linked in bacteria: ribosomes attach to mRNAs as they are being synthesized and begin translating them into protein. An mRNA that is being transcribed may already have one or more ribosomes attached to it. Binding of ribosomes to the *trp* mRNA influences which of the two stem-loop structures can form and so determines whether or not termination occurs. Immediately upstream of the stem-loop region is a short open reading frame containing 14 codons followed by a stop codon which is translated before the structural genes; two out of these 14 codons are for tryptophan. If levels of tryptophan are adequate the ribosome will translate the coding region following closely behind the RNA polymerase. In these circumstances the presence of the ribosome prevents formation of the larger stem-loop allowing the terminator loop to form and transcription ends. If tryptophan is lacking, the ribosome will be stalled as it translates the coding region. The RNA polymerase will move ahead and the first stem-loop will be free to form. Formation of the terminator loop is then blocked and transcription of the operon can proceed. The speed at which the ribosome translates the coding region will not be the same for each transcript. When tryptophan is present in the medium at intermediate levels some transcripts will terminate and others will not, thus allowing fine adjustments in the levels of transcription of the operon. Overall, the *trp* repressor determines whether the operon is switched on or off and attenuation determines how efficiently it is transcribed; both depend on the level of tryptophan in the cell. Attenuation allows the cell to synthesize tryptophan according to its exact requirements.

Attenuation is not restricted to the *trp* operon and occurs in at least six other operons that encode amino acid biosynthetic enzymes. Some operons such as the *trp* and *phe* operons are regulated by repressors and attenuation and others such as the *his*, *leu*, and *thr* operons rely only on attenuation.

Regulation by alternative sigma factors

Bacterial RNA polymerase is composed of five individual polypeptide subunits (two α, β, β' and ω). Another subunit, the **sigma (σ) factor,** is responsible for initiating transcription by recognizing bacterial promoter DNA sequences. Bacteria, including *E. coli*, make alternative sigma factors that recognize different sets of promoters and cause the RNA polymerase to transcribe different sets of genes. This is used as a way of regulating gene expression where environmental conditions dictate major alterations in the pattern of gene expression. The σ^{70} factor is the most common factor used by *E. coli*. Alternative σ factors come into play in a variety of situations including the response to heat shock in *E. coli* and sporulation in *Bacillus subtilis*.

Regulation of transcription in eukaryotes

Regulation of gene transcription is achieved by the interaction of gene promoters and DNA binding proteins called **transcription factors** (Figure 5). Transcription of a gene by RNA polymerase is initiated at the promoter and the efficiency of transcription initiation can be varied by interactions between short regulatory DNA sequences present in the promoter and transcription factor proteins. The regulatory sequences present in the promoter are on the same chromosome (DNA double helix) as the coding sequence and are said to be *cis*-**acting**.

In eukaryotic cells protein-coding genes are all transcribed by RNA polymerase II. Transcription is initiated by the formation of the transcription initiation complex (TIC) which involves binding of RNA polymerase II and a number of associated proteins called **basal transcription factors** to the DNA of the promoter at a characteristic sequence known as the **TATA box**. This has the sequence 5′TATA(A/T)A(A/T) 3′ and is present in most but not all eukaryotic genes located approximately 25 bp upstream of the transcription initiation site. Its function is to locate the RNA polymerase in the correct position to initiate transcription. Some genes, especially those expressed only in specific tissues or cells, do not have a TATA box but instead have an initiator sequence usually located over the transcription start site. Other genes, usually those expressed at low levels, have neither a TATA box nor an initiator element.

The efficiency of transcription initiation and hence the amount of mRNA produced is influenced by additional transcription factors that bind other DNA sequences present in the promoter and can interact with the proteins of the TIC affecting its stability.

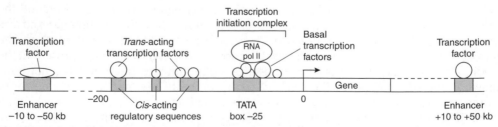

Figure 5.   Regulation of gene expression in eukaryotes.

Transcription factors can increase or decrease the rate of transcription. Many different transcription factors exist, each of which recognizes and binds a DNA sequence in the promoter. The sequence recognized can vary between promoters and the binding site is usually described as a consensus sequence which incorporates possible variations. Transcription factors are synthesized in the cytoplasm but exert their effects in the nucleus. As such, they are often referred to as ***trans*-acting** factors.

The rate of transcription of a gene can also be influenced by sequence elements called **enhancers** that may be located thousands of base-pairs distant from the transcription start site. Enhancers are typically 100–200 bp long and contain sequences that bind transcription factors and can stimulate transcription of the linked gene. The position of the enhancer relative to the gene it influences can vary, and may be upstream or downstream. Enhancers work independently of their orientation and are equally effective facing either forward or reverse. Interaction between the enhancer and its promoter occurs by looping of the intervening DNA to bring the two into close proximity. Some enhancers contain sequences that bind transcription factors that influence transcription negatively. These are known as **silencers** and they may be responsible for restricting expression to specific cell types. Enhancers and silencers generally operate by opening or closing the structure of chromatin (Section A8). Other distant sequences called **locus control elements** exist which influence expression of entire clusters of genes by controlling access of transcription proteins to the DNA. Examples of these are the locus control regions that regulate expression of the globin gene families (Section C11).

Transcription factors

These are a large number of proteins that regulate the expression of genes. They are distinct from the basal transcription factors that interact with the RNA polymerase II to form the TIC. Transcription factors have varied patterns of expression: some occur only in specific cell types whereas others occur in all cell types. Transcription factors act in a **combinatorial manner** where they interact with each other. Often this requires binding together to form homodimers with other molecules of the same transcription factor or heterodimers with different transcription factors.

Transcription factors have a modular structure composed of discrete protein domains with specific functions. Three types of domain occur commonly:

- DNA binding domains – examples include, helix-turn-helix, zinc finger, and basic domains;
- dimerization domains – examples include leucine zippers and HLH;
- transactivation domains – these are rich in acidic amino acids.

Some transcription factors can repress transcription. This may be achieved in a number of ways. Some interact directly with the transcription initiation complex. Others may act indirectly in a number of ways including: (i) blocking the DNA binding site of an activating transcription factor; (ii) formation of a dimer that lacks a DNA binding domain; or (iii) binding of a repressor protein to the activation domain of another transcription factor.

Regulation of gene expression by hormones and cytokines

Hormones are agents produced by cells that act on other cells influencing their function in a number of ways including altering the pattern of gene transcription. Hormones may be small molecules, often steroids such as estrogens and glucocorticoids, or polypeptides such as insulin. Cytokines are proteins that act in a similar way to hormones, often with blood cells as their targets. Hormones and cytokines modulate gene expression in target cells in different ways. Steroid hormones are lipid-soluble and so can pass through the cell membrane into the cytoplasm where they bind to a transcription factor called the **steroid hormone receptor**. Binding causes the steroid hormone receptor to be released from an inhibitory protein. It then dimerizes and is translocated to the nucleus where it activates transcription of target genes by binding to promoter sequences. Polypeptide hormones and cytokines act in a different way to steroid hormones by binding to receptor proteins on the surface of the target cell. Gene activation is triggered by a process called **signal transduction** in which a network of proteins is sequentially activated by protein phosphorylation or proteolysis. Ultimately this leads to stimulation of transcription of target genes by binding of transcription factors to gene promoter sequences.

Post-transcriptional regulation of gene expression by RNA interference

Gene expression can also be regulated by preventing translation of the mRNA. RNA interference (RNAi) occurs when an RNA molecule base-pairs with a mRNA molecule, making it double stranded. This can obstruct translation and/or cause degradation of the mRNA. The interfering RNA is known as **antisense** because it is complementary to the 'sense' mRNA which codes for amino acids. It was used to produce the first commercial genetically modified plant, Flavrsavr tomato, where the activity of the enzyme polygalacturonase was reduced. This enzyme softens ripe tomatoes, and without it, fruit can be allowed to ripen before picking (Section G3). The technique is widely used to reduce gene expression experimentally to observe the phenotypic changes.

Eukaryotes use RNAi mechanisms to regulate gene expression. Double-stranded RNA is formed by foldback pairing or by hybridization between mRNA and antisense RNA. This double-stranded RNA is cut into 21–28 bp-long short interfering RNAs (siRNA) by an enzyme, called 'Dicer' because it 'dices' the RNA. The siRNAs assemble with proteins including a member of the *argonaute* family and become single stranded, retaining the antisense strand. The RNA-protein complex is called a **RNA-induced silencing complex** (RISC). It pairs with the complementary mRNA. If the match is perfect, or near perfect, the mRNA is cleaved, leading to its degradation, and the RISC can go on to cleave other mRNA. If there are mismatches, the mRNA may not be cleaved but translation may be arrested. Perfectly pairing small RNAs are called **short interfering RNAs** (siRNAs) and small RNAs with some mismatches are called **microRNAs** (miRNAs). The target sequences in the mRNA are frequently, but not exclusively, in the untranslated regions at either end.

MicroRNAs are produced from *mir* genes and occur in plants, invertebrates (about 100 genes), vertebrates (about 250 genes), and some fungi. The genes have inverted sequences which pair as the stem of a loop. This stem-loop is cut off by *Drosha* then trimmed to 22 bases per strand (18 bases paired, 2-nucleotide 3′ single-strand ends) by *dicer*. Genome sequencing data are being used to design siRNAs to inhibit gene expression experimentally. Double-stranded (ds) RNA can be produced by inserting the sequence between promoters so both strands are transcribed. The dsRNA can be transfected into cultured cells, or injected into cells or organisms, and triggers RNAi. This allows the phenotypic effect of loss-of-gene product to be observed without the time and expense of producing inactivated mutants. It can be applied to cultured human cells or to organisms where genetic manipulation is difficult. It is especially useful for investigating genes of unknown function.

Some viruses produce dsRNA which triggers the inhibition system to destroy the viral RNA. The same mechanism is also involved in methylation of DNA, condensation of centromeric heterochromatin, and modification of chromatin containing transposable elements or repeated sequences. This appears to be caused by low levels of transcription of sequences in opposite orientation, possibly at different loci, producing complementary RNA strands which can pair. RNAi is involved in epigenetic regulation (Section A8).

Differentiation and development

Transcription of genes requires access to the DNA for RNA polymerase. Eukaryotes switch some genes off by modifying the chromatin so as to make them inaccessible. This involves methylation of DNA, normally cytosines, and methylation of specific amino acids in the histones. The chromatin becomes so compact it is inaccessible until a chromatin remodeling complex reverses the changes. The switching-off is semipermanent; it persists through somatic division, and in development lasts the lifetime of the organism. In some cases it persists in successive generations (Sections A8 and C11).

A8 Epigenetics and chromatin modification

Key Notes	
Overview	Epigenetic effects (by current usage) are the long-term changes in gene expression produced by modification of chromatin structure without change in DNA sequence. These may maintain differentiation in cellular lineages in a multicellular organism (eye tissue or bone cells) or affect expression from generation to generation. Many epigenetic effects are reset at gametogenesis or embryogenesis.
Chromatin modification and the histone code	Condensation (compaction/heterochromatinization) of a region of chromatin can prevent transcription of genes in that region, presumably by blocking successful access by the transcription machinery (transcriptional repression). This condensation is associated with modification of the DNA and histones which promote a cascade of binding of proteins which compact the chromatin. DNA is modified by methylation of cytosine to 5-methylcytosine at 5′CpG3′ dinucleotides in vertebrates or in 5′CNG3′ trinucleotides in plants. (Cytosines are not methylated in invertebrates.) Following DNA replication, maintenance methylases recognize hemimethylated metCpG·CpG palindromes (CpG reads the same on each strand) and methylate the new strand to match the old. *De novo* methylases methylate unmethylated DNA. Histones are mainly modified on the N-terminal tails which project from nucleosomes and bind other proteins. Control involves acetylation or methylation of lysine, and also phosphorylation of serine. A complex code of specific modifications exist which regulate specific genes. This is known as 'the histone code.' The inactivation is directed to specific sequences by double-stranded RNA in a mechanism related to RNA inhibition (Section A7).
Position effect variegation	Condensation can spread from inactivation centers until it is stopped by a boundary element or insulator. Genes moved closer to inactivation centers by chromosomal rearrangements may be inactivated by spreading heterochromatin reaching their new position. Growing clones of affected cells give a variegated expression pattern.
Deamination of methylcytosine	5-Methylcytosine may spontaneously deaminate to thymidine, producing a T·G mispairing. This may be misrepaired to T·A, so causing a mutation. It is estimated that 80% of the original CpGs in mammalian DNA have been lost by this mechanism. It follows that CpG only

survives where it is either unmethylated in germline cells, or where it is required for coding purposes or for control of gene expression so that selection removes mutants.

CpG islands

CpG-rich regions are found at the start of many mammalian genes, where they apparently have a regulatory function. CpGs in these regions are relatively unmethylated while undergoing transcription to RNA, but may be methylated in cells where those genes are inactive.

X-chromosome inactivation

Balanced gene expression requires a dosage compensation mechanism to make transcription from the two X chromosomes in females equal to the single X in males. In mammals, only one X is active, others are inactivated and condensed and the state of each X is maintained through somatic mitoses. In *Drosophila melanogaster*, the single X chromosome is transcribed at a higher rate to equal the two Xs in females (Section C9).

Imprinting

Imprinting has been found in mammals and plants, and results from embryonic identification of the parent of origin of each allele, followed by expression of only one allele (uniparental or monoallelic expression). Over 30 imprinted genes have been identified in mammals, with similar, but not identical, patterns in mice and humans. Loss of the active allele leads to deficiency and disease, despite the presence of a normal allele switched off on the other chromosome (e.g. Prader–Willi syndrome from loss of paternal 15q12 and Angelman syndrome from loss of maternal 15q12). The evolutionary-selective cause is thought to be competition between genes from male and female parents, or between different male parents. Genes switched on from the male tend to promote the individual offspring (e.g. placental growth collecting extra nutrients from the mother) and may affect behavior, promoting sibling competition. Genes switched on from the mother tend to counteract this, equalizing resource allocation between offspring.

Sporadic effects and cancer

Cancers frequently show abnormal methylation and loss of activity in tumor suppressor genes. Genes with a transposable element in their promoter region may have their expression modified by methylation of the transposon, often in a sex-specific way.

Origins of methylation

The methylation of DNA in mammals and plants is similar to the antiviral restriction systems of bacteria, which suggests that DNA methylation evolved to inactivate retroviruses and transposable elements. This accords with targeting directed by dsRNA against multicopy sequences, especially effective in plants

Related topics	(A7) Regulation of gene expression	(C9) Sex determination
	(B4) Chromosomes	(F3) Genes and cancer

Overview

The term 'epigenetic' has changed its meaning over time, but current usage refers to changes in genetic expression that are heritable to some extent but do not involve sequence changes in the DNA. It is more permanent than normal gene regulation by repressors (Section A7) and must involve a means of marking DNA in a manner that persists through DNA replication and cell division. There are three main manifestations: (i) differentiation – the tissue-specific regulation of genes that define cells as being a particular type for the lifespan of the organism; (ii) imprinting – the monoallelic expression of certain genes where the allele from parents of one sex is repressed; (iii) sporadic chromosomal rearrangement that puts an epigenetic control locus near a gene not normally under that control. The initial recognition of sites for epigenetic modification is directed by RNA-carrying proteins which then modify the chromatin in a long-lasting way.

Chromatin modification and the histone code

Epigenetic effects are apparent in multicellular animals and plants. Inactive chromatin has higher levels of 5-methylcytosine. In vertebrates the cytosine in 5'CpG3' dinucleotides may be enzymatically methylated while in plants the main target is 5'CNG3' triplets (N=any nucleotide). Invertebrates do not extensively methylate cytosine but have equivalent condensed heterochromatin.

Modification of histones was thought to be simple, with acetylation corresponding to active chromatin. There is, however, a complex range of modifications signaling alternative states. This is known as **'the histone code'** because specific modifications to histones cause specific proteins to bind and regulate specific genes. The best studied cases are the N-terminal tails of histones H3 and H4. H3 has the amino acid lysine at positions 4, 9, 14, 18, 23, and 27 which may be mono-, di- or tri-methylated or acetylated. Tri-methylation of lysine 4 and acetylation of lysine 9 seems to be activating, whereas methylation of lysine 9 promotes binding of heterochromatin protein 1 (HP1) and is inactivating. The Polycomb gene product is involved in maintaining tissue differentiation in *Drosophila*, where it binds to tri-methylated lysine 27 and excludes HP1. The serines at positions 10 and 28 may be phosphorylated. More work will discover the significance of other modifications.

Genes in heterochromatin are usually inactive, presumably because the structure prevents access by RNA polymerase. 5-Methylcytosine forms hydrogen bonds to guanine 1.8 times as tightly as cytosine does, stabilizing the double helix. Invertebrate DNA (including *Drosophila*) is not extensively methylated, if it is methylated at all. The marking of chromatin in invertebrates presumably functions through RNA and protein modifications alone.

Following DNA replication (Section A1), only the old strand is methylated. The hemimethylated mCpG·CpG palindromes in the DNA are prime targets for maintenance methylase enzymes that methylate the cytosines on the new strand, opposite the mCG on the old strand. Several methyltransferases exist in mammalian cells. DNMT1 is most abundant, has a strong preference for hemimethylated DNA, is concentrated at replication foci, and is the main maintenance methylase. DNMT3A and DNMT3B are the main *de novo* methyltransferases. Methyl-CpG binding proteins (MECPs) interact with methylated DNA. MECP2 directly inhibits transcription factor IIB, can displace histone H1 from nucleosomes, and interacts with other methyl-CpG binding-domain proteins (MBD1, MBD2, MBD3) and histone deacetylase enzymes to remodel and condense chromatin structure. Condensed inactive chromatin also contains methylated histones. The first indication of how chromatin modification is directed came in 2002 when it was discovered that RNA interference is involved. Double-stranded RNAs (dsRNA) are processed into fragments about 22 bp long. These interact with proteins, guiding them to homologous mRNA which is destroyed (Section A7). In a yeast, *Schizosaccharomyces pombe*, low levels of transcription through centromeric repeated sequences (Section B4) in both directions produced long dsRNA which, after processing, directed proteins to those DNA sequences and started the modifications producing heterochromatin. RNA is also involved in X-chromosome inactivation and antisense transcription is implicated in imprinting (below).

Position effect variegation

DNA condensation (heterochromatinization) tends to spread out from inactivation centers by cooperative binding of heterochromatin proteins and chromatin modifying enzymes until it encounters a boundary element or insulator. If genes are moved closer to inactivation centers by chromosomal rearrangement, or boundary elements are deleted, the spreading inactivation (heterochromatin) may reach those genes and turn them off. The pattern is inherited clonally, so some clones express the gene, others do not, giving a variegated appearance. The classic visible examples are split and white loci on the X chromosome of *Drosophila*. When they are moved close to the centromere by an inversion, the heterochromatin may extend over these genes, inactivating them in a wedge-shaped clone (Figure 1). The gene nearer the centromere (*split, spl*) is inactivated first, so white cells always form in a clone of split cells. A human disease, fascioscapulohumeral muscular dystrophy (FSHD), is caused by a deletion of repeated sequences (Section B3) near the telomere of the short arm of chromosome 13. The deletion does not appear to contain active coding sequences, but, following loss of the repeats, several nearby genes are hypomethylated and overexpressed, suggesting a reduction in chromatin condensation.

Deamination of methylcytosine

5-Methylcytosine deaminates to thymidine which does not hydrogen bond to guanine. Replication, or inappropriate mismatch repair of the guanine to adenine, will replace the

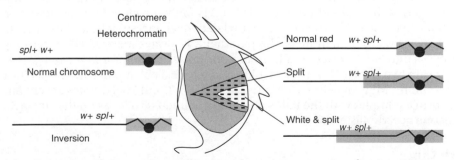

Figure 1.   Extension of chromatin condensation from the centromere first inactivates the split allele, then the white allele, to make a white wedge inside the clone with split phenotype.

original $C \cdot G$ pair with $T \cdot A$. This process is estimated to have removed 80% of the original CpG dinucleotides from mammalian DNA. They only persist where there is either no methylation in germline cells or where selection maintains them by removing mutations in essential codons and in control regions adjacent to coding sequences.

CpG islands

There are clusters of CpG dinucleotides, called **CpG islands**, near the start of many mammalian genes where they are involved in control of transcription. They are mainly found at the 5′ (mRNA) end of the gene in the promoter region, often into the first exon, but may occur throughout the coding sequence. Restriction nuclease HpaII cuts CCGG if the central C is unmethylated, and cuts CpG islands into small fragments in DNA from cells where a gene is active. Experiments with some genes show that their DNA is not cut as frequently, or not cut at all, in tissues where the gene is not transcribed, indicating that the inactive DNA is methylated. Sequence data show that about 40% of start codons are in a CpG island, and first exons have higher CpG densities than later exons. Sequencing also reveals that some genes have much more extensive and CG-rich islands than others, and there is a continuous gradation between genes with extensive islands and genes with no CpG islands, rather than a clear difference between genes with and genes without CpG islands.

X-chromosome inactivation

The presence of two X chromosomes in female mammals and *Drosophila* requires a mechanism to equalize their mRNA production to the single X in males to maintain a balance with the diploid autosomes. In mammals only one X per cell remains active, the second X in females, and all the extra copies in polysomic individuals, are inactivated (Sections B4, C9). X inactivation in mammals involves methylation of CpG islands. In *Drosophila* the single X in males is upregulated to double transcription of mRNAs. Both mammals and *Drosophila* use a mechanism based on mature RNA, transcribed from the affected X, binding to that same X (in *cis*). In mammals, the process is controlled by the X-inactivation center (*Xic*) which is about 1 Mb long and contains several elements: X-inactive specific transcript (*Xist*) codes for a 17 kb noncoding RNA which is transcribed from, and coats, the inactive X. *TsiX* (which is *Xist* backwards) is transcribed from the opposite strand to *Xist* so is complementary to it. *TsiX* starts near the short *DXPas34* gene. Deletion of DXPas34 causes that X always to be inactivated, so *Tsix* and *DXPas34* presumably counteract *Xist*. Another element, *Xce* (X-chromosome-controlling element) has a role in choice between the two chromosomes, and some alleles of *Xce* increase the chance of their X remaining active. After the inactive X is coated with stable *Xist* RNA it becomes late replicating, then histone variant macroH2A1.2 is found in some nucleosomes, histones are hypoacetylated and DNA is hypermethylated. In all marsupials, and in the extraembryonic membranes in mammals, the paternal X is always inactivated, suggesting that the process may have started as a form of imprinting. In *Drosophila melanogaster* (fruit fly) there are two small RNAs, roX2 (1.1 kb) and roXi (3.5 kb) and five male-specific lethal (MSL) proteins, MLE (maleless), MSL-1, MSL-2, MSL-3, and MOF (males absent on first). These form a complex with the RNAs and bind at hundreds of sites on the single X. MOF is a histone acetyltransferase.

Imprinting

Imprinting is a form of allele activation and inactivation that depends upon which parent a particular chromosome came from. Only the allele on one chromosome, either maternal or paternal in origin, is expressed. The inactive allele is generally methylated. This imprinting has so far been found to affect over 30 genes in mice and humans. It is also termed **monoallelic expression** or **allelic exclusion** and is consistently maintained in specific tissues. At an early stage of development, the allele from one parent is selectively inactivated, but the identification process is not clear. DNA is methylated in gametes, but, between 6 and 8 hours after fertilization, the pronucleus from the sperm is actively demethylated. The maternal DNA is not methylated after replication, so is demethylated by dilution at each mitosis (passive demethylation). The paternal and maternal chromatin must be different for this to occur, but it is not clear at which stage the alleles are recognized for imprinting. No signals have been identified, but the demethylation suggests that methylation of sperm DNA is not the primary signal for imprinting. Imprinting is controlled at imprinting centers. In some cases where there is competition between two adjacent promoters for transcription factor binding, methylation which suppresses one gene may permit expression of the other.

In humans, there are two large imprinted regions, one on chromosome 11p15.5, and the other on chromosome 15q12. Each carries a cluster of imprinted genes, but some genes in each cluster are oppositely imprinted or not imprinted at all (Figure 2). Loss of the single functional allele causes deficiency disease because the other allele is imprinted and inactive. In the case of 15q12, loss of the paternal chromosome segment causes Prader–Willi syndrome; loss of the maternal segment causes Angelman syndrome. These conditions have different symptoms because different sets of genes are inactive in the same region (Section F1).

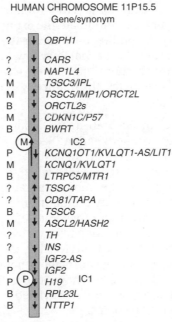

HUMAN CHROMOSOME 11P15.5
Gene/synonym

?	OBPH1
?	CARS
?	NAP1L4
M	TSSC3/IPL
M	TSSC5/IMP1/ORCT2L
B	ORCTL2s
M	CDKN1C/P57
B	BWRT
(M)	IC2
P	KCNQ1OT1/KVLQT1-AS/LIT1
M	KCNQ1/KVLQT1
B	LTRPC5/MTR1
?	TSSC4
?	CD81/TAPA
B	TSSC6
M	ASCL2/HASH2
?	TH
?	INS
P	IGF2-AS
P	IGF2
P (P)	H19 IC1
B	RPL23L
B	NTTP1

Figure 2.   The imprinted region of human chromosome 11p15.5 showing direction of transcription (arrows) and imprinting. Letters on the left indicate expression of paternal allele (P), maternal allele (M) or both (B). Circles show the position of maternal and paternal inactivation centers (IC2 and IC1, respectively).

There is a hypothesis that competition between alleles from male and female, or between different male parents, provides a selective advantage for imprinting. Genes switched on from the male tend to promote the individual offspring (e.g. placental growth collecting extra nutrients from the mother) and may affect behavior, promoting sibling competition. Genes switched on from the mother tend to counteract this, equalizing resource allocation between offspring.

In region 11p15.5, the gene for insulin-like growth factor type two (IGF2) is active on the paternal chromosome. A gene for insulin-like growth factor receptor (IGFR) which removes and counteracts IGF2 is active in the maternal genome. Diploid zygotes containing two paternally derived genomes develop masses of extraembryonic and placental material; those with two genomes from the mother develop embryonic tissue, but very little placenta. Studies in mice confirm occasional events in humans. Some imprinted genes are only differentially expressed in the brain, which controls behavior. They may promote sibling rivalry (paternal) or cooperation (maternal) after birth. Imprinting has been detected in plants, where both sets of alleles are active in embryo development (analogous to mammals) but not in *Drosophila*, *C. elegans*, or zebrafish, where the female's contribution to the embryo is completed before the egg is fertilized.

Sporadic effects and cancer

Abnormal DNA methylation patterns are common in cancer cells, causing abnormal expression of genes promoting cell division (oncogenes) or repression of cell-cycle-regulating genes (cancer suppressors). Methylation of the functional allele in a cell heterozygous for an inactive allele leads to total loss of function. This frequently occurs in tumor cells and contributes to development of cancers (Section F3). It may occur after rearrangement by position effect, or because chromatin modification has become inaccurate in the clone of cancerous cells. Methylation of a transposable element inserted into the promoter region of a gene can modulate the expression of that gene. The probability of inactivation may vary depending on whether the chromosome is transmitted maternally or paternally. For example, the viable yellow mutation in mice is dominant when active. When passed through sperm, it is inactive in about 19% of progeny, but, when passed through successive females, the frequency of methylation and inactivation increase each generation.

Normal development requires an accurate methylation system. Several human diseases are associated with defective methylation (Section F1). ICF syndrome (immuno-deficiency-centromeric instability-facial anomalies) is associated with loss of catalytic activity in DNA methyl transferase 3B (DNMT3B) which is involved in controlling DNA methylation. Rett syndrome is the most common sporadic cause of mental retardation in females (about 1/12000). It is caused by mutations in methyl-CpG binding protein 2 (MECP2) gene, which is located on the X chromosome. Affected females are heterozygous, and inactivate the functional allele in about half of their cells. Some MECP2 function is essential to survival, so no affected males survive. The most common cause of mental retardation in males is fragile X syndrome. The fragile X mental retardation 1 (FMR1) gene normally contains 29 repeats of CGG. Carriers have 50 or more copies, and in some mothers, the allele transmitted to offspring is hypermethylated in association with an increase of the number of repeats beyond 200 copies.

Origins of methylation

Eukaryotic methylation is similar to bacterial restriction systems used to counteract bacteriophages (viruses). The original function of methylation is thought to be inactivating genomic parasites such as retroviruses and transposable elements. About one-third of the human genome is of this type of element, and they contain about 90% of the methylcytosine. Different copies of the transposable elements are likely to be transcribed in different directions, so two complementary RNA molecules can pair. This produces dsDNA which triggers the RNA interference mechanism (Section A7). Plants and fungi (and presumably animals) can thereby identify, methylate and inactivate DNA elements that occur in multiple copies. Plants often inactivate transgenes inserted by genetic engineering techniques. Hypomethylation of transposable elements in cancer cells activates their recombinase genes leading to mobility, chromosome breakage and rearrangement, and further mutation.

B1 Concepts of genomics

Key Notes

Background

Since 1980 genomes of increasing size have been sequenced. Before beginning a sequencing project of a genome it is necessary to produce good framework maps. This can be done by physical or genetic mapping. Originally restriction fragment length polymorphisms (RFLPs) were used, but these have been superseded by variable number tandem repeats (VNTRs), and single nucleotide polymorphisms that are now available in large numbers.

Genetic maps

Genetic maps are based on recombination frequencies between markers. Genetic maps are good for ordering genes, but because the frequency of recombination is not constant throughout the genome, they do not give accurate measurements of physical distances between markers. High definition genetic maps are required to map genes.

Physical maps

Physical maps are constructed by subdividing the genome into smaller pieces. The genes or DNA markers on each of these are then determined. Maps can also be produced by cloning genomic human DNA in specialized vectors. Yeast artificial chromosomes were of great importance in the manufacture of earlier physical maps, but are being replaced by bacterial artificial chromosomes (BACs) and P1 artificial chromosomes (PACs). The cloned inserts are organized into continuous arrays of overlapping fragments (contigs). These are then anchored to the framework map by use of sequence tagged sites. Somatic cell hybrids have been very useful in developing physical maps of certain mammalian species.

Sequence data

Two methods have been used to determine the sequence of long stretches of human DNA. Both are based on BACs. The first uses fingerprinting to map BACs into contigs. Each element of the contig is then sequenced and the sequences produced are joined together on the basis of the contig map. The alternative method sequences the BACs first, and then aligns the sequences to deduce the complete sequence of the chromosome. This has some problems dealing with repeated sequences. Both methods have been successful in producing extensive chromosomal DNA sequences.

Placing genes on the map

Genes can be placed on the framework map through three processes. Their recombination frequencies to known markers in the genetic framework map can be used to place them between markers. Known cDNAs can be mapped directly by probing contigs' DNA. This allows genes to be ascribed to a specific clone in a mapped contig. The same approach can be used with anonymous cDNA. These are known as expressed sequence tags (ESTs). This allows the mapping of genes whose function is still unknown. Direct analysis of DNA sequence can be used to identify novel genes, by searching for sequences characteristic of gene structures. Gene mining is the process of looking for new members of a specific gene family.

Genome comparison	Comparison of genomes from different species reveals information about the genes required at various levels of complexity and the evolution of different taxa. Massively parallel sequencing has allowed the coding regions of genomes from over 1000 humans to be sequenced. The data show that individuals carry 250 to 300 recessive mutations, and children have about 60 new mutations which their parents did not have.
Environmental sequencing	Environmental sequencing takes DNA from environmental samples such as seawater. The presence of more than a thousand species may be identified, including many unknown microbial species.
Related topics	(B3) Eukaryote genomes (G1) Genetics in forensic science (C5) Linkage (G6) Ethics

Background

Deciphering the complete sequence of a genome is essentially mapping the genome to the highest limit of resolution. The first organism to have its genome sequenced was a bacterial virus, ϕX174 (Section E3). This work was published in 1980. The genome was very small, containing only 5368 bp. This was followed by small chloroplast and mito-chondrial genomes. These results raised the possibility that species with much larger genomes, such as humans, could also be sequenced. Currently almost 2000 genomes have either been sequenced or are under analysis.

The reasons for undertaking sequencing of whole genomes are numerous. Although it would be simpler to sequence only cDNA molecules and thus characterize the fraction of the genome that codes directly for protein (Section E2), it is now considered important to look at the complete sequence of all chromosomal DNA. This allows regulation of gene expression to be more clearly understood, includes genes with low-level expression that might not be found in cDNA libraries, and enables analysis of noncoding sequences that are important in evolution and variation within species.

Before sequencing technologies can have any significant impact a genome must be analyzed to produce a **framework map**. This is achieved by two related, but distinct, processes; **genetic and physical mapping**. The rate of progress of mapping depends on the availability of genetic markers. A marker is any defined segment of DNA, protein or phenotype, although DNA markers are now almost totally predominant. If a marker varies within the population it is termed polymorphic (Section D1). Ideally markers should be distributed approximately evenly throughout the genome. The first DNA polymorphisms to be utilized were restriction fragment length polymorphisms (RFLPs). These are detected as alterations to Southern blot (Section E1) patterns produced when a specific cloned DNA fragment is used to probe genomic DNA. Different patterns are produced if a mutation is present in the sequence of DNA recognized by the restriction endonuclease, or if there has been a large rearrangement of DNA within the region corresponding to the probe. Thus, any cloned DNA can be used to detect RFLPs even if nothing is known about its sequence, or its map position (an anonymous fragment).

The Donis-Keller map is an early example of a framework map of the human genome. Published in 1987 it used 180 RFLPs. Although useful, RFLPs had many drawbacks, including the fact that they are relatively uncommon, and tend to have only two different alleles (biallelic). They have been replaced by polymorphisms that result from variable *in situ* expansions of specific DNA sequences. These are of different types but are collectively referred to as variable number tandem repeats (VNTRs, Section B3). VNTRs show high levels of variation within the population. This is described as being highly **informative**. Microsatellites are the most frequently used VNTRs. Recently polymorphism due to alterations of a single nucleotide, SNPs (Section B3), have become a mapping tool of great potential, because very large numbers of these occur throughout genomes. The human genome contains over 15 million loci that display SNPs. That means that, on average, they are separated by only approximately 200 bp of DNA. SNPs are similar to RFLPs and hence have fewer heterozygotes. However, because of their high number, and the fact that they can be typed by DNA microchip technology (Section E1), they are likely to become the DNA markers of choice for mapping genomes in future. The incorporation of markers into framework maps is discussed next.

Genetic maps

Genetic maps are based on recombination frequencies and are constructed in a manner similar to the linkage analysis described in Section C5. One major advantage resulting from the use of molecular markers such as DNA polymorphisms is that they are codominant (Section C1). This allows the genotype of each individual to be determined directly rather than by inference from phenotypes.

Framework maps made in this way are extremely useful in ordering genes and markers along a chromosome although they do not give an accurate measurement of physical distance between loci. This is because the frequency of crossing-over is not constant throughout the genome (Section C5). It is high in regions close to the telomeres and reduced near centromeres. There are also occasional localized hot-spots for recombination.

A good example of a high definition genetic framework map is the Genethon map of the human genome constructed from analysis of microsatellite inheritance through several generations of a number of families. This uses over 5000 microsatellite loci. The average distance between markers is 1.6 cM (centiMorgan), and the largest gap between markers is 11 cM. This has the necessary definition to act as a framework map for accurate mapping of genes.

Physical maps

In parallel to the development of genetic maps, physical approaches have been used to produce framework maps. These employ processes in which the genome is subdivided into separate fragments and the genes and DNA markers present on each individual fragment determined. Such an approach is known as **physical mapping**. The major technique used in physical mapping is the production of maps by organizing cloned fragments of DNA into **contiguous arrays of overlapping fragments (contigs)**. This can be carried out over large regions of the genome for long-range mapping or for smaller areas for intense analysis leading directly to sequencing. In the former the original cloning vectors used were yeast artificial chromosomes (YACs). These can carry up to two megabases of DNA (Section E4). However, they have several drawbacks in that they are relatively unstable, and can lose all or part of the DNA insert. They are also notoriously chimeric; sequences from different parts of the genome can be incorporated into the same YAC. YAC contigs are anchored to the framework map by identification of sequence tagged sites (STSs). These are elements of defined sequence that have been mapped precisely onto framework maps and can be shown to be present on specific YACs by PCR or Southern blotting (Sections E1 and E2).

Subsequently two alternative cloning vectors, **bacterial artificial chromosomes (BACs)** and **P1 artificial chromosomes (PACs)** (Section E4) have been developed. These are much more stable than YACs, have low incidences of chimerism, and can accommodate up to 250 kb of DNA. BACs, in particular, have proved to be ideal vectors for use in producing contigs that can be subsequently utilized to produce sequence data, and have become the mainstay of most modern genomic analyses. In some, instances where genomes of mammals have been sequenced, very useful physical maps have been produced from somatic cell hybrids. These are produced in tissue cultures by fusing cells from different species. Human × mouse hybrid cells are a good example. When grown in culture these lose most of their human chromosomes. This system can be used to produce hybrids containing only a single human chromosome. Thus all human genes present in or DNA isolated from such a cell are known to map to that specific human chromosome.

An alternative to this approach of producing pure DNA from a single chromosome is to separate metaphase chromosomes in a Fluorescence Activated Cell Sorter. However the presence of some contamination from other chromosomes can cause problems with this method.

Sequence data

Two alternative approaches have been applied to generate and organize DNA sequence data. Both approaches are centered on BACs. The first represents a development of the physical mapping technique. BACs are used to make contigs. This is done by a process referred to as **fingerprinting**. A number of methods for fingerprinting exist but a simple approach is shown in Figure 1. In this, individual BACs are digested with a single restriction enzyme, and the products electrophoresed on gels. BACs that contain the same or overlapping regions of human DNA in their insert will give at least one restriction fragment of exactly the same size. In this way the cloned inserts can be ordered into contigs. The individual members of this contig can then be sequenced (Section E3). The nucleotide sequences are joined together from their known order in the contig. By combining contigs this eventually produces the entire sequence of bases along a specific chromosome.

The other approach is to sequence the individual BACs directly and then to use the sequence data produced to produce a contig map. This is referred to as the **shotgun** approach. It has the advantage of skipping out one step in the procedure, but has difficulties when interspersed repeated sequences are encountered, as these sequences are found in several different loci within the genome. Both methods have been used successfully to produce genome sequences.

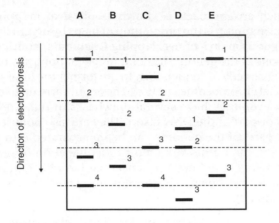

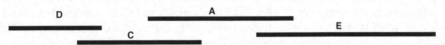

Figure 1. This shows the pattern of restriction fragments for five hypothetical BAC inserts, A, B, C, D, and E cloned from a region of the human genome. These have all been digested by the same restriction enzyme and the resulting fragments have been electrophoresed. Each insert produced three or four fragments of DNA after digestion. These are numbered in order of decreasing size for each insert. The horizontal broken lines indicate fragments of identical size. These represent the same stretch of the human genome.

It can be seen that insert A[1] is identical in size to fragment E[1]. This means that these two inserts overlap. Similarly A also shares fragments A[2] and A[4] with insert C. C does not share any fragments with E and therefore, in the human genome, C and E overlap with opposite ends of inserts A. By the same logic, insert D can be seen to overlap with C. However, because insert B has no fragments of identical size with any of the other inserts it is not part of this contiguous stretch of DNA in the human genome.

Typically the sequence is first produced as a draft. This will contain some sequencing errors and gaps that are not covered by the contigs. This is then refined and the final accuracy should be better than 99.99% which would still leave 300 000 errors in the human genome.

Placing genes on the map

Genes can be placed on the map by three different methods.

i.   Genes for a specific trait such as an inherited disorder are mapped to markers on genetic framework maps through pedigree analysis. This usually employs a procedure known as **Lod analysis** where the most likely recombination frequencies between a gene and a series of polymorphic loci are estimated. This provides an accurate map position showing that a gene lies between specific markers on the framework map.

ii. If the mRNA for a gene has been cloned as a cDNA this can be used as a probe to screen contigs of genomic DNA. This will allow placement on the physical map. cDNAs from anonymous mRNA molecules can also be used. These are known as **expressed sequence tags** (ESTs), and provide information on the location of novel genes.

iii. Genes can also be identified directly from the DNA sequence. This is done by analyzing the base sequence for indications of gene structure. These include open reading frames (i.e. stretches of bases with no stop codons), the presence of splice sites indicating exon/intron boundaries and CpG islands (Section A8) that are found at the 5' end of many genes. In this way novel genes can be identified. Unknown genes that are members of known gene families can be identified by comparing DNA sequences with the sequence of a previously characterized member of the family. Genes showing significant homology can be identified. This process is known as **gene mining**.

Genome comparison

Comparison of genomes from different species has revealed both a high level of conservation of genes through evolution and a reasonable number of novel genes restricted to particular groups. Three genes are thought to have evolved from noncoding sequences in humans since the split from chimpanzees. Only a fraction of the genome was screened and extrapolation predicts 18 such genes in the whole human genome. As the species diverged approximately six million years ago this represents about three new genes per million years. When the first draft of the human genome was published the number of human genes was estimated to be about 40 000. This number has been reduced to about 20 000 as better sequence data accumulated and each predicted gene was closely examined. It turns out that most metazoans have about this number of genes. Plants may have about 30–50% more. This may be an error, or due to recent polyploidy or they may be more tolerant of mutations or be selected to carry out more complex chemistry and so need more enzymes than heterotrophs. OMIN (Online Mendelian Inheritance in Man, a database of mapped human genes) currently lists 13 126 genes in humans, of which only 2692 are known to have mutations causing disease. *Drosophila* stand out with only 13 600 genes. They tend to use splicing variation (Section A4) extensively to generate many versions of a protein from one gene. Table 1 shows the genome sizes and gene numbers currently estimated for a range of species.

The development of **massively parallel sequencing** (Section E3) has allowed the collection of huge amounts of sequence data, which in turn facilitates comparisons. The data can only be used by aligning them with the 'gold standard' sequence painstakingly generated by the initial publically funded human genome project.

A single base mutation can profoundly affect the health of an individual. The '1000 genome project' now has extensive sequence data, concentrating on regions containing protein coding genes from 1092 individual humans representing over 25 ethnic populations around the world. They have identified 15 million single nucleotide polymorphisms (SNPs), about one million small insertions and deletions, and more than 20 000 structural changes. This should include 95% of the variations in any individual anywhere in the world. The principal current use of this data is to carry out genome-wide association studies (GWASs) to see whether some of these variants are found more frequently in people with particular disease conditions (Section F2). With so many sequence variants available they should include most mutations that cause disease. It is hoped that some show strong correlations with particular diseases and thus immediately identifying the cause of the disease. Their preliminary findings show that an individual human typically carries between 250 and 300 mutations which inactivate genes. Presumably these are recessive so do not affect the phenotype significantly. They also sequenced two mother-father-child trios in detail and estimated mutation rates of 1×10^{-8} and 1.2×10^{-8} per base pair per generation.

Table 1. Genome size in millions of base pairs (mega base pairs, Mb), currently estimated gene number and approximate percentage shared with humans for representative species

Organism (species)	DNA content Mb	Number of genes (estimated)	% similarity to humans
Bacteria			
Bacterium *Mycoplasma genitalium*	0.58	483	
Bacterium *Escherichia coli*	4.64	4452	
Fungus			
Yeast *Saccharomyces cerevisiae*	12.07	5800	30
Plants			
Rice *Oryza sativa*	389	41 000	
Thale cress *Arabidopsis thaliana*	140	27 500	
Animals			
Nematode *Caenorhabditis elegans*	97	19 099	40
Fruit fly *Drosophila melanogaster*	165	13 600	50
Human *Homo sapiens*	2900	20 000	
Chimpanzee *Pan troglodytes*	3000	20 000	>98
Mouse *Mus musculus*	2500	30 000	80
Rat *Rattus norvegicus*	2750	22 000	80
Sea urchin *Strongylocentrotus purpuratus*	814	23 500	60

Environmental sequencing

Massively parallel sequencing also allows all the genomes in an environmental sample to be examined at once. Most bacteria cannot be cultured, but sequencing DNA from bacteria filtered from the Sargasso Sea produced 1.045×10^9 bp of DNA sequence estimated to come from about 1800 species, including 148 previously unknown bacterial phylotypes. Specific indicative sequences such as ribosomal DNA can be selectively sequenced to reduce the sequencing costs and produce data to show the presence of species only identified by their DNA. This allows microbial communities to be compared from a global range of habitats.

B2 Prokaryote genomes

Key Notes

Organization of prokaryotic DNA

Prokaryotes are composed of eubacteria (including *Escherichia coli*) and archaebacteria. In prokaryotes, the genome consists of a supercoiled circular DNA molecule called the bacterial chromosome. Prokaryotic cells have a dense central area called the nucleoid composed of a protein core from which loops of supercoiled DNA radiate. Some nucleoid proteins may help to package the DNA. Separation of replicated bacterial chromosomes for cell division may be achieved by separate attachment points on the cell membrane.

Prokaryotic genes

Almost all bacterial genes occur on the chromosome. A few exist on plasmids. The *E. coli* chromosome has been sequenced and the position of the genes located. The genes are arranged as operons, or as single copies, and account for about 90% of the DNA sequence. The remainder is noncoding, intergenic DNA which includes important sequences such as the origin of replication. Bacterial chromosomes vary in size reflecting differences in the number of genes and probably the size of the intergenic DNA.

Plasmids

Plasmids carry genes that may confer useful properties to bacteria. They replicate independently. Some integrate into the bacterial chromosome. Plasmids may be stringent (low copy number) or relaxed (high copy number). Many different plasmids exist. These include: resistance (R), fertility (F), col (colicin), and virulence plasmids. Bacteria may contain several types of plasmid. Plasmid incompatibility restricts the types of plasmid that can coexist in a bacterium.

Bacterial transposons

These are DNA sequence elements that use recombination to move about the genome. They encode transposase enzymes that catalyze their own movement. Insertion sequences are *E. coli* transposons that have a short inverted repeat sequence at either end. When an insertion sequence transposes, a host DNA sequence at the site of insertion is duplicated such that the insertion sequence is flanked by a direct repeat. Transposons cause insertional mutation of genes. When they move to another site a duplication of the original target sequence is left behind and the gene remains mutated.

Archaebacteria

These are organisms found in extreme environments; they differ from eubacteria. They are thought to represent a group of organisms that is distinct from both prokaryotes and eukaryotes.

Related topics

(A2) Genes
(A7) Regulation of gene expression

(C4) Recombination

Organization of prokaryotic DNA

Living organisms are divided into primitive forms such as bacteria known as **prokaryotes** and higher organisms called **eukaryotes**. The structure and organization of cells in prokaryotes and eukaryotes are different. Eukaryotic cells have a complex internal structure with membrane-bound organelles and a separate nucleus. Prokaryotic cells lack this organization and have no distinct nuclear compartment. The prokaryotes are divided into **eubacteria** or true bacteria including *E. coli*, and an unusual group distinct from eubacteria called **archaebacteria**. Most of the information available refers to *E. coli* and other eubacteria. Prokaryotes have a single circular DNA molecule referred to as the **bacterial chromosome** that contains almost all of the genes. The DNA molecule is extremely long relative to the dimensions of the cell. To allow it to fit inside the cell, it is compacted by a process called **supercoiling**. Enzymes called **topoisomerases** introduce additional turns into the double helix that cause the DNA strand to wind up on itself and adopt a more compact form (Figure 1). The topoisomerases act by breaking the DNA polynucleotide and rotating the two ends relative to each other. The enzyme then rejoins the ends and the polynucleotide reacts by winding up on itself. This is called positive supercoiling. Topoisomerases can also remove coiling in a process called negative supercoiling by creating a turn in the opposite direction. Supercoiling also occurs in eukaryotic cells where it is involved in packaging of DNA in eukaryotic chromosomes.

Although prokaryotes lack a distinct nucleus, analysis of bacterial cells by electron microscopy shows a darker central area containing DNA and protein known as the **nucleoid** and an outer area called the cytoplasm. The exact structure of the nucleoid is uncertain but it is known to have a central protein core from which supercoiled loops of DNA radiate (Figure 2). Some of the proteins isolated from the nucleoid resemble the histone proteins found in eukaryotic chromosomes and it is thought that the proteins of the nucleoid may help to organize the folding of the DNA into its compact structure. The length of prokaryotic chromosomes relative to the dimensions of the cell means that replication and partitioning of DNA molecules during cell division is a potentially difficult task. This may be achieved by the DNA molecules having separate attachment points on the cell membrane which move away from each other as the cell divides.

Prokaryotic genes

Almost all the genes present in bacteria occur on the bacterial chromosome. A few other genes exist on small circular DNA molecules called **plasmids** that are present in bacteria in addition to the chromosomal DNA. The organization of the genes on the bacterial chromosome is best characterized for *E. coli* which has about 4300 genes carried by a chromosomal DNA molecule of 4.6 million base pairs. The entire sequence of

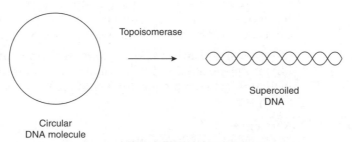

Topoisomerase

Supercoiled
DNA

Circular
DNA molecule

Figure 1. Supercoiling of circular DNA molecule.

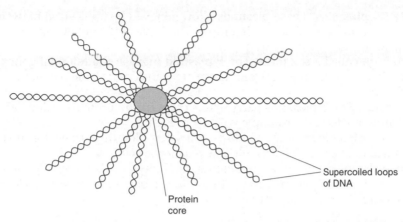

Figure 2. Structure of *E. coli* nucleoid.

the *E. coli* chromosome and the relative positions on the chromosome of the genes have now been determined. Some genes are arranged as families called **operons** that encode proteins with related functions and are regulated in a coordinated way. Other genes occur at random positions along the chromosome with no apparent organization. Most of the genes are present as single copies. The main exception is the ribosomal RNA genes that occur as a cluster of seven copies. A similar pattern of organization is seen in other eubacteria and gene maps are most similar in related species. About 90% of the DNA of the bacterial chromosome is accounted for by the genes. The remaining 10% is **intergenic DNA** which separates individual genes. Some parts of the intergenic DNA have important functions such as the location of the **origin of replication** of the bacterial chromosome. Other intergenic regions may be involved in interactions with DNA packaging proteins. The size of prokaryotic genomes varies between species of bacteria. The size variations reflect differences in the number of genes and possibly larger intergenic regions.

Plasmids

These are circular DNA molecules present in addition to the bacterial chromosome in almost all bacteria. Plasmids carry genes not found on the bacterial chromosome which often confer useful properties, such as resistance to antibiotics, to the bacterium. Plasmids have their own origin of replication and so can replicate independently of the chromosomal DNA. Some small plasmids use the cell's enzymes to replicate. Larger plasmids may carry genes that encode their own replicative enzymes. Other plasmids integrate into the host genome and are copied at the same time as the bacterial chromosome. This integrated form of plasmid is called an **episome** and may pass through many cell divisions before excising itself to exist as a separate plasmid again.

Many different plasmids are found in bacteria and individual species may contain several types. Plasmids can be classified according to the genes they carry and the characteristics they confer on the host cells. Five types have been identified:

- **Resistance (R) plasmids**. These carry genes that make bacteria resistant to antibiotics such as ampicillin and chloramphenicol. The way in which resistance is conferred varies. An example is the RP4 plasmid found in *Pseudomonas* and other bacteria. R plasmids have important consequences for the treatment of bacterial infections as they represent a way in which antibiotic resistance can spread between species of bacteria.

- **Fertility (F) plasmids**. These plasmids allow genes to be transferred between bacterial cells in a process called **conjugation**. The F plasmid contains genes that direct the transfer of the F plasmid from one bacterial cell to another by means of a tube-like structure called a **sex pilus**. The F plasmid may carry additional genes which it acquires from the chromosome and these are transferred to the recipient cell during conjugation.
- **Col plasmids**. These plasmids carry genes that encode proteins called **colicins** that can kill other bacteria. An example is ColE1 of *E. coli.*
- **Degradative plasmids**. These encode proteins that allow the host bacterium to metabolize unusual molecules such as toluene or salicylic acid.
- **Virulence plasmids**. These plasmids confer the ability to the bacterium to cause disease. An example is the **Ti plasmid** that is found in the bacterium *Agrobacterium tumefaciens* that causes crown gall disease in plants.

Plasmids vary in size with the smallest around 1 kb in length and the largest up to 250 kb. Individual plasmids vary according to the host cells in which they occur. Some are present in many different species of bacteria and others in just a few species. The number of plasmid molecules in a bacterial cell also varies. Plasmids that are present as just one or two copies are said to have a low copy number and are called **stringent** plasmids. Other plasmids have a high copy number with 10 or more plasmid molecules present and are known as **relaxed** plasmids. There are also restrictions on the types of plasmid that can coexist in bacterial cells. This feature is called **plasmid incompatibility**. Plasmids that occur in the same species of bacteria must belong to different incompatibility groups.

Bacterial transposons

These are DNA sequence elements that are capable of moving around in the genome. They occur in both eukaryotes and prokaryotes where they are present on bacterial chromosomes and on plasmids. The movement process is called **transposition** and depends on recombination between DNA sequences. Transposons are autonomous units and each encodes an enzyme called **transposase** that catalyzes its own transposition. Many different transposons are known. The first to be identified were the **insertion sequences** which occur in *E. coli* (Figure 3a). Several types of insertion sequence have been identified and as many as 10 copies of each may be present in bacterial genomes. Transposition is a relatively infrequent process occurring only every 10^3–10^4 cell divisions. Insertion sequences can transfer between bacteria during conjugation and can also transfer between related species. A characteristic feature of insertion sequences is that they have a **short inverted repeat** at either end. These are duplicated sequences in which the two copies point in opposite directions. This means that the same sequence is encountered moving from the flanking sequence at either end towards the insertion sequence. In addition, when an insertion sequence transposes, a host DNA sequence at the site of insertion is duplicated such that the insertion sequence is always flanked by a short duplication of the target sequence, known as a **direct repeat** (direct means that the two copies are in the same orientation). The ability of transposons to move around in the genome means that they can mutate a gene when they transpose into it. When they move to another site the short duplication of the original target sequence is left behind so that the gene remains mutated even after the transposon has left (Figure 3b).

Archaebacteria

These organisms are distinct from eubacteria. They are now recognized to be as different from other prokaryotes as they are from eukaryotes. The archaebacteria comprise

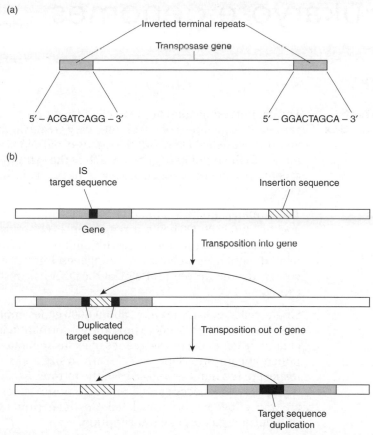

Figure 3. Insertion sequences. (a) Structure of an insertion sequence (IS) showing a possible inverted repeat sequence. (b) Gene mutation by transposition of insertion sequences.

three groups of related organisms: **methanogens, extreme thermophiles**, and **extreme halophiles**. Most archaebacteria live in extreme environments making them difficult to culture and so to study. Biochemically they differ from eubacteria with respect to the structure and composition of their cell wall. Genetically, several distinctions have been identified that are related to the ribosomal RNA genes including the presence of introns and a different structure and organization of the genes.

B3 Eukaryote genomes

Key Notes

DNA content, C-value paradox

The genome is the total of the nuclear DNA in a gamete. The genomes of eukaryotes vary greatly in the amount of nuclear DNA, but the quantity of DNA is not related to the number of genes. Much of the extra DNA is repeats of sequences which appear to be parasitic 'junk' DNA.

The human genome

The human genome is 3 billion base pairs long and contains approximately 20 000 genes arranged on 23 chromosomes. Less than 1.5% of the DNA codes for amino acids. Gene-related sequences including pseudogenes, introns, and control regions account for 25% of the DNA. The rest is extragenic DNA.

Genes

Single-celled eukaryotes have about 6000 genes; multicellular ones have 13 000 to 26 000. The coding information in eukaryote genes occurs as a series of exons separated by noncoding introns. Genes vary greatly in size and also with respect to the number and sizes of the introns. Leader and trailer sequences occur at the 5′ and 3′ ends of genes; these are transcribed but not translated. Upstream promoter sequences regulate gene transcription.

Gene families

Many genes occur as families containing multiple copies of genes with identical or related sequences. The genes in a family may be present at single or multiple loci. Gene families may also occur as individual clusters at multiple loci.

Pseudogenes

These are diverged members of gene families that have acquired one or more inactivating mutations. Processed pseudogenes are nontranscribed DNA copies of mRNAs, probably derived by a mechanism involving reverse transcription. Gene fragments are inactive genes that lack part of the parent gene. They are thought to have arisen by deletion or recombination of the original gene sequence.

Extragenic DNA

This is composed of sequences that are not genes, gene-related sequences, or pseudogenes and accounts for about 75% of the human genome. Most extragenic sequences (70–80%) are unique or exist as a small number of copies. The rest (20–30%) are moderately or highly repeated sequences present as tandem arrays or dispersed throughout the genome. Extragenic DNA has no known function.

Dispersed repetitive sequences	These consist of SINEs and LINEs (short and long interspersed nuclear elements, respectively). SINEs include human Alu sequences. These are a family of sequences about 250 bp long present as about 1 million highly dispersed copies. They are thought to be derived from processed pseudogenes that acquired the ability to move about the genome. LINEs are longer than SINEs. The human L1 LINE is 6500 bp and exists as 60 000 copies. LINEs are retroelements and have the ability to copy themselves using reverse transcriptase and to move about the genome.
Clustered repetitive sequences	Larger eukaryote genomes have extensive regions containing long tandem arrays of repetitive sequences. These are called satellite DNA. Short units are classified as micro (less than seven bases per repeat), or mini (seven to about 25 bases); the rest are just called satellite DNA and individual units may be thousands of base pairs long, usually with smaller repeats inside them. CA dinucleotide repeats and mononucleotide repeats account for 0.8% of the entire human genome.
Variable number tandem repeats (VNTRs)	VNTRs are repetitive sequences that vary according to the number of times the repeated sequence is present. Variation occurs at a given locus between individuals. Polymerase chain reaction (PCR) can be used to detect the variations. VNTRs are used in forensic science to identify individuals at the scene of a crime and in medical genetics to identify carriers of genetic diseases.
Related topics	(A2) Genes (F1) Genetic diseases (B1) Concepts of genomics (G1) Genetics in forensic science

DNA content, C-value paradox

The haploid DNA content of a eukaryote is its C value. This is low for single-celled organisms (*Saccharomyces cerevisiae*, a yeast, 12 156 590 bp) and higher for multicellular species. The range of DNA contents does not reflect the number of genes nor the complexity of the organism (Section B1). The variation is almost entirely due to noncoding sequences, much of it repetitive DNA (see below) and it serves no known useful function. It has been shown that, in plants, DNA content limits the minimum time required for a cell cycle, and that fast-growing annual weeds all have low DNA contents, as do many deciduous trees which replace their leaves each year. The best current explanation is that the excess DNA is parasitic, originating as transposable elements, related to retroviruses. The upper limit to DNA content occurs when the costs of replicating it cause a significant reduction in fitness. Species with long, slow life cycles, limited by factors other than the availability of phosphorus, nitrogen, and energy, can bear a larger burden of parasitic (junk) DNA.

The human genome

This term is used to describe the different types of sequence that together make up the DNA in a human cell. The DNA in the human genome is about **3 billion base pairs** long and is estimated to contain **20 000 to 21 000** genes. The DNA is arranged as a set of 23 chromosomes each of which is a single, double-stranded DNA molecule 55–250 million base pairs long. The genes and gene-related sequences account for about 25% of the DNA (Figure 1). The remainder is called extragenic DNA and has no known function. Mutations of a single base pair which show variation in the population are called **single nucleotide polymorphisms**. There are about 15 million, one every 200 bp on average, and they are very important for genetic mapping (Section B1).

Genes

The biggest surprise from genome sequencing has been the small number of genes. Initial guesses of 50 000 to 100 000 genes were clearly too big, because of the mutational load. There would have been around two new mutations per zygote, to be removed by selection. There are about 20 000 protein-coding genes identified in humans, and the number is now stabilizing. The coding information in a eukaryote gene is present as a series of segments of DNA sequence called **exons**, separated from each other by intervening noncoding sequences called **introns**. In genome searches, genes are detected as an 'open reading frame' (ORF): a region putatively encoding over 100 amino acids, between a start and stop codon, including introns identified by splicing sequences (Section A4). Genes vary greatly in size and also with respect to the number and sizes of the introns. Some genes such as the histone H4 gene are just a few hundred base pairs long. Others, such as the Factor VIII gene, are several hundred kilobase pairs (kbp) in length and contain many large introns such that the actual coding sequence accounts for just a few percent of the total gene sequence. Additional sequences are present which are associated with genes. **Leader** and **trailer sequences** occur at the 5′ and 3′ ends of the gene which are transcribed but not translated. **Promoter sequences** occur upstream of the point where transcription begins and regulate synthesis of mRNA from the gene. The promoter may extend up to about 1 kbp upstream but other regulatory sequences, called enhancers, that influence transcription may occur at sites much further away. Many genes are built from a relatively small set of about 1200 functional domains (e.g. ATP-binding site, transmembrane domain).

Gene families

Some genes exist as a number of copies with identical or related sequences that can be grouped into families. Some families are as old as eukaryotes. Gene families may be organized physically in a number of ways (examples are human):

i.   all of the genes in the family occur at the same chromosomal locus – an example of this is the growth hormone gene family whose five members are clustered on chromosome 17;
ii. the genes belonging to the family may occur at different loci – for example, the five members of the aldolase gene family are on different chromosomes;
iii. the genes of a family may exist as a series of clusters on different chromosomes – an example are the homeobox genes which occur as four clusters on separate chromosomes each containing about 10 individual genes.

In some multigene families all the genes are identical and may encode a protein required in large amounts by the cell such as histones. In other families, the genes are not identical but show some sequence divergence. In some cases the divergence is so great that

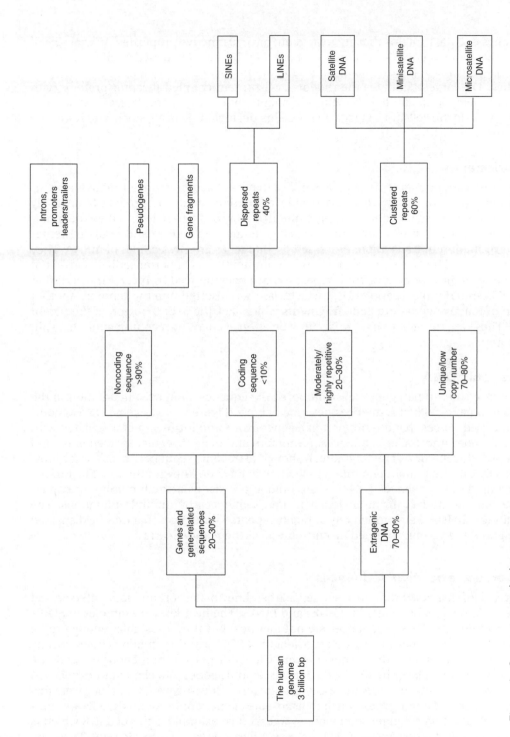

Figure 1. Sequences in the human genome.

the genes encode proteins that are related but have distinctive properties. An example of this is the α and β globin gene families whose members are expressed at different stages of embryonic development and in adults. In humans, 6177 genes belong to the 40 largest families. The largest family is the rhodopsin G-protein-coupled receptor proteins with 737 members. Some are light-sensitive pigments, but many more detect specific molecules outside the cell and trigger a response inside, including taste, smell, and hormone responses.

Pseudogenes

These are diverged members of gene families that have acquired one or more inactivating mutations so that they are no longer able to function and do not produce biologically active protein. A pseudogene is simply a mutated copy of a parent gene. Often the mutations present are nonsense mutations which generate stop codons and result in premature termination of translation. Processed pseudogenes are DNA copies of mRNA. They lack introns and because they do not have a promoter they are not transcribed and do not result in protein production. They appear to have been derived by reverse transcription of mRNA into double-stranded DNA which was then inserted into the genome. Another group of inactive genes are **gene fragments** which lack the 5′ or 3′ region of the parent gene. These are thought to have arisen by a deletion event or by recombination that split the parent gene.

Extragenic DNA

This part of the human genome is composed of sequences that exist in addition to the genes and gene-related sequences described above (Figure 1). Extragenic DNA is composed of sequences that are not part of a gene (exons and introns), not associated with a gene (leader and trailer sequences, promoters and distant regulatory elements) and not a pseudogene or a gene fragment. Although extragenic sequences account for most of the DNA in the human genome (70–80%), they have no known function. Most of the extragenic DNA sequences (70–80%) are unique or exist as a small number of copies. These may be derived from ancient repeated sequences by multiple mutations. The remainder (20–30%) are moderately or highly repeated sequences that may be dispersed throughout the genome or lined up end-on-end as long tandem arrays.

Dispersed repetitive sequences

Two types of dispersed repetitive sequence exist, known as short and long interspersed nuclear elements, abbreviated to **SINEs** and **LINEs**. The best-known examples of SINEs are **Alu elements**. These sequences are not identical but they are similar enough to be classed as a family. They have an average length of 250 bp and 1.3 million copies occupy 10.5% of the human genome. They are very widely dispersed throughout the genome occurring in most places including the introns of some genes. Alu elements are believed to have arisen as one or more processed pseudogenes. It is suggested that they were fortuitously inserted by enzymes from a transposable element or retrovirus. LINEs are similar to SINEs but have longer sequences. A well-known example is the **L1 LINE** which is 6500 bp long and is present as 60 000 copies. LINEs are a type of **retroelement**. These are sequences that are capable of moving through the genome by a process called **transposition** which allows them to copy themselves by reverse transcription and to insert the copy into the genome at a distant site (Figure 2). Most L1 elements are truncated, but the full-length version has two genes producing the reverse transcriptase and an enzyme for integration of the product into the chromosome.

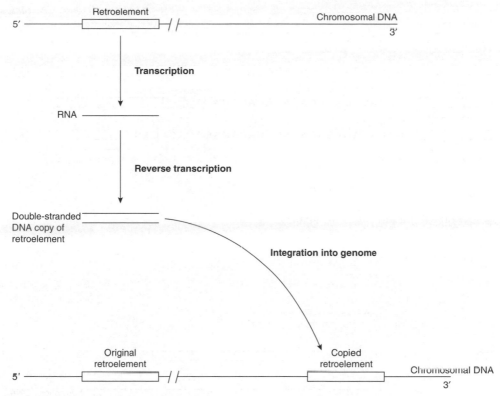

Figure 2. Transposition of retroelements.

Clustered repetitive sequences

Tandem repeated sequences are called **satellite DNA sequences** because the identical repeats produce a peak of a specific density after centrifugation of sheared DNA. Large eukaryote genomes contain extensive regions in which repetitive sequences are arranged end-on-end as long tandem arrays. Satellite DNA (Figure 1) occurs in various blocks each with a different repeat unit. The repeated units may be from one base up to thousands of bases long, but only simple repeats in small blocks can be studied easily. Although some satellite DNA is scattered around the genome, most is located around the chromosome centromeres where it may have a structural role. In simple sequence repeats (SSRs) the repeat units are small. **Minisatellite** DNA occurs in smaller clusters up to 20 kbp in length and has repeat units up to 25 bp; **microsatellite** DNA occurs in short clusters, usually less than 150 bp, and has a repeat unit which is usually 4 bp or less. Microsatellite DNA is very common. One type, which contains the dinucleotide CA as the repeat sequence, accounts for 0.5% of the entire genome. Mononucleotide repeats, consisting of a single repeated base, account for a further 0.3%.

Variable number tandem repeats

The length of microsatellite DNA sequences such as the CA repeats varies from person to person. At any given location in the genome, the number of repeat units present in an individual microsatellite DNA sequence may vary by as many as 10 or more. As such, these sequences are known as variable number tandem repeats or **VNTRs**. The polymerase chain reaction (PCR) (Section E2) can be used to analyse VNTRs (Figure 3).

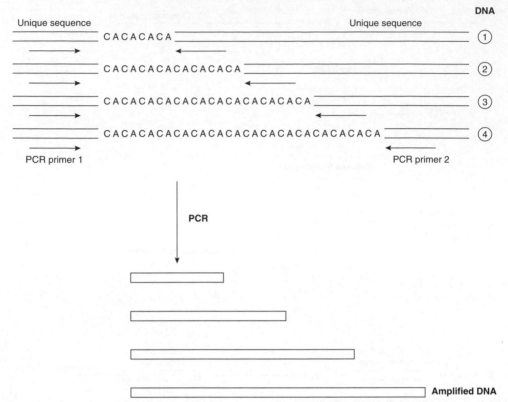

Figure 3. VNTRs. CA is repeated 4–16 times. PCR using primers specific for flanking unique sequence produces amplified DNA of varying lengths.

Using primers specific for unique sequences on either side of the repeat sequence, DNA molecules of varying lengths are amplified depending on how many repeats are present. By examining several VNTRs it is possible to build up a genetic profile that is unique to the individual being tested.

VNTRs have a number of very useful applications based on establishing the identity of an individual. They can be used in forensic science to link suspects to a crime using biological material recovered from the scene (Section G1). They also have a number of uses in medicine including matching patients undergoing transplants with donor organs, identifying carriers of genetic diseases, and establishing paternity. VNTRs have also been widely used in the human genome mapping project as gene markers to identify the location and order of genes on chromosomes (Section B1).

B4 Chromosomes

Key Notes

Prokaryote and eukaryote chromosomes

Prokaryote chromosomes consist of a single DNA molecule that is usually circular, with only a small amount of associated protein. Each chromosome has a single origin of DNA replication. Eukaryotes have several linear chromosomes and the DNA is tightly associated with large amounts of protein. Each eukaryote chromosome has multiple origins of DNA replication.

Chromosome morphology

Eukaryote chromosomes are visible by light microscopy only during cell division, after they have replicated. Chromosomes are placed into morphological groups, metacentric, submetacentric, acrocentric, and telocentric, according to the position of their centromere. Within a species each autosome is given a specific number in ascending order from the largest to the smallest. Chromosome banding aids chromosome identification and gives some information as to the underlying organization of the chromosome. G-banding gives a series of light and dark bands along the length of the chromosome. C-banding produces dark bands in regions of constitutive heterochromatin.

Specialized chromosome structures

Centromeres are points on chromosomes to which the spindle fibers are attached. This is mediated through specialized protein structures known as kinetochores. Centromeres are composed largely of highly repeated satellite DNA sequences. Specialized structures at chromosome ends are known as telomeres. These are also composed of short repeated DNA sequences. The number of repeats decreases with age in somatic cells, but is maintained in germ cells and tumor cells by the enzyme telomerase. Telomeres prevent recombination between the ends of chromosomes.

Nucleolar organizer regions (NORs) contain tandem repeats of the major ribosomal RNA genes and are located in secondary constrictions. When the region of the chromosome distal to the NOR is small it is referred to as a chromosomal satellite.

Molecular structure of chromosomes

Chromatin is the term given to the association of DNA and proteins that composes chromosomes. It contains basic proteins, histones, and nonhistone acidic proteins. The histones form nucleosomes around which the DNA is wound. Nucleosomes consist of two discs containing histones H2a, H2b, H3, and H4; 146 bp of DNA are associated

	with each nucleosome and linker DNA leads to the next nucleosome. A single molecule of histone H1 attaches outside the core. This molecule is responsible for further folding of the nucleosomes into solenoids and more complex structures. Acidic proteins are involved in the chromosomal scaffold and in gene regulation.
Functional and nonfunctional chromatin	Heterochromatin is inactive chromatin, whereas euchromatin is actively involved in RNA transcription. Heterochromatin appears denser than euchromatin under the electron microscope and stains darker under the light microscope. Some chromatin can exist as either hetero- or euchromatin. This is called facultative heterochromatin. One of the two X chromosomes in cells of female mammals is converted to heterochromatin. It forms a small dark body attached to the nuclear membrane. Chromatin that is permanently heterochromatic is called constitutive heterochromatin, and can be identified by C-banding.
Alteration to chromosome numbers	Polyploidy is where the altered chromosome number is a multiple of the haploid chromosome number. This is rare in animals but important in plants. Small changes in chromosome number are classed as aneuploidy. In humans up to 4% of conceptuses are aneuploid, but very few of these survive to birth. Those that do survive tend to involve smaller chromosomes or alterations of the sex chromosomes. Aneuploidy arises by nondisjunction of homologous chromosomes or chromatids. It is more common in older mothers. Translocations can cause inheritance of trisomy 21 within families. Loss of a chromosome has more severe effects than gain of an extra chromosome. Some aneuploidies are often found as mosaics.
Related topics	(A8) Epigenetics and chromatin modification (B3) Eukaryote genomes (B5) Cell division (C3) Meiosis and gametogenesis (D8) Polyploidy

Prokaryote and eukaryote chromosomes

All cellular life-forms have structures carrying genes, encoded in DNA, that are referred to as chromosomes. There are, however, major differences between these structures in prokaryotes and even the most simple of the eukaryotes. In prokaryotes the chromosome consists of a single DNA double helix that is usually circular and has relatively few proteins associated with it. DNA replication proceeds from a single origin. Eukaryotes have, in almost all cases, a number of different chromosomes that are linear and are contained within a membrane-bound organelle; the nucleus. The DNA molecules are intimately associated with large amounts of specific proteins. These may have functional or structural roles. The amount of DNA per chromosome is much greater in eukaryotes and because of this there are multiple origins of replication on each chromosome.

Chromosome morphology

Eukaryote chromosomes are usually only visible when a cell is in the process of dividing (Sections B5 and C3), after the chromosome has been replicated into identical double structures known as **chromatids** (daughter chromosomes). Chromosomes are classified on the basis of their morphology. This is determined by the position of the centromere (primary constriction). Figure 1 shows four typical morphologies for chromosomes. In **metacentric** (mediocentric) chromosomes the centromere is close to the midpoint of the chromosome. This divides the chromosome into two roughly equal halves (arms). Where the centromere is sufficiently far away from the midpoint for a long arm (q arm) and a short arm (p arm) to be distinguished the chromosome is referred to as **submetacentric**. The other two morphological classes relate to chromosomes in which the centromere is close to one end of the chromosome. In **telocentric** chromosomes the centromere is at the end of the chromosome and there is only one arm. If the centromere is so close to the end of the chromosome that the short arm is only just discernible then the chromosome is termed **acrocentric**. Students often confuse submetacentric and acrocentric morphologies. The differences between the two are easily recognized in the chromosomes of humans (Figure 2).

In any species the complete diploid set of chromosomes is referred to as the **karyotype**. The autosomes are numbered in order of decreasing size, and the sex chromosomes are referred to as X or Y (Section C9). A chromosome preparation can be photographed, the homologous chromosomes paired, and set out in order. This is known as an **ideogram**, and is the conventional way to display karyotypes.

By grouping chromosomes by relative size and morphology it is usually possible to individually identify each chromosome in a species. This process was made much simpler by the development of treatments which, when applied to chromosomes prior to staining, produce a pattern of dark and light bands unique to each chromosome. Although there are several different banding techniques available, only the two major processes, G-banding and C-banding, will be described here.

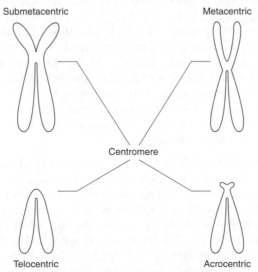

Figure 1. Chromosome morphology. The morphological class is based on the position of the centromere.

G-banding patterns can be produced by a wide range of different treatments but the most commonly used is a mild pretreatment of chromosome preparation slides with protease enzymes such as trypsin. The treated slides are then stained with Giemsa (hence the term G-banding). The process causes chromosomes to stain as a series of dark G-bands and pale interbands. The pattern is unique to each homologous chromosome pair and greatly aids the process of chromosome identification. These patterns have been used to create cytological maps of each chromosome in many species so that subchromosomal regions can be accurately identified. This is important in various processes such as gene mapping and medical genetics (Section F1).

Figure 2 sets out the G-band pattern for each of the human chromosomes diagrammatically. The pattern of bands allows each individual chromosome to be divided into a series of regions and subregions. Using chromosome 1 as an example the first division is between the long (q), and the short (p) arm. Figure 2 shows that the q arm is further divided into regions 1, 2, 3, and 4. There are two further levels of subdivision so that a specific region may be accurately defined as 1q42.1. Note that not all chromosomes or chromosome regions are equally subdivided. These regions are arbitrary, and are decided by international conventions of cytogeneticists. As techniques have allowed the banding of longer, less-condensed chromosomes so the number of subregions has increased. G-banding not only provides us with a convenient method for identifying chromosomes, but it also gives us information as to the overall organization of DNA and genes within eukaryote chromosomes. In general the darkly stained G-bands are rich in the bases adenine and thymine, whereas the pale-staining interbands are richer in guanine and cytosine. More genes are located in the interbands than in the G-bands. C-banding also gives us an insight into the organization of chromosomes. C-banding produces a number of dark bands. These are largely confined to areas around centromeres. These indicate regions of **constitutive heterochromatin** and are discussed in greater detail later in this section.

Specialized chromosomal structures

All eukaryote chromosomes contain two different areas which have specific structural importance. These are the **centromeres** and **telomeres**. In addition some chromosomes contain **nucleolar organizer regions** (NORs). Centromeres are the sites at which the spindle attaches during cell division and functional centromeres are essential to this process. Any chromosome fragment which loses its connection to a centromere will not segregate to daughter cells at the end of cell division. The best studied centromeres are those of yeast where some are as short as 200 bp. Most centromeres are much larger than this. Normally the centromere consists of highly repeated satellite DNA (Section B3). In humans different chromosomes can be distinguished by the presence of specific alphoid satellite DNAs within their centromeres. Connection of the chromosome to the microtubular spindle fibers is effected by proteins that attach to the centromere forming a multi-layered structure known as a **kinetochore**.

Telomeres are not simply the ends of chromosomes and DNA molecules, but are specialized structures. They contain multiple repeats of simple, short DNA sequences. In humans the repeat sequence is TTAGGG, but there is little variation between eukaryotes: similar sequences are found in plant and protist species. Specific proteins bind to the telomere region and the resulting nucleoprotein structures are thought to prevent recombination between the ends of different chromosomes. The number of repeats per telomere is high in germ cells but decreases with age in somatic tissues; this is a molecular marker of the aging process. Telomere length is maintained by the enzyme **telomerase**, a protein that contains RNA complementary to the telomere repeat DNA sequence, which acts as a template for extension of the telomere. Telomerase is absent from somatic cells

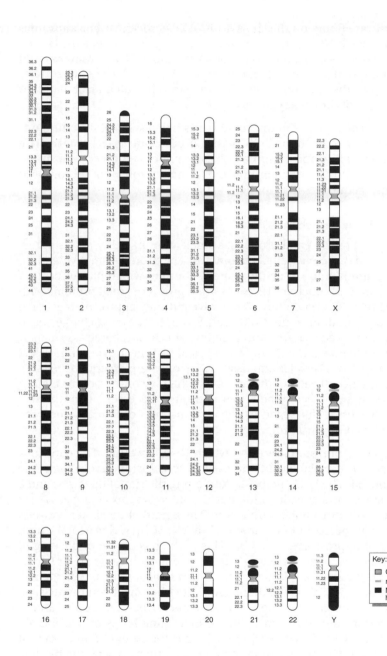

Figure 2. The human karyotype. This shows one member of each pair of homologous chromosomes, drawn to show their individual pattern of G-bands (G-banding is described later in this section). The chromosomes are arranged in order starting with the longest, chromosome 1. The X chromosome is placed in order of its length between autosomes numbers 7 and 8. The Y chromosome is placed at the end. Each chromosome arm is subdivided into regions. Only chromosome 2 and chromosome 19 appear to be metacentric. Chromosomes 13, 14, 15, 21, 22, and Y are acrocentric. The remainder are submetacentric, showing a variety of different arm ratios. Chromosomes 13, 14, 15, 21, and 22 are drawn to show chromosomal satellites distal to the region of ribosomal DNA. Reproduced from Strachan and Read (2010) *Human Molecular Genetics*, 4th Edn. Garland Science, Abingdon.

Key:

▨ Centromere

= rDNA

■ Noncentromeric heterochromatin

but reappears in tumor cells, where telomere length is stabilized (Section F3). NORs are usually found at secondary constrictions. They consist of tandemly repeated 5.8S, 18S, and 28S rRNA genes. In most species the 5S rRNA genes are clustered elsewhere in the genome. In humans NORs are found on the short arms of all acrocentric chromosomes except the Y chromosome. Each NOR consists of approximately 80–100 repeats. During interphase the NOR decondenses and a nucleolus forms around it; NORs from different chromosomes can be incorporated into a single nucleolus. When the cell enters metaphase of mitosis (Section B5) the chromosomes may appear to be still attached at their short arms. This is known as **satellite association**.

The secondary constriction can be so pronounced that the small distal region of the chromosome appears to be unconnected to the body of the chromosome; this has given rise to the term **chromosomal satellite** and in humans these can be seen on chromosomes 13, 14, 15, 21, and 22 (Figure 2). The term chromosomal satellite must not be confused with the term satellite DNA sequences (Section B3).

Molecular structure of chromosomes

Chromosomes are composed of DNA and proteins; a small amount of RNA is also present. The mixture of DNA and protein is called **chromatin**. The proteins are divided into two classes, **histones** and **nonhistone** or **acidic** proteins, both of which play important roles in chromatin structure and function. Histones are a group of small proteins with molecular masses of less than 23 kDa. In terms of dry weight they approximately equal DNA in the composition of chromatin. At physiological pH they have a basic charge due to the high frequency of the amino acids lysine and arginine. This basic charge assists their intimate interactions with the polyanion DNA. Five types of histone are found, H1 H2a, H2b, H3, and H4. This is true for all species and tissues with only rare exceptions relating to H1. The amino acid sequence of each of the histones is highly conserved throughout evolution, suggesting that these molecules have an important role, essential for the survival of eukaryotes. This has been elucidated by the identification of **nucleosomes**, the basic building blocks of chromatin structure.

Nucleosomes consist of a core of histones around which DNA is wound. The core consists of two discs arranged in parallel each composed of four histone molecules, one each of H2a, H2b, H3, and H4. The DNA molecule runs along the rim of the discs, and a molecule of histone H1 sits on the outside of the nucleosome complex acting as a seal; 146 bp of DNA are associated with a nucleosome core. The length of the linker between nucleosomes varies between species but in humans it is about 60 bp giving a total length of DNA per nucleosome of approximately 200 bp. This is the basic level of packing of DNA in chromatin. Further packing depends to a great extent on histone H1. H1 molecules can interact to hold the individual nucleosomes in a helical structure giving rise to a **solenoid** of 30 nm diameter. This is the diameter of the fiber most commonly seen in electron micrographs of chromatin, but more densely coiled structures are also found. Simple nucleosome structures are shown in Figure 3. Increasing levels of packing are observed within the nucleus. The highest level of packing is found in chromosomes at the metaphase of cell division. The organization of these structures involves the binding of chromatin fibers on to a chromosomal scaffold. This is made up largely of the acidic (nonhistone) nuclear protein, **topoisomerase II**. Specific regions of the DNA which run for several hundred base pairs and are rich in the bases adenine and thymine, known as **scaffold attachment regions** (SARs), link the DNA molecule to the chromosomal scaffold. The intervening material is arranged as loops of different lengths. The DNA in these is shown when histones are removed from preparations of metaphase chromosomes; electron micrographs show long lengths of DNA spooling out from the chromosomal scaffold.

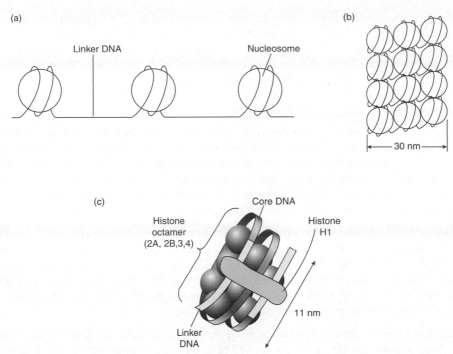

Figure 3. Nucleosomes. (a) Diagrammatic representation of a series of nucleosomes. (b) Nucleosomes coiled together to form a solenoid. (c) Basic structure of a nucleosome showing histones and DNA.

Nonhistone nuclear proteins are involved in a number of aspects of chromatin function, including the regulation of gene expression, where transcription factors are of major importance. These are described in more detail in Section A7.

Functional and nonfunctional chromatin

Not all areas of chromatin are equally involved in gene transcription. Some chromatin is effectively inert. This is known as **heterochromatin** as compared with active chromatin or **euchromatin**. As seen by electron microscopy heterochromatin has a denser pattern of chromatin fibrils. It can also be differentiated by staining at the light microscopy level. Some chromatin is heterochromatic in all tissues and at all stages of development; this is **constitutive heterochromatin**, and is detectable by C-banding. As noted above centromeres are often detected by C-banding. There is a strong association between protein and satellite DNA sequences in these regions that prevents loss of DNA during the C-banding procedure; these regions therefore stain strongly. Other regions of constitutive heterochromatin include the long arm of the human Y chromosome and regions of sex chromosomes in other animals. Interstitial C-bands are more frequently encountered in plants, where they can be used to identify chromosomes because plant chromosomes do not display G-bands.

Some regions of chromatin can exist in either the heterochromatic or euchromatic state; these are known as **facultative heterochromatin**. Female mammals have two X chromosomes, one of which is largely inactive as far as transcription is concerned. It is converted into heterochromatin and can be observed as a small dense spot on the side of the interphase nucleus, known as a **Barr body** or **X chromatin**. In this way there is a dosage compensation (Section A8) between males and females because in the male there is only one X chromosome and the Y chromosome is composed largely of constitutive hetero-chromatin. Which X chromosome is inactivated is a random event; in approximately half of the cells of a female mammal the paternally inherited X chromosome is inactivated and in the other half of the cells the maternally inherited X is inactivated; the inactivated X chromosome is reactivated during gametogenesis.

Alteration to chromosome numbers

Chromosome numbers may be altered by errors that take place in meiosis, the division process associated with gametogenesis (Section C3). These result in the formation of gametes with gains or losses of genetic material. If the gametes are viable they will give rise to progeny in which each cell has the same, abnormal, chromosome complement. Errors that occur during division of somatic cells, **mitosis** (Section B5), result in an individual with more than one karyotype. Such an individual is termed a **mosaic**.

Polyploidy occurs when the chromosome number is increased or decreased by a multiple of the haploid chromosome number. The euploid series describes different ploidy levels. An example using humans is given in Table 1. Although triploids and tetraploids make up approximately 10% of spontaneously aborted human fetuses, they are not observed in live-born infants. Polyploidy is rare in the animal kingdom, but frequently observed in plants, where it is important in evolution (Section D8).

Aneuploidy refers to small deviations from the chromosome numbers of the euploid series. **Nullisomy** is the absence of any copy of a specific chromosome in a cell. One copy of a chromosome per cell is referred to as **monosomy**, two as **disomy** and three as **trisomy**. Monosomy is the norm in gametes and disomy in somatic tissue. Table 2 contains examples of aneuploid syndromes found in humans.

Aneuploids for other chromosomes are also found, but usually only in fetuses that undergo spontaneous abortion. In total, 4% of human conceptuses are thought to be trisomic. In those aneuploid syndromes that survive to term there is a marked reduction in life expectancy. Individuals who have only part of a chromosome present three times express the associated syndrome less severely. In this context it is relevant to note that the autosomal trisomies listed all involve small chromosomes. The presence of an extra copy of a large chromosome would probably affect the expression of too many genes to allow development of the fetus.

Sex chromosome aneuploids are relatively common, despite the fact that the X is a large chromosome known to carry many genes. How can this be explained? As noted above only one X chromosome is active in any mammalian cell so the presence of extra copies has less effect than would otherwise be predicted. Turner's syndrome (XO) is rarer than other sex chromosome aneuploidies; it is often found as a mosaic, where it is mixed with XX cells, and this may aid survival.

Table 1. The human euploid series

Chromosome number	Ploidy	Normally found in
23	Haploid	Gametes
46	Diploid	Somatic cells
69	Triploid	
96	Tetraploid	

Table 2. Aneuploidies in human populations

Condition	Chromosome involved	Approximate frequency
Involving gain of an autosomal chromosome		
Edward's syndrome	18	1 in 5000
Pateau's syndrome	13	1 in 5000
Down's syndrome	21	1 in 750
Sex chromosome aneuploidies		
Turner's syndrome	XO	1 in 10000
Klinfelter's syndrome	XXY	1 in 2000
Triple X syndrome	XXX	1 in 2000

Aneuploid gametes arise from errors at both first and second division of meiosis (Section C3) through a process known as **nondisjunction**. This is where either two homologous chromosomes fail to separate at meiosis I or a centromere fails to split at metaphase of meiosis II (Section C3). Nondisjunction involving the splitting of centromeres can also arise at mitosis. Nondisjunction is thought to arise as random events similar to mutations in the DNA sequence; however, they are affected by external factors. This is shown clearly for trisomy 21, where the incidence increases with the age of the mother. In some cases there is a familial predisposition to trisomy 21, due to the presence of carriers in the family who have a normal phenotype but carry a translocation involving chromosome 21. They can transmit both a normal copy of chromosome 21 and the translocation chromosome, which also bears chromosome 21, to their offspring. An affected child will also inherit a normal copy of chromosome 21 from their other parent and thus will have three copies of the chromosome. This is described in Figure 4.

All of the syndromes mentioned, with the exception of Turner's syndrome (XO), have involved the presence of an extra chromosome. Individuals with loss of a single autosome, monosomy, are very rare and have a very short life-span. In humans nullisomics are never viable.

Gametes		Zygote
Carrier	Normal	
21,14	21,14	Viable normal
21/14	21,14	Viable carrier
21,21/14	21,14	Trisomy 21
14	21,14	Lethal nullisomic
14,21/14	21,14	Nonviable
21	21,14	Lethal nullisomic

Figure 4. Trisomy 21 inheritance. If an individual is a carrier for trisomy 21 (Down's syndrome) because of the presence of a translocation between chromosome 21 and chromosome 14 they will produce a number of genetically different gametes due to irregularities in segregation of the translocation at meiosis. These gametes will be fertilized by normal gametes monosomic for both chromosome 21 and 14. The two sets of gametes and the outcome of their fusion are shown above. The translocation is denoted as 21/14. Note that the gametes will not all be produced at equal frequency.

B5 Cell division

Key Notes

Eukaryote cell cycle

The cell cycle is the period between two divisions. It consists of four stages, G1, S (the DNA synthesis phase), G2, and mitosis. Cells in different phases of the cell cycle can be identified using a fluorescence-activated cell sorter. Cells in S phase can be identified by labeling with analogs of thymidine.

Mitosis and cytokinesis

Nondividing cells are in interphase. Mitosis is divided into four phases. In prophase the chromosomes condense. At metaphase they are aligned on the equator of the cell. When the centromeres split, the two chromatids separate towards opposite poles. This is anaphase. As the chromatids near the poles the cell moves into telophase. Nuclear membranes form around the two nuclei and the chromosomes begin to decondense. At the same time, cytokinesis is initiated at the equator of the cell.

Regulation of the cell cycle

Cells in multinucleate organisms are often not traversing the cell cycle. Cells are stimulated to grow and divide by growth hormones. Once a cell passes the restriction point it is committed to divide. Growth hormones act through the signal transduction pathway. Progress through the cycle is dependent on complexes between cyclins and cyclin-dependent kinases. Checkpoints exist at several crucial stages in the cycle to ensure that the process of cell division is highly regulated. Mutations in cell cycle genes are usually found in tumor cells.

Related topics

(A1) DNA structure (F3) Genes and cancer
(C3) Meiosis and gametogenesis

Eukaryote cell cycle

The progression from one cell division to the next can be regarded as a cyclic process, the **cell cycle**. During this time the cell must replicate its contents and then organize the distribution of its components equally between two daughter cells. Except in the production of gametes (Section C3) the nuclei of eukaryote cells divide by **mitosis** and in parallel with this the cytoplasm divides by **cytokinesis**. These processes are easily visualized in fixed or living cells. The period between two consecutive divisions is referred to as a **cell cycle**. The replication of DNA is accomplished during a period in the cell cycle known as the **synthetic** or **S phase**. The period preceding S phase is called G_1 (gap 1), and the period between S phase and division is known as G_2 (gap 2).

Where suspensions of single cells can be produced, **fluorescence-activated cell sorters** can identify cells as being in G_1, S, or G_2 phases of the cell cycle. Apart from this, cells in G_1 and G_2 are not easily identified, but cells in S phase can be detected if cells are allowed to incorporate a labeled precursor of DNA. Analogs of thymidine are best suited to this purpose as this nucleoside is specific to DNA. The proportion of cells labeled after a short exposure to labeled precursor is equal to the average proportion of the cell cycle that the cells spend in S phase.

Mitosis and cytokinesis

Mitosis has many similarities with meiosis (Section C3), the reduction division in game-togenesis, and care must be taken to avoid confusion. The stages of mitosis are set out in Figure 1. G_1, S, and G_2 phases are collectively referred to as **interphase**. As a cell leaves G_2 it enters **prophase** of mitosis. The individual chromosomes gradually become apparent as their chromatin structure begins to condense. They are now double structures, **chromatids**, as DNA has been replicated in S phase. Towards the end of prophase the nuclear membrane and nucleolus begin to break down. Released nucleolar proteins attach to the surfaces of condensing chromosomes. The cell moves from prophase to **metaphase**. At this point the nuclear membrane and nucleolus are absent and the chromosomes are aligned on the equator of the cell. This is achieved by the action of tublin-containing spindle fibers which run from both poles of the cell and attach to the centromeres of the chromosomes (Section B4). The subsequent phase, **anaphase**, commences as soon as the centromeres are cleaved allowing the chromatids to separate. The individual chromatids, better described as **daughter chromosomes**, are pulled to opposite poles of the cell by the mitotic spindle. When the two sets of chromosomes are well separated toward the poles of the cell they begin to decondense, nuclear membranes form around the decondensing chromatin and the cell is transiently binucleate; this is **telophase**. At the same time, in animal cells, a **cleavage furrow** forms across the equator of the cell. This structure, which is composed of actin-containing microfilaments, progressively tightens

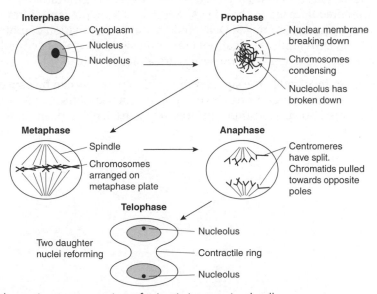

Figure 1. Schematic representation of mitosis in an animal cell.

and eventually divides the cell into two daughter cells; this is **cytokinesis**. In plant cells cytokinesis is achieved by the formation of a new cell wall across the equator of the cell.

Regulation of the cell cycle

This refers to the rate at which cells traverse the cell cycle. The essential point to grasp is that in multicellular organisms most cells do not divide regularly. These quiescent cells require stimulation before they divide and are referred to as being in the G_0 phase. Only specific tissues such as bone marrow in animals or root tips in plants have high proportions of cycling cells. Cells are triggered to leave G_0 by specific hormonal growth factors. They then move through G_1 phase. This can vary in length from cell to cell, but once a cell passes a specific point, the **R** or **restriction point**, it is committed to attempt to complete the cycle, pass through division and reach G_1 again. The subsequent portions of the cycle, S phase, G_2, and mitosis have relatively fixed lengths.

Progress through the cell cycle as a response to the effects of growth stimulating hormones is mediated by a process known as **signal transduction** (Section F3). This comprises a series of effector molecules that translocate and amplify the initial signal produced by the growth hormone, eventually causing transcription of genes involved in cell growth and division. At the same time progress through the cell cycle is negatively regulated by a series of **checkpoints** which the cell cannot pass until certain criteria have been fulfilled. Thus growth and division of cells remains a highly regulated process as is necessary for the development of multicellular organisms. Key players in the progression of the cell cycle are a group of kinases, enzymes that add phosphate groups to their substrates. Phosphorylation of specific target proteins at different stages of the cell cycle is necessary for the cycle to proceed. The kinases bind to a second group of proteins, **cyclins**. These regulate the kinase activity. In the absence of cyclin the kinases are inactive and are therefore called **cyclin-dependent kinases** (**cdks**). Different members of the cyclin family appear at different points of the cell cycle.

Checkpoints have been identified at three stages of the cell cycle, the boundary between G_1 and S phase, the boundary between G_2 and mitosis and within mitosis itself. These ensure that the various processes involved in cell growth are coordinated. For instance, cells with unrepaired damage to their DNA (Section A6) are prevented from entering S phase, where replication of damaged DNA could result in the creation of mutations. Cells with uncompleted DNA replication cannot enter mitosis; mitosis is not completed before the spindle is organized and all chromosomes are attached to it.

Mutations of genes involved in the signal transduction pathway can be cancer inducing; these are known as **oncogenes**. Genes involved in the checkpoints are also often mutant in tumors; they are known as **tumor suppressor genes**. This is dealt with in detail in Section F3.

B6 Bacteriophages

Key Notes

Bacteriophages

Bacteriophages (phages) are viruses that infect bacteria. They are composed of a nucleic acid genome of single- or double-stranded DNA or RNA surrounded by a protective protein coat. Three shapes occur: icosahedral (MS2); filamentous or helical (M13); head and tail (T4 and λ). Bacteriophage genomes vary in size and complexity with the largest genomes associated with the more complex phage structures such as head and tail. The phage genes encode capsid proteins and proteins involved in DNA replication. All phages have some requirement for host cell enzymes.

Phage life cycles

Phage infection follows lytic or lysogenic pathways. Lytic infection results in cell lysis. In lysogenic infection, the phage remains quiescent for many generations before inducing cell lysis.

T4 lytic pathway

Phage T4 undergoes lytic infection of *E. coli*. The phage attaches to the cell surface and injects its DNA into the cell. Phage DNA is replicated and phage genes are expressed. Expression of host cell genes and replication of host DNA is arrested. Phage DNA is packaged into capsids and new phage particles are released following cell lysis.

Lambda lysogenic life cycle

Phage λ undergoes lytic and lysogenic infection of *E. coli* but lysogenic infection is more common. The phage attaches to the cell surface and injects its DNA into the cell. The DNA integrates into the host cell chromosome where it remains quiescent. Eventually a switch occurs to lytic infection triggered by chemical or physical stimuli associated with DNA damage. Phage DNA is excised from the chromosome and is replicated. Phage genes are expressed and new phage particles are produced which are released following cell lysis.

M13 phage

M13 infection of *E. coli* is intermediate between the lytic and lysogenic pathways. The M13 genome is a single-stranded circular DNA molecule which replicates via a double-stranded replicative form. New phage particles are released without cell lysis or are passed on to daughter cells following division of infected cells. M13 is used as a cloning vector to produce single-stranded DNA for sequencing.

Gene expression in lytic infection	Phages regulate expression of their genes to different degrees. Usually genome replication precedes the synthesis of capsid proteins. Lysozyme is produced at the end of the lytic cycle. Simple phages such as φX174 express all their genes at the same time. Other phages have early and late phases of expression. Early gene expression is associated with replication of the phage genome and late gene expression with the synthesis of structural proteins. Expression of early genes activates the expression of later genes by modifying the specificity of the host RNA polymerase.	
Gene expression in lysogenic infection	During lysogenic infection of *E. coli* by phage λ, gene expression is inhibited by binding of the cI repressor protein to P_L and P_R promoters that control expression of early λ genes. Binding of cI also stimulates its own expression. The λ cro protein acts as a repressor of cI gene transcription. The choice between the lytic and lysogenic pathways depends on the relative levels of cI and cro. If cro levels are high, cI is repressed and the block on λ gene expression is removed resulting in lytic infection. If cI levels are high lysogeny is maintained. The switch to lytic infection following lysogeny can be induced by ionizing radiation which induces expression of the RecA protein which cleaves cI.	
Related topics	(A7) Regulation of gene expression (B7) Eukaryotic viruses	(E1) Using sequence specificity to study nucleic acids (E2) PCR and related technology

Bacteriophages

Bacteriophages (phages) are viruses that infect bacteria. Many different types of bacteriophage exist which tend to be species specific infecting only certain bacterial hosts. Like other viruses, their structure consists of an inner nucleic acid genome surrounded by an outer protective protein coat. Three types of phage structure are seen (Figure 1): (i) **icosahedral**, in which the individual polypeptide molecules form a geometrical structure that surrounds the nucleic acid – an example is the phage MS2 that infects *E. coli*; (ii) **filamentous or helical**, in which the polypeptide units are arranged as a helix to form a rod-like structure – an example is the *E. coli* phage M13; (iii) **head and tail**, the most complex phage structure consisting of an icosahedral head and a filamentous tail that allows the phage nucleic acid to enter the cell – examples of this type are the *E. coli* phages T4 and λ. The nucleic acid in phages may be DNA or RNA which may be single-stranded or double-stranded. Usually all the genes are present on a single nucleic acid molecule; however, in some RNA phages the genes are present on more than one molecule and these phages are said to be **segmented**. Phage genomes vary greatly in size from just a few kilobase pairs (kbp) to about 150 kbp. The number of genes varies roughly in proportion to genome size. The genome of the phage M13 is about 6 kbp and contains just 10 genes. Larger phages, especially those with complex structures (e.g. head and tail), have many more. For example, phage T4 at 166 kbp has 150 genes. Phage genes encode proteins required

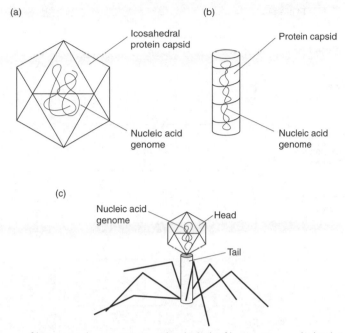

Figure 1. Structure of bacteriophages. (a) Icosahedral, (b) filamentous or helical, (c) head and tail.

for the construction of the capsid and enzymes involved in phage DNA replication. All phages also require at least some host proteins and RNAs. A feature of some phages (e.g. φX174), also seen in viruses of eukaryotic cells, is that they have overlapping genes translated in different reading frames.

Phage life cycles

The events that follow infection of a cell by a phage vary but essentially two patterns, known as **lytic** and **lysogenic infection**, are seen. In the lytic pathway, the phage causes lysis of the host cell soon after infection. In the lysogenic pathway, the phage causes cell lysis only after an extended period during which it remains quiescent. Different types of phage may undergo either type of infection. Phages that undergo lytic infection are called **virulent**. Those undergoing lysogenic infection are said to be **temperate**.

T4 lytic pathway

Phage T4 undergoes lytic infection of *E. coli* (Figure 2). Phage particles attach to a receptor protein on the surface of the cell called *ompC* and phage DNA is injected into the cell via the T4 tail structure. Inside the cell, transcription of the phage genes begins and synthesis of host cell DNA, RNA, and protein is arrested. Host cell DNA is depolymerized and the nucleotides are used to replicate phage DNA. Phage capsid proteins are synthesized and new phage particles are assembled. Finally, the host cell is lysed and 200–300 new phage particles are released.

Lambda lysogenic life cycle

Temperate phages such as λ can undergo lytic infection of *E. coli* but more commonly follow the alternative lysogenic pathway (Figure 3). After the phage injects its DNA into

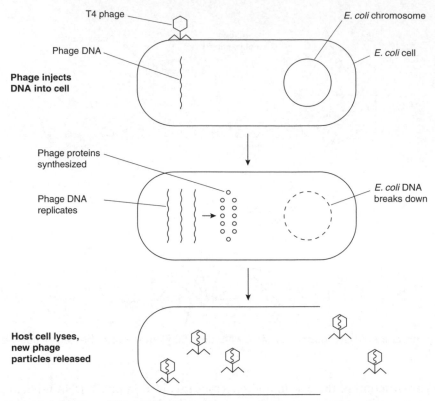

Figure 2. Lytic infection of *E. coli* by phage T4.

the cell, the DNA circularizes then integrates into the host cell chromosome. Integration occurs by recombination and involves a 15 bp phage sequence homologous to a sequence on the *E. coli* chromosome. Integration always occurs at the same position and the integrated form is known as a **prophage**. It remains undisturbed for many generations and is replicated with the host cell chromosome. Eventually, after numerous cell divisions, a switch occurs to the lytic mode of infection. This is induced by a number of physical or chemical stimuli, all of which are linked to DNA damage and possibly signal the imminent death of the cell. In response to these stimuli, the phage DNA is excised from the chromosome by a reverse recombination event. Phage genes are expressed and capsid proteins are synthesized. Phage DNA is replicated and is packaged into the capsids. Eventually the cell lyses and new phage particles are released.

M13 phage

Some phage life cycles show variations on the lytic or lysogenic pathways. An example is M13 which infects *E. coli*. Its genome is a single-stranded circular DNA molecule. On infection, a double-stranded **replicative form (RF)** is produced by host cell enzymes. Multiple copies of the RF are synthesized which act as templates for the production of single-stranded forms which are packaged into capsids and released from the cell without cell lysis. Phage particles are also passed to other cells following cell division and cells

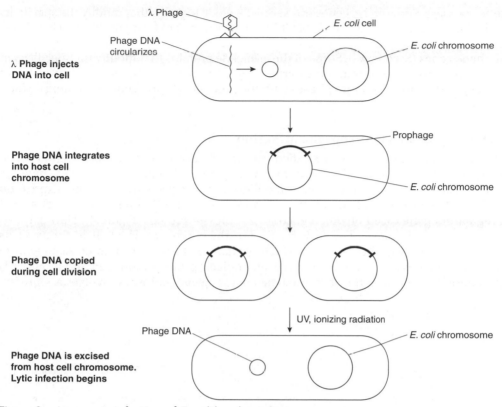

Figure 3.   Lysogenic infection of *E. coli* by phage λ.

infected by M13 continue to grow and divide giving rise to infected daughter cells. Several features of the M13 life cycle have made it a useful cloning vector for molecular biology. The double-stranded circular RF can be easily purified like other plasmids and the single-stranded phage DNA is an ideal template for DNA sequencing (Section E3).

Gene expression in lytic infection

Phages show different degrees of regulation of gene expression. Usually genome replication precedes the synthesis of capsid proteins and lysozyme, the enzyme that causes cell lysis, is produced at the end of the lytic cycle. For simple phages such as φX174, regulation of gene expression is minimal. All of its 11 genes are transcribed by the host RNA polymerase upon infection. However, lysozyme production, which is required last, is delayed by being translated more slowly than the other transcripts. With most other phages there are two phases of gene expression known as **early** and **late**. Early gene expression is usually associated with replication of the phage genome and late gene expression with the production of structural proteins. The early genes are transcribed first and are responsible for activating expression of the late genes. In T4, the earliest genes are transcribed by the host RNA polymerase using phage promoter sequences that normal *E. coli* genes possess. Some of the genes transcribed encode proteins that modify the specificity of the host RNA polymerase such that it no longer recognizes host promoters. This causes host gene expression to be switched off but also leads to the transcription of a second set of phage genes (Section A7). Some of these encode proteins that modify the specificity of the polymerase again so that it transcribes a third set of genes. In this way expression of phage genes is organized into phases with the products of early genes switching on later genes.

Gene expression in lysogenic infection

The bacteriophage λ has been studied extensively as a model of phage gene expression. On infection of *E. coli*, λ may follow the lytic or lysogenic pathways. There are three phases of λ gene expression: immediate early, delayed early, and late. Expression of the early genes is regulated from two promoters, P_L and P_R, present on either side of a regulatory gene called cI which encodes a repressor protein (Figure 4). During lysogeny transcription from P_L and P_R is blocked by the cI protein, which also stimulates its own expression when bound to P_R which overlaps with the cI promoter. As long as cI is present, expression of the early genes and of later genes is repressed and lysogeny is maintained. The choice between the lytic and lysogenic pathways involves a λ protein

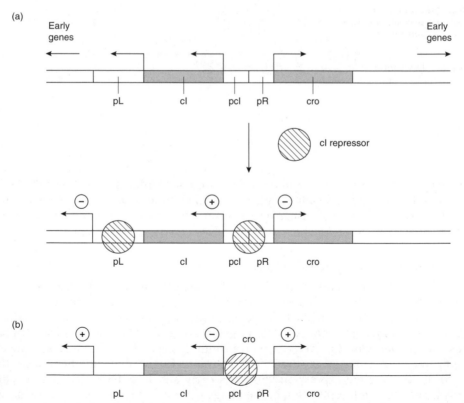

Figure 4.   Control of λ phage gene expression. (a) cI repressor blocks transcription of early genes and stimulates cI gene transcription. (b) Cro repressor blocks transcription of the cI gene allowing early genes to be expressed.

called **cro** which acts as a repressor of cI gene transcription. After infection, the host RNA polymerase starts transcribing λ genes from a number of λ promoters. cI is expressed and prevents early gene transcription thus blocking progression of the lytic pathway. Cro is also expressed and if sufficient cro protein accumulates cI is repressed and the block to early gene transcription is removed allowing progression to the lytic pathway. In host cells with good nutrition and low levels of cAMP (Section A7), which will produce large numbers of bacteriophage, the lytic cycle is favored. In a starved host with high cAMP levels, cI production continues and the bacteriophage follow the lysogenic path. As long as enough cI protein is present lysogeny will be maintained. If cI levels fall, however, lysis will be induced spontaneously. The induction of lysis following lysogeny can be triggered by agents such as ionizing radiation that activate a general protective mechanism in *E. coli* called the SOS response. This involves the ***RecA*** gene whose protein inactivates the cI repressor by cleaving it. When cI is inactivated in this way, the early genes are activated, lysogeny ends, and the lytic pathway is induced.

B7 Eukaryotic viruses

Key Notes

Viruses
Viruses are infectious agents composed of a nucleic acid genome surrounded by protein. They reproduce by using the host cell's ability to synthesize nucleic acids and protein. Many viruses are pathogenic.

Viral genomes
These may be single- or double-stranded, DNA or RNA, linear or circular. Genomes may be positive or negative or both senses. Viral genomes may be segmented. Some viral particles have incomplete genomes and only replicate in the presence of wild-type virus or by complementation.

Structure of viruses
The protein coat surrounding the viral nucleic acid is the capsid; this may be icosahedral or filamentous in shape. Some viruses have a host-cell-derived lipid envelope containing host cell proteins and viral glycoproteins that play a role in infection. Structural viral proteins are antigenic and are useful in the development of vaccines.

Replication strategies
Viruses use many different strategies for replication. All depend on the host cell completely for translation and to different extents for transcription and replication. DNA viruses are more dependent on host cell enzymes than RNA viruses, and small DNA viruses are more dependent than large DNA viruses. RNA viruses require virus-encoded RNA-dependent RNA polymerases. Retroviruses are RNA viruses that replicate via a double-stranded DNA intermediate produced by virus-encoded reverse transcriptase that copies RNA into DNA. Host cells that support infection are said to be permissive. The capacity of a virus to cause disease is called virulence.

DNA viruses
The genomes of DNA viruses vary greatly in size (5–270 kbp) and may be linear or circular, single- or double-stranded DNA (ssDNA, or dsDNA). Large DNA viruses have complex life cycles with coordinated patterns of gene expression and genome replication. Herpes simplex virus type 1 has a 150 kbp genome with over 70 genes present on both strands with some genes overlapping. The viral genes are expressed in three phases (immediate early, early, and late), and expression is coordinated with viral genome replication. Small DNA viruses have fewer genes and are more dependent on the host cell for replication and gene expression. The SV40 virus has a 5 kbp genome containing five overlapping genes present on both strands of the DNA. Viral gene expression

makes use of overlapping reading frames and alternative splicing. The genes are expressed in two phases (early and late). Early gene transcription produces the large and small T antigens which stimulate viral and host cell transcription and replication, and are responsible for the tumorigenic properties of the virus.

RNA viruses

The genomes of RNA viruses may be single- or double-stranded. Single-stranded genomes may be positive or negative or both senses. The viral *pol* gene encodes an RNA-dependent RNA polymerase required for the transcription of viral genes. RNA viruses are not dependent on host polymerases and so may replicate in the cytoplasm. Due to the lack of proofreading ability of the RNA polymerase, RNA viruses have a high mutation rate allowing them to evolve rapidly causing changes in antigenicity and virulence. Some RNA viruses exist as a collection of mutant forms that can only replicate by complementation. The high mutation rate also imposes an effective limit on the viable size for an RNA viral genome of about 10 000 nucleotides.

Retroviruses

This group includes human immunodeficiency virus (HIV). Retroviruses have a single-stranded, positive-sense RNA genome. Two copies are present in each viral particle. On infection, viral reverse transcriptase synthesizes a dsDNA copy of the viral RNA which integrates into the host cell genome and acts as a template for viral replication and gene expression. Some retroviruses cause cancer and carry genes originally derived from the host cell called oncogenes. Retroviruses carry three genes called *gag*, *pol*, and *env*: *gag* encodes capsid proteins; *pol* encodes enzymes involved in viral replication; and *env* encodes proteins found in the lipid envelope. The viral genome has repeat sequences at either end called long terminal repeats (LTRs). Host cell RNA polymerase II binds to the 5′ LTR and transcribes viral genes. Viral mRNAs are translated as polyproteins that are cleaved proteolytically to give mature proteins. HIV creates extra proteins by differential splicing of mRNAs. The reverse transcriptase of some retroviruses has a high error rate producing defective copies of the viral genome. Replication depends on complementation between the two genome copies present in each viral particle. The high turnover rate of HIV and its high mutation rate allow the virus to adapt to selective pressures.

Related topics

(B6) Bacteriophages   (F3) Genes and cancer

Viruses

Viruses are submicroscopic infectious agents composed of a nucleic acid genome surrounded by a protein coat. They are parasitic and reproduce themselves by infecting cells
and making use of the cell's ability to replicate DNA and synthesize protein. It is difficult
to define them as living organisms because they are incapable of existing independently.
Many viruses are pathogenic and cause destruction of the host cell leading to disease in
humans and other organisms. In addition to eukaryotes, prokaryotic organisms such as
bacteria can be infected by viruses called **bacteriophages** (Section B6).

Viral genomes

There are many different types of virus that infect eukaryotic cells. These differ with
respect to their nucleic acid genomes which may be single- or double-stranded DNA or
RNA. The nucleic acid may also be linear or circular. In some cases the viral genome is a
single molecule of nucleic acid but in others the viral genes exist on more than one molecule and the genome is said to be **segmented**. Single-stranded genomes are replicated
via a double-stranded intermediate and may be **positive sense** or **negative sense**. The
sense refers to the sequence of the genome and the mRNAs transcribed from it. If the
genome is positive sense, its sequence will be the same as its transcribed mRNAs. Conversely, if it is negative sense then the genome sequence is complementary to the mRNAs.
In some cases the genome will encode mRNAs which are of either sense. Many viral particles do not have a complete or functional genome and so are only capable of replication
when they are rescued by coinfection of the cell with a wild-type replication-competent
helper virus, or by a process known as **complementation,** in which coinfection occurs by
a defective virus with a different mutation.

Structure of viruses

The viral genome is surrounded by a protective protein coat known as the capsid which
is assembled from individual virus-encoded polypeptides. The combination of the
genome and the capsid is known as the **nucleocapsid**. Two basic shapes occur (Figure 1).
(i) **Icosahedral** – in this case the individual polypeptide molecules form a geometrical
structure that surrounds the nucleic acid; an example is poliovirus. (ii) **Filamentous** or
helical – in this case the polypeptide units are arranged as a helix to form a rod-like structure surrounding the nucleic acid genome. In many viruses the capsid is surrounded by
a **lipid envelope** derived from the host cell membrane as the virus is released from the
cell. Virus-encoded glycoproteins and proteins derived from the host cell may be inserted
into the membrane. The viral glycoproteins play an important role in facilitating infection by interacting with receptor proteins on the surface of the host cell. **Matrix proteins**
also occur which allow interaction between the nucleocapsid and the lipid envelope. The
matrix and the capsid proteins also have other roles associated with virus replication and
transcription of viral genes. The structural proteins of a virus are often antigenic and are
of interest in the development of vaccines.

Replication strategies

Viruses make copies of themselves by exploiting the ability of cells to replicate DNA
and to synthesize proteins. The strategies adopted by viruses for replication of the viral
genome and transcription of the viral genes vary greatly. All viruses rely on the host
cell for translation and to differing extents for transcription and replication. DNA
viruses are more dependent on the host cell enzymes than RNA viruses and small
DNA viruses more dependent than large DNA viruses on the host cell's replication and

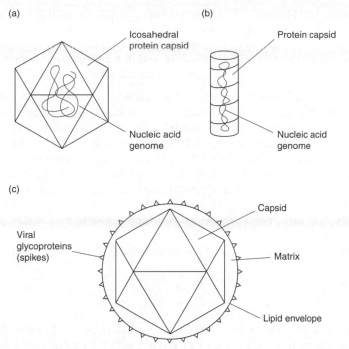

Figure 1. Structure of eukaryotic viruses. (a) Icosahedral. (b) Filamentous or helical. (c) Enveloped virus.

transcription machinery. RNA viruses require RNA-dependent RNA polymerases for replication which are not present in host cells but must be encoded by the virus. Other RNA viruses called **retroviruses** replicate via a dsDNA intermediate produced from the viral RNA by a unique virus-encoded enzyme called **reverse transcriptase** that copies RNA into DNA.

The susceptibility of a cell to infection by a virus is determined by its ability to support viral replication and the presence of cell-surface receptors specific for the virus. Cells capable of supporting replication are called **permissive**.

The consequence of viral infection is not always disease. The primary function of the virus is to replicate itself, which may coincidentally cause cell damage. If enough cells are damaged, this may result in disease. The capacity of viruses to cause disease is called **virulence** and may be disadvantageous to the virus because it may reduce the capacity for viral replication.

DNA viruses

The genomes of DNA viruses vary greatly in size from 5 kbp to as large as 270 kbp. The DNA may be single-stranded or double-stranded and may be linear or circular. Almost all DNA viruses replicate in the nucleus and make use of the host cell machinery for genome replication and viral gene expression. Large DNA viruses encode as many as 200 genes and have a complex life cycle with coordinated patterns of gene expression and genome replication. An example of large DNA viruses is the **herpes viruses** which infect a range of vertebrates, including humans, where they cause diseases such as chicken pox, shingles, and glandular fever. **Herpes simplex virus 1** is a well-studied herpes virus that causes cold sores. It has a 150 kbp dsDNA genome which contains over 70 genes present on both DNA strands, some overlapping with each other. The transcription and replication of the viral genome are tightly regulated. The viral genes fall into three groups called immediate early (α), early (β), and late (γ) which are expressed in a defined sequence following infection of a host cell. Expression is also coordinated with replication of the viral genome.

DNA viruses with small genomes have many fewer genes and consequently are more dependent on the host cell for replication and gene expression. Examples include polyoma virus and the monkey virus **SV40** which has been studied intensively due to its ability to cause tumors. Both are members of the papovavirus family. The SV40 virus has a small 5 kbp genome which contains five genes. The genes are accommodated in the small genome by being present on both strands of the DNA and by having overlapping sequences. Viral proteins are produced by a combination of the use of overlapping reading frames and alternative splicing. The genes are expressed in two phases after infection known as **early** and **late**. The early genes produce proteins that activate transcription called the **large T antigen** and the **small T antigen**. These stimulate both viral and host cell transcription and replication, and are responsible for the tumorigenic properties of the virus.

RNA viruses

The genomes of RNA viruses may be single-stranded or double-stranded. If single-stranded, the genome may be positive or negative sense. In some cases the genome will encode mRNAs of both senses. Transcription of viral genes from an RNA genome requires enzymes called **RNA-dependent RNA polymerases** which do not occur in host cells but are encoded by a viral gene called *pol*. Because RNA viruses do not require host cell polymerases, transcription and replication need not take place in the nucleus as is the case for DNA viruses. Consequently, many RNA viruses replicate in the cytoplasm. Unlike host cell polymerases, RNA-dependent RNA polymerases are not capable of proofreading and replicate their templates with a much higher error rate. The mutation rate is one base in 10^3–10^4 per replication cycle which is much higher than the rate for DNA viruses of one base in 10^8–10^{11}. The consequences of the much higher mutation rate for RNA viruses are that they are capable of evolving rapidly and can develop changes in antigenicity and virulence, quickly allowing them to adapt to changing environments and attempts to eliminate them by the host's immune system.

Some RNA viruses mutate so rapidly that they exist as a population of genome sequences which replicate by complementation and are known as **quasi-species**. Many of the mutations that occur in RNA viruses will be deleterious to viral replication. This effectively creates an upper limit to the genome size of about 10^4 nucleotides, equivalent to the size that, on average, would produce one mutation per genome.

Retroviruses

These are an important group of RNA viruses with single-stranded positive-sense RNA genomes. The group includes human immunodeficiency virus (**HIV**), the virus that causes the acquired immune deficiency syndrome (**AIDS**). Retroviruses contain two copies of the genome in each viral particle. On infection of a host cell, the ssRNA enters the cell, is converted to a dsDNA copy by **reverse transcriptase**, and is integrated into the host cell genome by a viral **integrase** enzyme. The integrated form is known as a **provirus** and acts as the template for replication of the viral genome and expression of the viral genes (Figure 2). The genomes of retroviruses have structural similarities and some homology

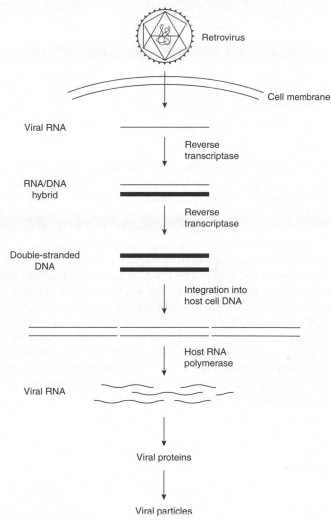

Figure 2. Life cycles of retroviruses.

with sequence elements present in the human genome called **retrotransposons**. Retroviruses vary in complexity. Some are extremely simple and differ from retrotransposons essentially only by having an *env* gene that encodes envelope glycoproteins required for infectivity. Others, such as HIV, have larger genomes with more complex life cycles, encode proteins that are active at different stages of the replication cycle, and are involved in regulating both viral and cellular functions.

Some retroviruses are known to cause cancer in animals although they are linked only rarely to human cancers. These retroviruses are said to be **oncogenic** and frequently carry a gene called an **oncogene** which was originally acquired from the host cell genome by recombination with the viral genome. Oncogenes encode proteins involved in regulating cell growth and have cellular counterparts called **proto-oncogenes**. Expression of the oncogene following viral infection results in uncontrolled cell division associated with the formation of a tumor (Section F3).

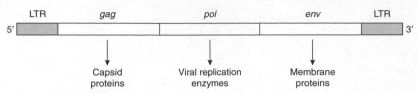

Figure 3. Retrovirus genome.

The genomes of retroviruses all have the same basic structure (Figure 3). Most of the genome is taken up by three genes called *gag*, *pol*, and *env*. *Gag* encodes the capsid proteins, *pol* encodes the enzymes involved in viral replication (reverse transcriptase, RNAse H, integrase, and protease), and *env* encodes the proteins present in the lipid envelope. At either end of the genome are duplicate sequences called **LTRs** (long terminal repeats) that are involved in viral replication, integration, and gene expression. Transcription of the viral genes from the integrated provirus depends on host-cell RNA polymerase II which binds to sequences in the 5′ LTR. The mRNAs are translated into **polyproteins** that are cleaved proteolytically to give individual viral proteins. Some retroviruses such as HIV produce additional mRNAs by differential splicing. The reverse transcriptase of some retroviruses has a high error rate. This means many of the genome copies produced are incapable of replicating themselves. This can be overcome by complementation between the two genome copies present in each viral particle, which can also recombine with each other during reverse transcription. These features, combined with a high turnover rate (10^9–10^{10} new viral particles per day), enable HIV-1 to adapt to new environments such as selective pressure from antibodies and drug treatments.

C1 Basic Mendelian genetics

Key Notes

Basic concepts	Any inherited trait such as eye color is referred to as a phenotype. All phenotypes result from the presence of a specific gene or combination of genes, the genotype. In hybrids one phenotype may be dominant to another. Pure-breeding lines are strains of a species which have been bred for many generations and have maintained the same phenotype.
The monohybrid cross	Crosses between pure-breeding (homozygous) lines differing in one inherited character yield progeny that all have the same phenotype. This is termed the F1 generation and is composed of heterozygous individuals. When these are inter-crossed, the next generation, the F2, shows both of the original phenotypes in the ratio 3:1 with the dominant phenotype being the majority class. Each individual carries two copies (alleles) of each gene. Homozygous individuals carry two identical alleles, heterozygous individuals carry two different alleles. Differences between alleles are due to mutation.
Detection of heterozygotes	When an individual that is heterozygous for one gene is crossed to a recessive homozygote only two classes of progeny are observed. The dominant and recessive phenotypes arise at equal frequency. This is a test cross.
Variations of the 3:1 ratio	The 3:1 ratio depends on complete dominance of one phenotype over the other. If the phenotypes under study show partial or codominance, a 1:2:1 ratio will be obtained. If either allele has a negative effect on viability this will also distort the ratio. Alleles that can cause lethality when homozygous are called lethal alleles. Semilethal alleles have a quantitative effect on viability.
Multiple alleles	Most genes exist in several different forms, multiple alleles. This is caused by mutations of bases at different sites within the same gene, thus affecting different amino acids in the encoded protein. These arise at random within the population.
Modern developments	DNA technology has developed a number of polymorphic DNA markers. These have a great advantage in that they are codominant and can be analyzed directly without the complications of dominance.
Related topic	(C2) More Mendelian genetics

Basic concepts

The first clear evidence pointing to what we now call genes came from the work of Gregor Mendel who carried out experiments on inheritance in pea plants in the middle of the nineteenth century. Before we examine his results it is necessary to establish an understanding of some of the basic terms that are used in the study of inheritance in higher organisms.

Phenotype

Any character (trait) which can be shown to be inherited, such as eye color, leaf shape, or an inherited disease, such as cystic fibrosis, is referred to as a phenotype. A fly may be described as having a red-eyed phenotype or a child as displaying the cystic fibrosis phenotype. The pattern of genes that are responsible for a particular phenotype in an individual is referred to as the **genotype.**

Pure-breeding lines

This refers to organisms which have been inbred for many generations in which a certain phenotype remains the same. Pedigree breeds of dogs or cats are commonplace examples of pure-breeding lines.

Dominance

Within a species there may be differences in the phenotype for one inherited character. In hybrids between two individuals displaying different phenotypes only one phenotype may be observed. For instance, in crosses of pure-bred fruit flies with short wings with pure-bred long-winged flies the progeny will all have long wings. The phenotype expressed in the hybrids is said to be dominant and the other recessive. In the example above long wings are dominant to short wings.

The key ingredients for success in Mendel's experiments were the use of pure-breeding strains of pea plants and the fact that he subjected his results to simple mathematical analysis.

The monohybrid cross

Mendel studied inheritance of several phenotypes in pea plants, but we will concentrate on only one of these: petal color. He made a cross between two pure-breeding lines of plants, one of which had violet petals and the other white petals. The hybrids produced in this cross were referred to as the **F1 (first filial)** generation. These all had violet flowers. Thus violet was dominant to white. He then allowed these plants to self-fertilize to produce the **F2 (second filial)** generation. Some plants had white flowers and others violet flowers. The ratio of violet to white flowered plants was close to 3:1. In crosses between plants differing in seed color, pod shape, or other phenotypes, the same pattern was observed. The recessive phenotype always reappeared in the F2 generation and made up approximately one-quarter of the plants.

These experimental data led Mendel to suggest that heredity was due to the action of specific factors, which we now call genes. This apparently simple conclusion was, however, in complete opposition to the conventional view that heredity was due to blending of fluids from both parents. Clearly no blending had occurred in Mendel's experiments, neither in the F1, where only one phenotype was expressed, nor in the F2, where both were expressed separately.

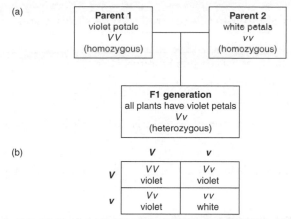

Figure 1.   (a) Gametes produced by the F1 plants carry *V* or *v* alleles. These fuse at random to give offspring with the genotypes *VV*, *Vv*, or *vv* as shown in (b). (b) Each of the four fusions arise with equal frequency. Hence the phenotype ratio in the F2 generation is 3 purple:1 white and the genotype ratio is 1*VV*:2*Vv*:1*vv*.

From our modern knowledge of genes and gene structure it is easy for us to appreciate how Mendel explained what was happening in his experiment. This is set out diagrammatically in Figure 1. He suggested that the pure-breeding violet-flowered plants carried two copies of a gene for violet pigment, *V*. The white flowered plants carried two copies of a variant of this gene that codes for white flowers, *v*. We refer to individuals with two identical copies of a gene as being **homozygous**. The F1 hybrids inherited two different copies of the pigment gene, *Vv* and are referred to as **heterozygous**. As violet is dominant to white these plants had violet flowers. When F1 plants self-fertilize three different classes of genotype, *VV*, *Vv* and *vv* are possible. These arise in the ratio 1:2:1 (Table 1). This is how the 3:1 phenotypic ratio is established (i.e. 3 violet:1 white).

The validity of this hypothesis was strengthened when individual plants from each of the F2 classes were self-fertilized. All white-flowered plants were found to be pure-breeding. One-third of the violet-flowered plants were pure-breeding and two-thirds gave purple- and white-flowered plants in the ratio 3:1. There have subsequently been many examples of the 3:1 ratio for the inheritance of characters controlled by a single gene in many different species. A few of these are listed in Table 2.

One new term needs to be defined at this stage. Genes become altered through the process of mutation. The different variants of a gene are referred to as **alleles**. Students often are confused between the terms genes and alleles. In the previous example it is better to refer to *V* and *v* as two alleles of a petal color gene. It is conventional to denote dominant alleles with uppercase and recessive alleles with lowercase letters. The 3:1 ratio is

Table 1. Products of F1 self-fertilization in pea plants

Genotype	Phenotype (petal color)	Ratio
VV	Violet	1
Vv	Violet	2
vv	White	1

Table 2. Examples of inheritance controlled by a single gene

Species	Character (phenotype)
Mice	Albino/normal coat, pale/normal ears
Red clover	Red/white flowers
Fruit flies	Normal/vestigial wings
Humans	Blue/brown eyes, cystic fibrosis, sickle cell anemia, phenylketonuria

referred to as the **monohybrid ratio**, and is the basis for all patterns of inheritance in higher organisms.

Detection of heterozygotes

One simple extension of the 3:1 phenotype ratio is a 1:1 ratio produced when an F1 individual is crossed to the homozygous recessive parent. As shown in Figure 2 the heterozygous F1 can produce only two classes of gamete, carrying either the dominant or the recessive allele. The parent with the recessive phenotype can only produce gametes with recessive alleles, and so the progeny of the cross have the dominant and recessive phenotypes in equal numbers, a 1:1 phenotype ratio. This type of cross is termed a **test cross**, and is useful in any situation where it is necessary to determine if an individual is heterozygous. It is also the expected phenotype ratio in families where one parent carries a rare dominant allele, such as Huntington's disease. Because the dominant allele is rare, the affected individual is unlikely to be homozygous.

Variations of the 3:1 ratio

The simple 3:1 monohybrid ratio is not always observed in instances where only one gene is responsible for a particular phenotype. This may be due to a number of factors.

Partial or incomplete dominance

In the preceding section the example used showed complete dominance. In other words the phenotype of the F1 generation was identical to that of one of the parents (the dominant phenotype). This is not always the case. Often the F1 is clearly intermediate

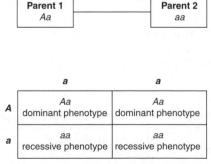

Figure 2.   The genotype and phenotype ratios are both 1:1 because the alleles contributed by the doubly recessive parent do not affect the phenotype.

between two parents. A simple example of this is the inheritance of petal color in snap-dragons. When pure-bred white- and pure-bred red-flowered plants are crossed, the F1 generation has pink rather than white or red petals. The F2 comprises three classes of plants (Table 3). This 1:2:1 ratio of three phenotypes is clearly different from the 3:1 but it is easy to see how the two relate. The cross is set out in Figure 3. The genotype of the red-flowered plants is *RR* and that of the white-flowered plants is *rr*. The F1 generation consists of *Rr* heterozygotes. In the F2 generation the ratio of genotypes is the same as that described previously. The difference here is that the *Rr* heterozygotes are pink, thus altering the ratio of phenotypes.

The red- and white-flowered F2 classes are homozygous and are therefore pure-breeding, whilst the pink heterozygotes when self-fertilized will always produce a 1:2:1 ratio of white-, pink- and red-flowered plants.

Codominance is similar to incomplete dominance, but here the heterozygote displays both alleles. Examples of this are found frequently in the inheritance of blood groups. In humans the MN blood group is controlled by a single gene. Only two alleles exist, *M* and *N*. Children whose father is an *NN* homozygote with group N blood and whose mother is a *MM* homozygote with group M blood are *MN* heterozygotes and have group MN blood. Both phenotypes are identifiable in the hybrid. Codominance also modifies 3:1 ratios to 1:2:1 ratios. Alleles that are differentiated by molecular methods such as polymerase chain reaction (PCR), or Southern blotting (Sections E1 and E2) are also codominant.

Table 3. Inheritance of petal color in the snapdragon

Genotype	Phenotype	Ratio
rr	White petals	1
Rr	Pink petals	2
RR	Red petals	1

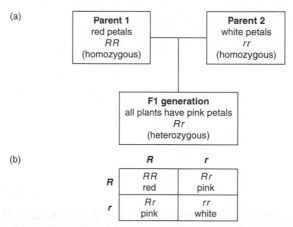

Figure 3.   (a) Gametes produced by the F1 plants carry *R* or *r* and these fuse at random to give offspring with the genotypes *RR*, *Rr*, or *rr* as shown in (b). (b) The genotype ratios are the same as in Figure 1 but the phenotype ratio is 1 red:2 pink:1 white.

Lethal alleles

Some alleles affect the viability of individuals that carry them. In most cases the homozygous recessive does not survive but the heterozygotes may have a normal life span. To be detected the heterozygotes need some observable phenotype. The best-known example of this is the inheritance of yellow coat color in mice. Yellow varieties can arise in strains of mice with different coat colors, for instance, black mice. Yellow coat color is dominant to black coat color: BB mice are black, BB^y mice are yellow. When two yellow mice are mated the progeny would be expected to be in the proportion shown in Table 4. However, B^yB^y is lethal and any mice with this genotype die *in utero*. Hence live-born progeny from this cross are in the ratio given in Table 5. The 3:1 phenotypic ratio has been distorted to a 2:1 ratio.

There is a concept here which sometimes causes problems to students. The allele B^y is recessive in relation to its effect on viability, but dominant in relation to coat color. It is important that you recognize this difference. It is quite common for a gene to be involved in two different phenotypes and an allele which is dominant for one phenotype may be recessive for another. Other examples where alleles are lethal when homozygous, but which have a dominant effect when heterozygous, include tailless Manx cats and short-legged Creeper chickens. Genes that are involved in developmental processes are often found to have lethal alleles. The presence of one mutant allele alters development so as to produce characteristic changes to the animal, but, when two are present, development is so aberrant as to cause death. This may occur *in utero* as described above or result in shortened life expectancy as found in several examples in humans, such as Tay-Sach's disease, Huntington's syndrome, or sickle cell anemia.

In other instances a homozygote may not be absent from a cross, but appear in reduced numbers. An example of this is vestigial wings in fruit flies, a condition that is caused by a recessive allele (*vg*). Alleles with this effect are referred to as **semilethals**.

These examples of alteration to the 3:1 ratio may at first appear complicated. It is essential that you realize that, in these cases, the behavior of the genes remains the same as in the 3:1 monohybrid ratio but the phenotypic ratio may change. Phenotypes depend on how genes act through protein synthesis, and how specific proteins interact in the cells and tissues of an organism. You know a great deal about this from Section A. You will see further examples of interaction between gene products in the following sections. However, if you concentrate on genotypes and the inheritance of alleles it will be much easier to understand how specific genes are influencing the inheritance of phenotypes.

Table 4. Expected inheritance of yellow coat color in a cross of $BB^y \times BB^y$ mice

Genotype	Phenotype	Ratio
BB	Black fur	1
BB^y	Yellow	2
B^yB^y	Yellow	1

Table 5. Actual inheritance of yellow coat color in a cross of $BB^y \times BB^y$ mice

Genotype	Phenotype	Ratio
BB	Black fur	1
BB^y	Yellow fur	2

Multiple alleles

All the examples used so far have employed genes with only two alternative alleles. For some genes only two alleles have been identified, but, for the majority of genes, a large number of alleles have been found. Examples of this include the human β-globin gene where a specific mutation at codon 6 results in an allele responsible for the hereditary syndrome sickle cell anemia, whilst mutations at several other sites in the gene cause a different syndrome, β-thalassemia. Although they are alterations of the same gene, the changes are to different codons. The resulting proteins have variant β-globins with discrete differences in amino acid sequence and so behave differently.

In rabbits, multiple alleles of one gene are responsible for a number of different coat-color phenotypes. There are four members of this allelic series: agouti, chinchilla, Himalayan, and albino. When homozygous each produces a distinct coat pattern. In heterozygotes there is a clear pattern of dominance. Agouti is dominant over all the other alleles, chinchilla is dominant over Himalayan and albino, while Himalayan is dominant only over albino, which fails to produce any pigment and hence is recessive to all the others.

Another well-known example of multiple alleles is the human ABO blood group system. Here a single gene codes for an enzyme that is responsible for the addition of sugar residues to a specific glycoprotein on the membrane of red blood cells. Three different alleles of the gene are known. One form of the enzyme adds a molecule of *N*-acetyl-galactosamine to the glycoprotein resulting in blood group A. A second allele codes for a variant enzyme that adds galactose instead of *N*-acetyl galactosamine resulting in blood group B. A third allele codes for a nonfunctional enzyme that cannot add any sugar to the glycoprotein, resulting in blood group O. All three alleles have arisen by mutation from a single ancestor.

The major histocompatability complex which determines the suitability of donor organs for transplantation is an example of a complex multiple allele system.

Modern developments

The traditional procedures of Mendelian genetics described above have altered very little since their development during the early part of the twentieth century. However, advances in our knowledge of genomes (Section B3) have produced huge numbers of polymorphic markers that have revolutionized our approach to genetic analysis. Polymorphisms identified as alterations of DNA sequences now dominate most genetical procedures. DNA polymorphisms are grouped in various classes (Section B3). Those whose variation depends on gain or loss of repeats are highly polymorphic (informative) and are used in forensic genetics (Section G1). Others such as RFLPs and SNPs have only limited variation in alleles, and are often described as biallelic. However, all DNA sequence markers have the great advantage that when analyzed using methods utilizing DNA hybridization or PCR the alleles are codominant and so dominance does not cause problems of interpretation.

C2 More Mendelian genetics

The dihybrid cross

Section C1 showed how the 3:1 monohybrid ratio could be used to explain the inheritance of a phenotype where only a single gene was involved. This is the basic Mendelian ratio and everything that follows depends upon it.

The obvious next step is to look at a situation where the inheritance of two different inherited characters are studied at the same time, a **dihybrid cross**. The simplest experimental system for this is to cross two pure-breeding strains of a species, one of which is homozygous for recessive alleles of two genes and the other homozygous for dominant alleles of the same two genes. Again Mendel was the first to carry out such experiments. In one of these he used pea plants that differed in two properties of the seed.

The choice of seed characteristics aided his work for two important reasons.

- It was easy to analyze large numbers of seeds accurately, thus improving the statistical basis of the work.
- By using seed characters it was possible to reduce the length of time the experiments would take. The F1 generation became the seeds produced by crossing the parental plants. When these seeds were planted out, grown on, and allowed to self-fertilize, the F2 generation was the seeds present in the pods. Thus he was saved the time (1 year) and effort of having to plant out these seeds and assess the characters in the plants that they produced.

Seed shape is determined by a single gene that has alleles for round, *R* (smooth), or wrinkled seeds, *r*. The round phenotype is completely dominant over wrinkled. Seed color can be yellow or green. Again this is determined by two alleles of a single gene. Yellow, *Y*, is completely dominant over green, *y*. The experiment is set out in Figure 1a. A cross between the two pure-breeding (homozygous) parental lines yielded an F1 generation which consisted only of round yellow seeds. As with the monohybrid cross (Section C1), the F2 generation showed considerable diversity. Four different phenotypes could be identified. Of 556 seeds analyzed, he found 315 round yellow seeds, 108 round green seeds, 101 wrinkled yellow seeds, and 32 round green seeds. This is close to a ratio of 9:3:3:1, which is referred to as the **dihybrid ratio**. Mendel obtained ratios close to this for several different combinations of pairs of genes, and since that time a great many other examples of this ratio have been demonstrated in crosses in plants, animals, and fungi.

The predicted ratios of phenotypes and genotypes can be determined graphically using what is known as a Punnett square. Simple examples of this were given in Section C1 to show the possible classes of progeny in the monohybrid cross. A Punnett square for the dihybrid cross is set out in Figure 1b. The 9:3:3:1 ratio is simply two 3:1 ratios combined, and shows that the alleles of the two genes behave (segregate) independently of each other. This is demonstrated more easily by looking at one gene at a time (Figure 2). Take the seed shape gene first: in the F2 generation a ratio of 3 round to 1 wrinkled would be expected. Now look at the seed color gene in those seeds that have the wrinkled phenotype. These should have a ratio of 3 yellow to 1 green. The same is true for wrinkled seeds, these should also have a ratio of 3 yellow to 1 green. If the ratios for the two phenotypes are multiplied across, the 9:3:3:1 ratio is obtained.

As noted earlier (Section C1) the 3:1 phenotypic monohybrid ratio can be distorted by factors such as incomplete dominance or lethal effects of certain alleles. These also affect the 9:3:3:1 ratio, but other factors can also modify this ratio. The 9:3:3:1 ratio depends on two conditions.

- The two different genes must not act on the same character. For instance if the proteins encoded by the two genes are involved in the same biochemical pathway then the ratios of phenotypes resulting from the genotypes in the F2 generation will be altered. This is discussed in detail later in this section, under **epistasis**.
- If the two genes lie close together on the same chromosome the four classes of gamete are not produced at equal frequencies. This is the basis of gene mapping studies and is examined under **linkage** (Section C5).

As with the monohybrid cross it is also possible to conduct a test cross with the F1 generation of the dihybrid cross. If, in the round/wrinkled, yellow/green example of the 9:3:3:1 ratio, an F1 plant is crossed with the homozygous recessive parent for both genes the progeny fall into four phenotypic classes: round yellow, round green, wrinkled yellow and wrinkled green. These classes occur with equal frequencies.

(a)

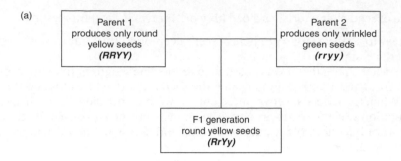

(b)

Gametes	RY	Ry	rY	ry
RY	**RRYY** Round Yellow	**RRYy** Round Yellow	**RrYY** Round Yellow	**RrYy** Round Yellow
Ry	**RRYy** Round Yellow	**RRyy** Round Green	**RrYy** Round Yellow	**Rryy** Round Green
rY	**RrYY** Round Yellow	**RrYy** Round Yellow	**rrYY** Wrinkled Yellow	**rrYy** Wrinkled Yellow
ry	**RrYy** Round Yellow	**Rryy** Round Green	**rrYy** Wrinkled Yellow	**rryy** Wrinkled Green

Phenotype class   *Genotype class*

Round, yellow seeds   *RRYY* (1), *RRYy* (2), *RrYY* (2), *RrYy* (4)
Round green seeds   *RRyy* (1), *Rryy* (2)
Wrinkled yellow seeds   *rrYY* (1), *rrYy* (2)
Wrinkled green seeds   *rryy* (1)

Figure 1.   (a) The production of F1 plants. These can produce four different gametes: *RY, Ry, rY* and *ry*. (b) The matrix (Punnett square) shows all the possible genotypes and phenotypes that can arise when these plants are self-fertilized. The figures in brackets indicate the number of times each genotype appears in the Punnett square.

It would be a useful exercise for you to work this out for yourself. However, this type of cross is most important in studies of gene mapping, and linkage and is covered in detail in Section C5.

At this point it is necessary to make a brief comment about systems where more than two genes are studied simultaneously. In the case of three genes, a **trihybrid cross**, in which complete dominance is observed for alleles of all three genes, the phenotypes observed fall into the ratio 27:9:9:9:3:3:3:1. The predicted ratios can be determined for crosses involving any number of genes by use of the expression $(3:1)^n$, where n is the number of genes.

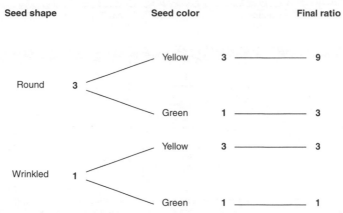

Figure 2.   Derivation of the 9:3:3:1 ratio. See text for explanation.

Epistasis

Epistasis refers to situations where the expected ratio of phenotypes is not observed due to some form of physiological interaction between the genes involved. This is usually seen as a distortion of the 9:3:3:1 ratio with a reduction in the number of different phenotypes observed. Many different ratios can be derived from the original 9:3:3:1 ratio. Three of these are shown in Figures 3, 4, and 5. Although each can be explained by specific examples from the literature, it is simpler if the different ratios are described here by reference to fictitious biochemical pathways in an unnamed plant species in which two enzymes, that are coded for by separate genes, are both involved in the production of petal pigments. The genes are denoted *A* and *B* with capital letters representing a functional dominant allele and lowercase letters representing nonfunctional recessive alleles. All of the examples involve the F2 generation produced from a doubly heterozygous *AaBb* F1 generation. In working through these you may find it useful to construct a Punnett square for the genotypes, as shown in Figure 1, and determine for yourself how the phenotypes are distributed.

In a strict sense 12:3:1 is the only ratio which was originally referred to as epistasis, because the presence of the A allele can completely mask the genotype of the *B* gene, but the term is used now wherever genes interact to alter the expected ratios.

There are several other variations of the 9:3:3:1 ratio caused by interaction between the gene products. These include 9:6:1, 15:1, and 13:3. You should attempt to think of biochemical pathways that would yield these ratios. Remember that in every case the ratios are derived by summing together the four phenotype classes 9, 3, 3, or 1 of the basic ratio.

The examples described here are deliberately made simplistic but illustrate the basic principles. You should realize that the term phenotype is capable of different interpretations. A plant breeder may be happy to simply use flower color as we have done to describe phenotype. On the other hand a plant biochemist might wish to interpret the results differently and assay enzymes A and B *in vitro*. The phenotypic ratios determined in this way would be different from those given above. If this is not obvious to you, work them out for yourself. The genotypes would, of course, not change.

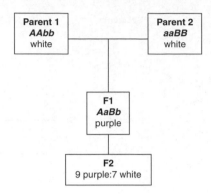

In this example the biochemical pathway would be a simple chain where enzyme A converts its substrate into a white product which is, in turn, the substrate for enzyme B which converts it to a purple product.

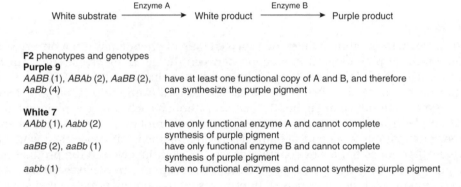

F2 phenotypes and genotypes
Purple 9

AABB (1), ABAb (2), AaBB (2), AaBb (4)	have at least one functional copy of A and B, and therefore can synthesize the purple pigment

White 7

AAbb (1), Aabb (2)	have only functional enzyme A and cannot complete synthesis of purple pigment
aaBB (2), aaBb (1)	have only functional enzyme B and cannot complete synthesis of purple pigment
aabb (1)	have no functional enzymes and cannot synthesize purple pigment

This is known as **complementation**, because the genes in the two white strains each complement the deficiency in the other strain, producing the purple product in the F1.

Figure 3.   9:7 ratio.

Mendel's laws

The monohybrid and dihybrid ratios come directly from the work of Mendel. Remember that this was carried out without any knowledge of chromosomes or DNA. Mendel's work is often expressed as the two laws or principles which he inferred to explain the inheritance of phenotypes.

*   The first of these states that in gamete formation the two alleles of the same gene **segregate** (separate) so that each gamete receives only one allele. This is clearly demonstrated in the monohybrid 3:1 ratio.
*   The second law states that alleles at any one gene segregate independently of alleles at any other gene. This derives from the dihybrid cross data where, in the example described in Figure 1, the four alleles *R, r, Y,* and *y* must act independently of each other so that the four different classes of gametes *RY, Ry, rY,* and *ry* arise in equal numbers.

These two principles form the basis of our knowledge of transmission genetics.

Handling problems

As either a student in an examination, or a geneticist carrying out research, you may be faced with data obtained from F1 and F2 generations of crosses. You would need to be

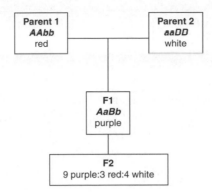

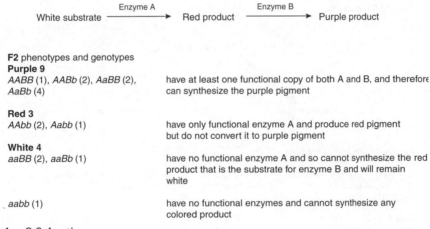

Figure 4.   9:3:4 ratio.

able to recognize ratios in order to decide how many genes are involved and whether or not epistasis is taking place. You may for instance be faced with the following example.

A cross between pure-breeding white-fruited and purple-fruited tomato plants produced an F1 generation in which all plants had purple fruit. In the subsequent F2 generation 160 plants were obtained; of these 99 had purple fruit, 25 had red fruit, and 36 had white fruit. How would you approach such a problem?

As you know nothing about the genes controlling fruit color in tomato you must first ask yourself the question, 'does the data fit any of the known Mendelian ratios?' The simplest way to proceed is to exclude those ratios which obviously do not apply. Clearly because there are three different phenotypes in the F2 generation any ratio with only two classes such as 9:7 or 3:1 are excluded. On examination of ratios with three phenotypic classes 9:3:4 looks a possible candidate, but a 1:2:1 ratio may also apply. How do you decide which ratio is the best fit to your data?

This is done using the χ^2 statistic, which is well suited to determining the goodness-of-fit to ratios. Detailed explanation of this test is given later (Section C14), but this example

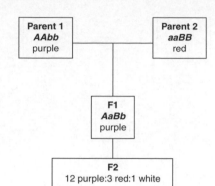

In this case we need a more complex biochemical pathway. Here two enzymes compete for the same substrate. Enzyme A which converts the substrate to a purple product has much higher affinity for the substrate than enzyme B which converts the substrate to a red product. The difference in affinity for the substrate is so marked that enzyme B can only work effectively if no enzyme A is present.

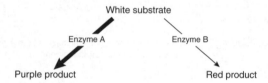

F2 phenotypes and genotypes
Purple 12
AABB (1), *ABAb* (2), *AaBB* (2),   these have at least one functional allele A and convert
AaBb (4), *AAbb* (1),   all the substrate to purple product
Aabb (2)

Red 3
aaBB (2), *aaBb* (1)   lack any functional enzyme A, but have functional
                                               enzyme B which converts the substrate to a red product

White 1
aabb (1)   have no functional enzymes and cannot synthesize any
                                               colored pigment

Figure 5.   12:3:1 ratio.

will be worked through here. The observed data are compared with that which would have been predicted by the ratio. This is set out for both possible ratios.

Observed result: 99 purple   25 red   36 white   160 total
Result predicted by a 9:3:4 ratio:

$$\tfrac{9}{16}\times160 \qquad \tfrac{3}{16}\times160 \qquad \tfrac{4}{16}\times160$$
$$90 \qquad\qquad 30 \qquad\qquad 40$$

$$\chi^2 = \frac{(99-90)^2}{90} + \frac{(30-25)^2}{30} + \frac{(40-36)^2}{40}$$
$$= 2.13 \text{ with two degrees of freedom}$$

This does not differ from the 9:3:4 ratio at the 5% level of probability. Result predicted by a 1:2:1 ratio:

$$\frac{1}{4}\times160 \qquad \frac{2}{4}\times160 \qquad \frac{1}{4}\times160$$
$$40 \qquad\qquad 80 \qquad\qquad 40$$

$$\chi^2 = \frac{(40-25)^2}{40} + \frac{(99-80)^2}{80} + \frac{(40-39)^2}{40}$$
$$= 10.53 \text{ with two degrees of freedom}$$

This is different from the 1:2:1 ratio at the 5% level of probability.

On this basis the 1:2:1 ratio is rejected in favor of the 9:3:4 ratio.

C3 Meiosis and gametogenesis

Key Notes

Reproduction

Reproduction can take place by asexual or sexual processes. In asexual reproduction only one parent is required and the genetically identical progeny form a clone. In unicellular organisms this is often accomplished by binary fission, and is the basis of vegetative propagation in plants. Sexual reproduction requires the fusion of two gametes to form a zygote. In the process of gametogenesis a specialized form of cell division, meiosis, causes halving of chromosome numbers and recombination of alleles. The latter is a result of both independent assortment of homologous chromosomes and crossing-over.

Meiosis

Meiosis is a two-stage division process that yields four daughter cells, each of which is haploid. In the first division (meiosis I), homologous chromosomes pair to form bivalents. As the cell progresses through prophase I the chromosome pairs shorten and individual chromatids and chiasmata become visible. At metaphase I the paired chromosomes separate. One member of each pair moves to the opposite poles of the cell. Anaphase I and telophase I follow resulting in the formation of two haploid cells. The second division (meiosis II) is divided into prophase II, metaphase II, anaphase II and telophase II. In meiosis II chromatids separate at metaphase II resulting in the formation of four haploid cells. This is often referred to as a tetrad. Pairing of homologous chromosomes in meiosis I is associated with the formation of the synaptonemal complex. Paired chromosomes cross over by reciprocally exchanging portions of their chromatids. In this process DNA double helices are broken and rejoined. This takes place at structures known as chiasmata.

Meiosis and recombination

Recombination, the rearrangement of alleles to new groupings, takes place by two processes: independent assortment of chromosomes at anaphase I, and by crossing-over at chiasmata. Both of these change the patterns of alleles that are passed on to the next generation.

Gametogenesis

Spermeogenesis in males and oogenesis in females have many similarities. In the gonads of both species, cells of the germline go through a series of mitotic divisions. These are followed by meiosis. Unequal division is seen in female

	meioses and only a single gamete is produced per germ cell. The other three cells produced are termed polar bodies. Males produce four gametes per germ cell. The process extends over a much longer period in females.
Production of aneuploid gametes	Nondisjunction at either first or second divisions of meiosis can result in aneuploid gametes. Nondisjunction in mitotic divisions after fertilization can result in mosaic individuals.
Related topics	(B4) Chromosomes (C4) Recombination (B5) Cell division (C5) Linkage

Reproduction

Reproduction takes place by one of two methods: asexual or sexual. Asexual reproduction involves the production of a new individual(s) from cells or tissues of a pre-existing organism. This process is common in plants and in many microorganisms. It can involve simple binary fission (splitting into two) in unicellular microbes, or the production of specialized asexual spores. These processes may be exploited for commercial purposes, as in the vegetative propagation of plants. More recently it has been possible to artificially regenerate whole organisms from a single cell. This was first shown in carrots and frogs, but it has now been reported in mammals, and, by implication, is possible in all mammals including humans. Asexually reproduced organisms are genetically identical to the individual from which they were derived. A group of such genetically identical organisms is known as a clone.

Sexual reproduction differs in that it involves fusion of cells (**gametes**), one derived from each parent, to form a **zygote**. The genetic processes involved in the production of gametes allow for some genetic changes in offspring. Sexual reproduction is limited to species that are diploid or have a period of their life cycle in the diploid state. The production of gametes is referred to as **gametogenesis**. This may be a complex process involving sexual differentiation and the production of highly differentiated male and female gametes, or in lower eukaryotes identical cells may fuse (isogamy). Whatever the biology of the process, one fact is obvious: gametogenesis must involve a halving of the chromosome number otherwise each succeeding generation would have double the chromosome number of its parents. Halving of chromosome numbers is achieved in a specialized form of cell division, **meiosis**, that is only observed in gametogenesis. During the process of reducing the number of chromosomes by half, the combinations of alleles are rearranged to give **recombinant gametes**. Two distinct processes are involved. These are **independent assortment** of chromosomes and **crossing-over**. These are described below.

Meiosis

The pattern of the meiotic cell cycle is very different from the typical mitotic cycle (Section B5). Meiosis involves two successive divisions resulting in the production of four cells each of which has half the number of chromosomes of the mother cell. There is no replication of DNA between the two divisions, which are known as **meiosis I** and **meiosis II**.

Meiosis I is divided into **prophase**, **metaphase**, **anaphase** and **telophase**. These are the same phases that are used to describe mitosis (Section B5), but beware, behavior of the chromosomes in the first division of meiosis is very different from that in mitosis.

In prophase each chromosome pairs with its **homolog** (copy of the same chromosome inherited from the other parent). The paired chromosomes are called bivalents (the term tetrad is sometimes used to describe these structures but should be reserved for the four cells produced at the end of meiosis). Each pair is held together by **chiasmata** (singular chiasma). These are structures where homologous chromosomes exchange reciprocal portions of chromatids. The process involves breaking of the DNA double helix in each chromatid at corresponding sites and then joining the DNA of one chromatid to the DNA of the other. This exchange of genetic material is known as **crossing-over**. The molecular basis of this process is outlined in Figure 1. See also Section C4.

The cytological progression of meiosis is set out schematically in Figure 2. Prophase of meiosis I is subdivided into five stages. In the first of these, **leptotene**, the chromatin is seen to condense into very long thin strands that appear tangled in the nucleus. As prophase proceeds the chromosomes become shorter and thicker. At **zygotene**, homologous chromosomes are seen as partially paired structures. The chromosomes are still very elongated at this stage and chromosome pairs may overlap or intertwine. By the third stage, **pachytene**, pairing is complete. It is still not possible to identify clearly the individual chromatids in each bivalent.

The pairing or **synapsis** of homologous chromosomes is a very precise process. The molecular mechanisms of this are still not completely resolved, but examination of paired chromosomes under the electron microscope reveals a structure known as the **synaptonemal complex**. This is a proteinaceous structure which is found only between homologs where they are synapsed (paired). Under the electron microscope it has a zipper-like appearance and is thought to be a major factor in the pairing process. The transition from pachytene to **diplotene** occurs as the homologous chromosomes begin to separate. This process begins at the centromeres and the bivalents are seen to be held

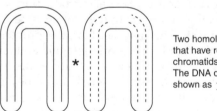

Two homologous chromosomes that have replicated into chromatids are paired together. The DNA double helices are shown as ——— or - - - - -

The chromatids break at the asterisk, and reciprocally exchange resulting in transfer of DNA and alleles of the genes that it encodes

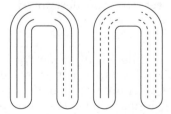

Figure 1.   The molecular basis of crossing-over.

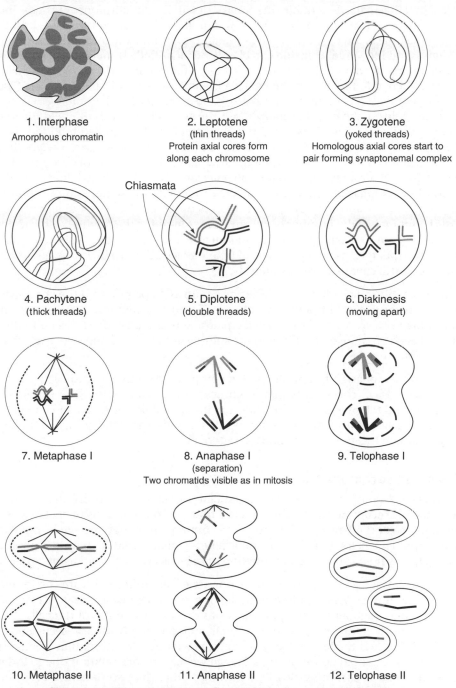

1. Interphase
Amorphous chromatin

2. Leptotene
(thin threads)
Protein axial cores form
along each chromosome

3. Zygotene
(yoked threads)
Homologous axial cores start to
pair forming synaptonemal complex

Chiasmata

4. Pachytene
(thick threads)

5. Diplotene
(double threads)

6. Diakinesis
(moving apart)

7. Metaphase I

8. Anaphase I
(separation)
Two chromatids visible as in mitosis

9. Telophase I

10. Metaphase II

11. Anaphase II

12. Telophase II

Sister chromatids separate to give four haploid cells – potential gametes.

Figure 2.   Diagrams 1–6 outline the changes in a nucleus undergoing meiosis I. Only two pairs of homologous chromosomes are shown. One of each pair is darker than the other. The terms in brackets under each of the prophase stages are literal translations of the meaning of the stage names. Steps 7–9 represent the remaining stages of meiosis I. After this the cells go into a transient interphase, and prophase of meiosis II. Steps 10–12 outline the major stages of meiosis II. The result is the formation of four haploid nuclei each of which carries recombined chromosomes.

together by chiasmata. The nucleolus breaks down. As the chromosomes continue to condense, the cell moves into **diakinesis**, the final subdivision of prophase I. In diplotene and diakinesis the four chromatids (daughter chromosomes) in each bivalent are identifiable and individual chiasma can clearly be identified. At **metaphase I** the nuclear envelope breaks down and the bivalents lie across the equator of the cell with their centromeres attached to the microtubules. This is similar to the spindle in mitosis (Section B5). The dynamic action of the spindle causes one member of each homolog pair to move to opposite poles of the cell. At **metaphase I** the sister chromatids are held together by proteins called cohesions. These hold the chiasmata in place and so hold the chromosomes together. At **anaphase I**, the cohesions in the chromosome arms are cut (but not in the centromeres) allowing the homologs to separate. The two sets of chromosomes are seen with their centromeres moving to the poles and the telomeres trailing behind. This is followed by **telophase I**, where two nuclei form around the segregating chromosomes and a degree of chromosome decondensation is observed. The two nuclei produced at meiosis I contain half the number of chromosomes found in the original cell. The diploid has been reduced to a haploid. The second meiotic division proceeds shortly after telophase I. Note that there is no replication of DNA between the two divisions, and that each chromosome still contains two chromatids.

The second meiotic division closely resembles mitosis. In **prophase II** the chromosomes are seen to recondense within the two nuclei. The nuclear membrane breaks down and the chromosomes are arrayed on the cell equator at **metaphase II**. At this point the centromere of each chromosome splits and the spindle pulls one chromatid of each chromosome to opposite poles.

Note that in metaphase I the spindle pulled complete chromosomes to each pole, while in metaphase II it is chromatids that are moved to opposite poles. This is followed by **anaphase II** and **telophase II**. At this stage the initial diploid cell has divided to give four haploid cells, each with a different genotype. In many instances the group of four haploid cells may remain together and is known as a **tetrad**.

Meiosis and recombination

From the stand-point of inheritance the crucial events in meiosis are those which are responsible for **recombination**, which means that the combinations of alleles passed by individuals to their offspring differ from the combinations they received from each of their parents. This helps to maintain the level of genetic variation within a population (Section D4). As noted above, recombination in meiosis arises by two distinct processes.

Each pair of homologous chromosomes consists of one inherited from the father and one from the mother. When paired homologous chromosomes separate (segregate) at anaphase I, one member of each pair moves to the opposite poles of the cell. The process does not differentiate between maternally and paternally inherited chromosomes and so both haploid nuclei are likely to contain a random combination of maternally and paternally inherited chromosomes. Hence we have recombination due to **independent assortment** of chromosomes. The likelihood that the two daughter cells contain new combinations of maternally and paternally inherited chromosomes depends on the chromosome number of the species, as set out in Table 1.

In the dihybrid ratios obtained by Mendel (Section C2) the pairs of genes showed independent assortment. Hence all four classes of gamete were produced in equal numbers, resulting in the 9:3:3:1 ratios he observed. The frequency of recombination in gametes is

Table 1. The likelihood of recombination according to chromosome number

Haploid chromosome number	Likelihood of recombination
1	0
2	0.5
3	0.75
n	$(2^n-2)/2^n$

further increased due to crossing-over. This only affects recombination between genes located on the same chromosome (synteny). The exchange of regions of chromatids inherited from different parents (observed at meiosis as chiasmata) will result in a new combination of maternally and paternally inherited alleles being inherited on the same chromosome by the next generation. This is discussed in great detail in Section C5. The breakage and rejoining of DNA double helices involved in this process is generally referred to as the mechanism of recombination, and is fundamental to many genetic phenomena in both eukaryotes and prokaryotes. For greater detail see Section C4.

Gametogenesis

Meiosis in animals is found only in ovaries and testes, and even in these tissues is restricted to cells that are destined to form gametes (the germline). Although the mechanisms of gametogenesis differ somewhat between organisms, the steps involved in gametogenesis in mammals outlined below are generally representative. These are summarized in Figure 3. In male gametogenesis (**spermeogenesis**), precursors of germ cells go through many rounds of mitotic divisions in order to maintain a pool of **spermatagonia**, which subsequently differentiate into **primary spermatocytes**. It is in these cells that meiosis takes place. After the first meiotic division the cells are referred to as **secondary spermatocytes**. These are haploid. The products of the second meiotic division are spermatids which differentiate into motile, tailed **spermatozoa**. The final activation of spermatozoa takes place after copulation.

In female mammals the pattern of **oogenesis** is superficially similar. Here **oogonia** go through mitotic divisions before differentiating into **primary oocytes**. These then undergo meiosis. Both daughter cells of the first meiotic division are haploid, but differ greatly in size. The larger daughter cell is the **secondary oocyte**, the smaller the **first polar body**. The two cells remain attached. Both undergo a second meiotic division. The secondary oocyte undergoes an unequal division producing a large **ovum** that contains almost all the cytoplasm, and a small **secondary polar body**. The primary polar body also divides to form two further secondary polar bodies. Only the ovum will transmit genes into the next generation. Thus oogenesis differs from spermeogenesis in that only one of the four meiotic products functions as a gamete. Another difference involves the timescale during which the processes take place. Spermeogenesis is a relatively continuous process beginning at puberty. In oogenesis, production of primary oocytes is completed in the fetus. Cells are held in prophase of meiosis I until puberty. The first meiotic division is only completed at ovulation, and the second division occurs after fertilization. In human females, primary oocytes can be held in meiotic arrest for up to 45 years. This may be important in the increased frequency of aneuploid births observed in older mothers (Section B4).

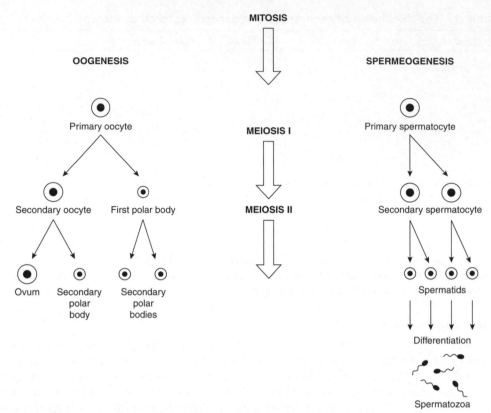

Figure 3.   Pattern of divisions in gametogenesis. For details see text.

Production of aneuploid gametes

Aneuploidy was described in Section B4. The major cause of aneuploidy is aberrant chromosome behavior at meiosis. Errors can arise at either of the two meiotic divisions. Failure of chromosomes to segregate properly is known as **nondisjunction**.

First division nondisjunction occurs when two homologous chromosomes move to the same pole at anaphase I. The end result of this is that, of the four cells arising from meiosis, two will be disomic (contain two copies of the chromosome) and two nullisomic (contain no copy of the chromosome) for the chromosome in question. Where two chromatids remain together at anaphase II the tetrad contains two normal monosomic cells (contain one copy of the chromosome), one nullisomic cell and one disomic cell. Aneuploidy can also arise due to nondisjunction at an early mitosis in the embryo. This results in two populations of cytogenetically different cells in the individual, which is known as a **mosaic**. This is common in Turner's syndrome (Section B4).

C4 Recombination

Key Notes

Overview

Recombination is a process which makes new combinations of genetic information on a DNA molecule (a chromosome) by cutting and splicing between existing DNA molecules (chromosomes) to form a crossover called a chiasma in meiosis.

General recombination

General recombination is a process of exchange between chromosomes (or DNA double helices) which share a long region (usually hundreds of base pairs) of homology. Two homologous double-stranded DNA molecules align, break, and rejoin by splicing to exchange their ends.

Holliday structure

A Holliday structure, or half chiasma, is an important intermediate in recombination. Two homologous (same sequence) double helices exchange nicked single strands of the same polarity, so that there is a partner switch. If the double helices rotate (rotary diffusion) the exchanged strands can wind across between the helices, and the branch point migrates. The four ends around the exchange point are structurally identical, and the Holliday structure can isomerize so that either the two 5′→3′ strands or the two 3′→5′ strands cross over. The two helices are eventually separated (resolved) by cutting and differently rejoining the crossed-over single-stranded DNA strands, either the two 3′→5′ or the two 5′→3′ strands. The ends of the helices are only exchanged (recombined) if the strands cut to resolve the structure are the ones which were not cut to create the Holliday structure in the first place (i.e. to achieve recombination, all four strands are eventually cut and swapped). The favored model for generating a Holliday structure by nicking single strands was the Aviemore (Meselson–Radding) model, where the strands to be exchanged are not nicked at identical points. Instead, one single nicked strand may transfer first and displace the identical strand from the other double helix. The displaced strand is then nicked and transferred further along. Degradation by exonuclease and synthesis by DNA polymerase can then make the ends meet. Ligation then restores continuity.

The double-strand break model

Recombination in the yeast *Saccharomyces cerevisiae* is initiated by a double-strand break in one chromatid (double helix). The 5′ ends are cut back by about 500 bases by exonuclease, and the remaining 3′ ends invade the homologous chromatid (double helix), generating two half-Holliday structures, one each side of the break. The gap is filled by synthesis, and further strand exchange completes the two Holliday structures. If these are resolved differently (one by cutting 5′→3′ strands, the other by cutting 3′→5′ strands) then recombination (exchange) of the ends of the chromatids will occur.

Conversion	The splicing of DNA molecules between chromatids with sequence differences (say one has AT where the other has CG) can generate hybrid (heteroduplex) DNA with mismatches (AG, CT). These may be corrected by mismatch repair pathways, converting the sequence on one chromosome to be like the sequence on the other. This gives a non-Mendelian ratio, three chromatids can have one sequence, one chromatid still has the other sequence. If the mismatch is not repaired before the next DNA replication phase, the mismatched bases will act as templates, generating different sequences in the two daughter cells after mitosis. This is called postmeiotic segregation (pms). Gap-filling synthesis in the double-strand break-repair model also converts the sequence in the gap.
Recombination enzymes	Most of the enzymes involved in the repair of DNA damage, both excision repair (ultraviolet-light-induced damage) and double-strand break repair (ionizing radiation, X-ray- and γ-ray-induced damage) also function in recombination in meiosis. There are several enzymes specific to meiosis as well. Deficiencies in some recombination-related enzymes are associated with human disease (e.g. breast cancer 1 gene product and ataxia telangiectasia).
Site-specific recombination	Bacteriophage lambda integrates into the host (*E. coli*) genome at a specific 15 bp sequence (*att* or 'O') found on both the bacteriophage and host chromosome. The bacteriophage gene *int* coding for integrase is necessary for this. Excision is also by recombination catalyzed by excisionase.
Related topics	(A1) DNA replication (A6) Mutagens and DNA repair (B6) Bacteriophages (C3) Meiosis and gametogenesis (C5) Linkage

Overview

Recombination means making new combinations of genetic information on a DNA molecule (a chromosome). (This is different from reassortment of whole chromosomes, which are separate DNA molecules, into gametes and zygotes.) Recombination mechanisms involve an interaction between two DNA double helices (= two double-stranded (ds) DNAs). The effect is that both dsDNAs break, and different ends rejoin, but the mechanisms are more like splicing events. The event is also described as a **crossover**, visible as a **chiasma** (plural **chiasmata**) at meiosis. A chromatid is the same as one double helix from a genetic viewpoint. Meiotic crossing-over (Sections C3 and C5) involves one double helix (= chromatid) from each of two homologous chromosomes.

General recombination

General recombination is the type of mechanism which eukaryotes use to recombine chromatids in meiosis. Bacteria also use it to incorporate homologous regions of foreign DNA into their chromosome after the DNA enters the cell by conjugation (sex), by transduction (inside a bacteriophage), or by transformation (simply DNA brought into the cell from its surroundings). It is a reciprocal process of exchange between two chromosomes, and in eukaryotes is normally conservative (nothing is gained or lost except for conversions; see below). The effect is that two homologous dsDNA molecules align, break at identical sites, and each broken end rejoins with the matching end from the other molecule, to exchange (recombine) the entire ends of the dsDNA. The molecular event is a splice, not a blunt cut and butt-end join, and sequence information is conserved. In bacteria the DNA which does not end up in a stable circular chromosome will be lost. General recombination requires two dsDNA molecules with a large region of homology (nearly identical sequence), usually hundreds of base-pairs long. The minimum length of homology probably varies between different species.

Holliday structure

The most important intermediate in recombination is a **Holliday structure** (named after Robin Holliday, and also called **Holliday junction** or **half chiasma**). This can be formed (theoretically) by making single-strand nicks in two strands with the same polarity, one in each double helix, and exchanging one single strand from each double helix, again with the same polarity, so that they pair with the intact strand in the other double helix (Figure 1). 3′ ends are shown being exchanged because these could prime DNA polymerase, but there is no evidence as to which ends cross first. This exchange is effectively a partner switch. In the original model, Holliday proposed that the nicks occurred at identical sites in both double helices (chromatids) as in Figure 1, but genetic evidence suggests that some asymmetrical exchange occurs.

To take account of this asymmetry the model was revised and called the **Aviemore model** after the location of the workshop where it was discussed, or the **Meselson–Radding model** after the publishers of the revision. The major difference from Holliday's model is that one single strand is nicked first and invades the other duplex (double helix).

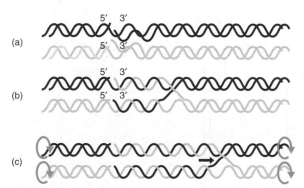

Figure 1.   (a) Formation of a Holliday structure by nicking of single strands and exchange of ends of the same polarity. (b, c) Branch migration by rotary diffusion moving exchange point.

The displaced strand is partially degraded, then transfers into the original donor duplex to complete the Holliday structure (Figure 2a). The single-strand gap on the donor is filled by synthesis of DNA by DNA polymerase (dashed line Figure 2a). In this way the transfer is asymmetrical near the original nick. This model is very versatile because any ends that are too long can be digested back by exonuclease, any gaps can be filled by polymerase, and free ends can be joined by ligase.

Holliday structures have interesting properties. If the double helices rotate (**rotary diffusion**) in the same direction then both the exchanged single strands can unwind from one helix and wind onto the other. In this way, the exact point where the strands cross over (the branch point) can move along the chromatids. This is called **branch migration** (Figure 1c). Another feature is that the four double helices going away from the site of the exchanged single-stranded (ss) DNA are structurally identical (Figure 2). The arrangement can undergo **strand isomerization** so that either the two 3'→5' strands are exchanged or the two 5'→3' strands (the strands going the other way) are exchanged. Please refer to Figure 2 while reading this. At the start (Figure 2a) the single-strand ends are transferred.

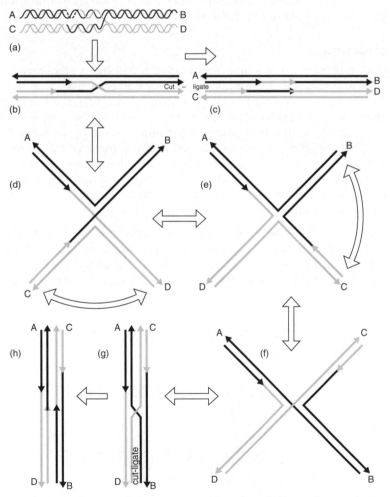

Figure 2. Formation (a, b), strand isomerization (d, e, f) and alternative resolution (b, c and g, h) of a Holliday junction.

This is more easily drawn as straight lines (Figure 2b). In the original arrangement the ends A and B are on one chromatid, ends C and D are on another. The structure can open (Figure 2d) then rotate to interchange ends C and D, uncrossing the cut strands B–C and D–A (Figure 2e). The structure is now open with all four ends equivalent. Interchanging ends B and C now makes the continuous (uncut) strands (A–B and D–C) cross over (Figure 2f). Cutting these strands and ligating their ends gives recombination (a crossover) (Figure 2g, h). Now the ends A and D are on one chromatid and the ends C and B on another.

Thus the structure can be **resolved** by cutting two DNA strands of the same polarity and rejoining them differently to separate the two dsDNA helices. If the single strands which were exchanged first are cut and the ends swapped and ligated, then the original dsDNA molecules will be restored without any recombination and the same ends will still be joined together (Figure 2c). If the other two strands are cut and swapped, both single strands in both helices will have been cut and swapped. This is a staggered double-strand break and will cause **recombination** of the genes on either side, because the ends of the DNA molecules will have been swapped over. That is what recombination is: a re-combination. This can only be understood by following the diagrams in Figure 2, or by making a three-dimensional model from rope or tubing. All the functions are carried out by proteins which encircle the crossover point, force rotation by unwinding and rewinding opposite helices, and cut/ligate the crossed single strands.

The double-strand break model

The double-strand break model of recombination was developed to explain the observed features of recombination in *S. cerevisiae*, bread yeast. It is quite possible that this model operates in all eukaryotes, and that the Aviemore model never actually occurs in nature. However, there is little evidence relating to the mechanism actually operating in other eukaryotes with larger genomes. Some scientists believe that the greater complexity and size of the genomes of most eukaryotes require a more cautious mechanism of recombination than double-strand breaks. Recombination in *S. cerevisiae* is initiated by the generation of a double-strand break in one double helix (Figure 3a). The protein which does this, SPO11, is related to a type of topoisomerase II, and forms covalent DNA-protein links across the break. The 5' ends are digested back from the break to leave about 500 bp of 3' ends, detectable biochemically. The model proposes that these 3' ends invade the intact homologous DNA double helix (Figure 3b), and can prime extension synthesis across the gap by DNA polymerase using the complementary strand of the intact dsDNA as a template (Figure 3c). When the gap is bridged, further strand exchange generates two Holliday structures, one on each side of the original break (Figure 3d). If these are both resolved in the same way (the same single strands are cut/rejoined in each) then there will be no recombination (Figure 3e) but the sequence in the gap in the broken chromatid will have been **converted** to be the same sequence as the other chromosome. If the two Holliday structures are resolved differently, there will be an exchange (crossover) which may be either side of the gap, and the ends of the two double helices will be recombined (Figure 3f).

Conversion

Conversion is a phenomenon which results in non-Mendelian ratios in the four haploid cells produced by one meiotic cell. If all genetic information is conserved, then two of the four haploid cells will have copies of one allele and the other two cells will have copies of the other allele, in accordance with Mendel's laws. Sometimes this rule is broken and

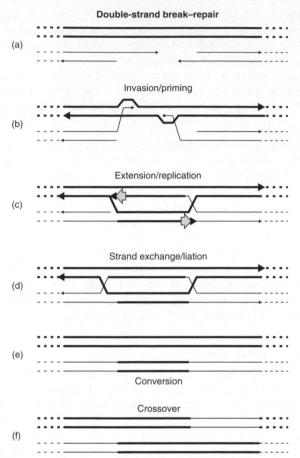

Figure 3.   The double-strand break model of recombination. (a) There is a double-strand break, then (b) the free ends invade an intact homologue. They are extended across the gap (c, open arrows), and one Holliday structure forms at each side (d). These can resolve to give conversion (e) or recombination (f).

a **3:1 ratio** occurs. In some fungi there is a second mitotic division which duplicates the haploid products of meiosis, forming eight haploid spores, changing the 3:1 ratio into 6:2. These can also give a **5:3 ratio**, **aberrant 4:4 ratio** or rarely a **7:1 ratio**. In each case the sister cells from one mitosis are genetically different, an event called **postmeiotic segregation** or **pms** (described below). The spliced DNA double helix formed during recombination must have contained a mismatched base pair (Figure 4). (Note that in the asymmetric exchange region only one chromatid of the four can have a mismatch.)

Suppose that one parent chromosome of a pair has an AT base pair, whereas the other (its homolog) has CG. If single strands are exchanged in this region then the **heteroduplexes** (made by one ssDNA strand from each chromosome) will have an AG base pair and a CT base pair. If these **mismatches** are recognized by repair enzymes they may be changed to an AT or a CG base pair, and either mismatch may be repaired either way (or not repaired at all), so three of the four chromatids may contain the same sequence. If a mismatch is not repaired before the next DNA synthesis then the two strands will be used as templates for replication into new duplexes. These will separate at the next mitosis giving

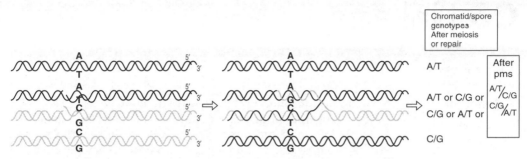

Figure 4.   Gene conversion by mismatch correction in heteroduplex DNA.

daughter cells with different sequences, one AT and the other CG. Because this happens at the mitosis after meiosis it is called **postmeiotic segregation**. There are three possibilities on each chromatid, two directions of repair, or no repair, so there are nine possible combinations, one of which restores the parental arrangement. A heteroduplex which could contain mismatches is also shown in Figure 2 and Figure 3. Symmetrical exchange without mismatch repair gives the aberrant 4:4 ratio after pms.

Conversion, particularly a 3:1 ratio after meiosis, is easily explained by the double-strand break model because the base sequence in the gap in the broken chromatid is always replaced by sequence information from the other homologous chromosome.

Recombination enzymes

Recombination involves many proteins. Several are similar to *E.coli* RecA protein; they bind ssDNA and insert it into homologous dsDNA. The proteins in bacteria, yeast, and humans are very similar. Most enzymes involved in DNA repair also function in meiotic recombination. Repair of double-strand breaks caused by ionizing radiation is very efficient in yeasts and involves high levels of recombination as the double-strand break-repair model predicts. Excision repair enzymes are also widely active in meiotic recombination. Mismatch recognition complexes, which locate errors such as thymine dimers after ultraviolet irradiation, also seem to function in meiosis. Some mismatch repair proteins are present at the stages of pairing and synapsis of the entire homologous chromosomes, and may help to restrict recombination to homologous DNA. Rejection of mismatches may prevent recombination between chromosomes in hybrids between related species, giving post-mating isolation of those species (Section D7). Other repair proteins localize at recombination events. The genes required for recombination are often essential for normal repair of damage to DNA. When they malfunction, the DNA damage causes mutations which may cause cancer. The normal product of the human breast cancer gene, *Brca1*, is a component of the synaptonemal complex, which has a role in pairing and recombination of chromosomes in meiosis (Section C3). The genes which are defective in ataxia telangiectasia are also important in meiotic recombination.

Site-specific recombination

Bacteriophage lambda can integrate its DNA into the chromosome of its host *E. coli* staying dormant and being replicated with the host DNA (a state called lysogeny). It does this by recombination between a 15 bp sequence called *att* (for attachment) in the host chromosome and an identical sequence in the circular bacteriophage chromosome (*att* may also be called 'O'). This reaction requires an integrase enzyme (called Int) encoded by the bacteriophage. Excision also occurs by recombination, this time catalyzed by excisionase (coded by the bacteriophage *xis* gene).

C5 Linkage

Key Notes

Definition

Linkage is the tendency for alleles of different genes to be passed together from one generation to the next. Only genes situated on the same chromosome can show linkage. A series of alleles on a single homologous chromosome is referred to as a haplotype. Genes on nonhomologous chromosomes are, by definition, unlinked and always show 50% recombination. Parental gametes carry the same set of alleles as were inherited together from one parent. Recombinant gametes carry alleles derived from both parents. The degree of linkage between two genes depends on the frequency of crossovers that occur between them during meiosis. The closer they are together the less likely a crossover will occur between them. Groups of genes that are linked are called linkage groups.

Simple measurement of linkage

Two-factor crosses involve crossing a double heterozygote to a double recessive. The frequencies of different phenotypes in the progeny equal the frequencies of different gametes in the double heterozygote. The proportion of recombinant gametes is the recombination frequency. This is used to give a measure of the distance between two genes. The percentage recombination between two genes is taken as the distance they are apart. One percent recombination equals one map unit or centimorgan (cM). In linkage analysis in humans it is more common to utilize molecular techniques to analyze genotypes than to study phenotypes.

Three-factor crosses

These are more accurate than two-factor crosses in that they identify and utilize many of the double crossovers that are missed by the two-factor cross. They allow ordering of genes and generate additive map distances.

Interference

It is difficult for chiasmata to form close to one another. For this reason the number of double crossovers observed in three-factor crosses may differ from that predicted from the frequency of single crossovers. If fewer crossovers are observed this is termed positive chromatid interference. This is measured as the coefficient of coincidence.

Linkage analysis in fungi

In ascomycete fungi that have ordered asci, first and second division segregation patterns can be used to determine distance between a gene and the centromere. A similar approach can be used to estimate distances between genes.

Recombination frequency and physical distance

The distance between genes as measured by recombination frequencies is not a precise measure of the physical distance because the frequency of crossing-over varies in different parts of the genome. It is useful for deciding on gene order. Physical maps provide distance between genes in absolute terms.

| Related topics | (B1) Concepts of genomics | (F1) Genetic diseases |
| | (C3) Meiosis and gametogenesis | |

Definition

Linkage is the tendency for alleles of two or more genes to pass from one generation to the next in the same combination. This usually means that the closer together any two genes lie on the same chromosome the more likely they are to show linkage.

A single homologous chromosome or chromosomal region can be described in terms of the alleles it possesses. These are referred to as a **haplotype**. The term can also be used to describe a complete haploid genome such as those of mitochondria or chloroplasts.

Genes located on different (nonhomologous) chromosomes cannot, by definition, show any linkage. As discussed in Section C3, genes on different chromosomes assort independently at the first division of meiosis.

Consider a double heterozygote *Aa, Bb* where the two genes involved are on different chromosomes. Assume that this individual has inherited the dominant alleles *A* and *B* from one parent, and the recessive alleles *a* and *b* from the other. In meiosis four genetically different gametes, *AB, Ab, aB*, and *ab* will be produced in equal proportions, Those gametes carrying *AB* or *ab* are referred to as **parental** and those carrying *aB* or *Ab* as **recombinant**. The recombination frequency is obviously 50%. This is the maximum recombination frequency that can be obtained between any two genes, and genes that show this recombination frequency are said to be **unlinked**.

What would have happened had the two genes been located on the same chromosome? Now the parental allele combinations *AB* or *ab* can only be rearranged (recombined) if a crossover occurs at a chiasma between the two genes (Section C3). The closer the two genes are together the less likely it is that a crossover will take place between them. The frequency of recombinants observed will therefore give an indirect measure of how close the two genes lie to each other.

In practice the frequency of crossovers on most chromosomes is high, and this means that genes that lie far apart on the same chromosome show 50% recombination. If two genes show linkage and a third shows linkage to only one of the original two, by definition they must all be on the same chromosome. They are said to constitute a **linkage group**. Thus linkage maps can be built up even though some genes in the group may not show linkage to all other members of the group.

Simple measurement of linkage

Except in certain circumstances where linkage can be measured to the centromere it is necessary to have heterozygosity for at least two genes in any study of linkage. Such a system is called a two-factor cross. An example is set out below.

An F1 hybrid was made by crossing two pure-breeding strains of tomato, one of which was homozygous for purple fruit and hairy stems, *PP HH*, and the other was homozygous for red fruit and smooth stems, *pp hh*. The hybrid had purple fruit and hairy stems. Thus purple fruit is dominant to red fruit, and hairy stems is dominant to smooth stems. To determine if the genes for fruit color and hairiness of stem are linked, the doubly heterozygous hybrid was crossed to the double recessive (red-fruited and smooth-stemmed) parent. This is a test cross (Sections C1 and C2). Four classes of progeny were obtained (Table 1): classes 1 and 2 are the same phenotypes as the original parental strains and are therefore *parental*; classes 3 and 4 represent new combinations of the alleles at the two genes and are therefore *recombinant*.

If the genes were unlinked a ratio of 1:1:1:1 would be expected between the four possible phenotypes (Section C2). Clearly, this is not the case. The total number of progeny is 500 of which only 70 are recombinant. The recombination frequency is 70/500 (14%). This is strong evidence that the two genes are linked. They have a map distance of 14 map units between them (1% recombination is equivalent to 1 map unit). Map units are referred to as centimorgans (cM) in memory of Thomas Hunt Morgan who was the first geneticist to explain linkage.

Is it necessary to use a test cross to determine linkage? It is not essential, and other crosses can be used; for instance, crossing the double heterozygotes together as in a dihybrid cross (Section C2) will give data on linkage, as the numbers in recombinant classes of the 9:3:3:1 ratio will be reduced. However, because the double-recessive parent contributes nothing to the phenotypes of the progeny in the test cross, the ratios obtained represent exactly the ratios of gametes and so this cross gives maximum information on linkage. In current studies of linkage in humans (Section B1) recombination frequencies are usually determined by the use of molecular methods to identify differences in DNA sequence (the genotype) rather than the phenotype as described in the examples given here. Working with genotypes allows more information to be obtained from crosses. The two-factor cross has limited use in determination of linkage and gene mapping studies. To map the order of genes along a chromosome and to give more accurate estimates of the distances between these genes at least three genes should be studied in the same cross.

Three-factor crosses

A major advantage of a three-factor cross is that genes may be placed in order and that the map distances, at least over relatively short distances, are additive. The worked example given below again involves a test cross. For simplicity, in this example alleles denoted by capital letters are completely dominant over the lowercase alleles, and the three genes are simply named after their dominant alleles.

Table 1. Phenotypes produced in a test cross between F1 generation purple-fruited, hairy-stemmed tomatoes (*Pp*, *Hh*), and the double recessive red-fruited, smooth-stemmed parent (pp, hh)

Phenotypic class	Frequency	
(1) Purple-fruited, hairy-stemmed	220	} parental
(2) Red-fruited, smooth-stemmed	210	
(3) Purple-fruited, smooth-stemmed	32	} recombinant
(4) Red-fruited, hairy-stemmed	38	
Total	500	

Table 2. Phenotypes of progeny produced from the cross between an individual heterozygous for three genes, and one homozygous recessive for the same three genes

Phenotypes of progeny	Frequency	Class
Parental		
ANR	347	(1)
anr	357	(2)
Recombinant		
ANr	52	(3)
anR	49	(4)
Anr	90	(5)
aNR	92	(6)
AnR	6	(7)
aNr	7	(8)
Total	1000	

An individual heterozygous at three genes *Aa, Nn,* and *Rr* is crossed with the homozygous recessive parent *aa, nn,* and *rr.* The frequency of progeny with different phenotypes is given in Table 2.

If there were no linkage between these genes each of the eight classes of progeny should have arisen with equal frequency. This is clearly not the case: classes (1) and (2) represent progeny from gametes where no recombination had taken place between the three genes. These are parental gametes because they retain the parental combination of alleles of the three genes: classes (3) and (4) represent progeny from gametes where recombination has occurred between the genes *N* and *R*; classes (5) and (6) represent recombination between the genes *A* and *N*; and classes (7) and (8) are the most important because they represent gametes where recombination has taken place both in the interval between *A* and *N* and also between *N* and *R*. These are referred to as double crossovers; they are less frequent than any of the other classes because they require two independent events to take place. The two least frequent classes of progeny in a test cross of this nature can be used to identify which gene lies between the other two. In this case the central gene is *N*.

To determine the map distance between the genes it is necessary to quantify all the recombination events that have occurred.

(i) To determine the map distance between *A* and *N*:
add progeny in classes (5) and (6)   (recombination between *A* and *N*)
and progeny in classes (7) and (8)   (double crossover, one of which is between *A* and *N*)
Express the total as a percentage of all progeny.

$$\frac{(90+92+6+7)\times100}{1000} = 19.5\% \text{ recombination, or } 19.5 \text{ cM.}$$

(ii) To determine the map distance between N and R:
by the same logic as used for the map distance between A and N the distance between genes N and R is

$$\frac{(52+49+6+7)\times100}{1000}=11.4\% \text{ recombination, or } 11.4 \text{ cM.}$$

(iii) To determine the map distance between A and R:
this requires a summation of all recombinants.

$$\frac{(52+49+90+92+6+7+6+7)\times100}{1000}=30.9\% \text{ recombination, or } 30.9 \text{ cM.}$$

Note that the distances are additive.

$$A \longleftarrow \text{ 30.9 cM } \longrightarrow R$$
$$A \xleftarrow{\text{19.5 cM}} N \xleftarrow{\text{11.4 cM}} R$$

Three-factor crosses are more accurate than two-factor crosses because they detect and utilize double crossovers that would go undetected in two-factor crosses. Because more recombination events are detected the calculated recombination frequencies are greater. This is not a problem as recombination frequencies merely reflect the frequency of crossovers detected rather than act as a physical measurement of actual distance between genes. This point is referred to again in Section G1.

Interference

Crossing-over takes place at chiasmata. These are physical structures involving two chromatids. Not surprisingly the presence of one chiasmata in a particular chromosome region can reduce the frequency of others forming close to it. This can result in a reduction in the number of double crossovers observed. In the previous example the observed frequency of double crossovers is 13/1000 (0.013). By using the data for the observed single crossover between A and N and between N and R we can predict the expected number of double crossovers.

Single crossovers between A and $N=195$
Single crossovers between N and $R=114$
Predicted number of double crossovers $(195/1000) \times (114/1000) \times 1000 = 22.3$

This is greater than the observed number, 13, suggesting that **positive chromosome interference** is observed in this region. The extent of interference is calculated as the **coefficient of coincidence** (**S**), the observed number of double crossovers divided by the expected number of double crossovers. In this case it would be:

$$S = \frac{13}{23.3} = 0.58$$

Linkage analysis in fungi

Several ascomycte fungi produce asci which hold the haploid ascospores produced after meiosis in a specific linear order, an **ordered tetrad**. This order reflects the organization of the bivalents at meiosis. The distal and proximal pair of ascospores each contain the products of one bivalent after the completion of both meiotic divisions. In some species the ascus contains eight spores. This is simply due to each spore having duplicated by mitosis (Section B5) so that the original pattern is displayed by pairs rather than single spores. The example shown in Figure 1 represents the products of meiosis in a fungus heterozygous for pale and dark spores. The asci contain eight spores.

Where the four asci of the same color are found together there have been no crossovers between the spore color gene and the centromere. This pattern is known as **first division segregation**, because the two phenotypes are physically separated at the first meiotic division. In order to show other patterns a crossover must have taken place between the centromere and the gene for spore color. The color phenotypes are now separated after the second meiotic division, **second division segregation**. Depending on which chromatids are involved in the crossover, different patterns of light and dark spores can be observed in asci showing second division segregation (recombinant asci). These are shown in Figure 2. The percentage recombination is determined as:

$$\frac{1/2(\text{number of second division segregating asci}) \times 100}{(\text{total number of asci})}$$

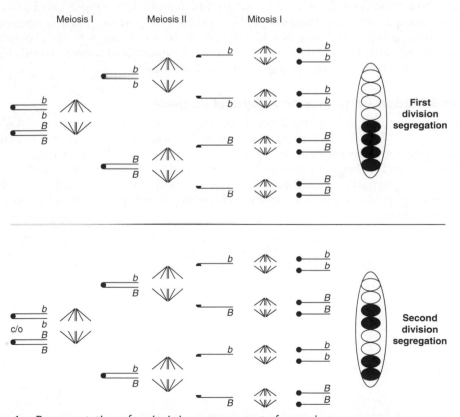

Figure 1. Representation of meiosis in an ascomycete fungus heterozygous at a gene for spore pigment. The allele *B* codes for dark spores, and allele *b* for pale spores. In the upper panel there is no crossover between the spore pigment gene and the centromere. This results in first division segregation. In the lower panel there is a single crossover (c/o) between the gene and the centromere. The color phenotypes do not separate until the second meiotic division; this is second division segregation.

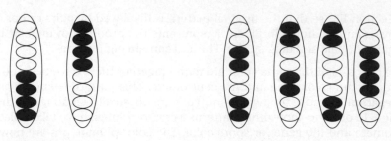

First division segregation   Second division segregation

Figure 2.   Different arrangements of asci in ascospores showing first or second division segregation. As shown each pattern is present twice depending on which member of the bivalent is distal at the time of meiosis. The two different patterns of second division segregation asci depend on which pairs of chromatids were involved in the crossover.

The number of second division segregating asci is halved because in each of these asci only half of the chromatids have had a crossover between the spore color gene and the centromere. As well as the distance to the centromere the distance between two genes can also be mapped in ordered tetrads using a similar procedure. Map distances can also be determined by counting phenotypes of individual ascospores in fungal species which do not produce ordered tetrads, or when the geneticist is too lazy to micro-dissect each ascospore individually from the ascus. However, only analysis of ordered tetrads allows mapping of the distance between a gene and the centromere.

Recombination frequency and physical distance

We have assumed that recombination frequency is a measure of the actual distance between genes. This is only true in a semiquantitative sense. It is now known that the frequency of crossovers varies in different regions of the genome. It tends to be lower near centromeres and higher near telomeres. It can also differ between males and females. However gene maps based on linkage data give a reasonably accurate indication of distance between genes and they are most useful in determining the order of genes along the chromosome.

C6 Transfer of genes between bacteria

Key Notes

Overview

Bacteria do not show Mendelian inheritance, because they are haploid and lack meiosis. However they can undergo recombination. An artificial process for achieving this is transformation, where DNA from one bacterial strain is mixed with another. DNA is taken up only by competent cells, and can alter their phenotype. Frequencies of co-transduction can be used to map closely linked genes. Processes involving bacterial viruses and plasmids also bring about recombination.

Plasmid-mediated gene transfer

Plasmids are chromosome-like structures found in bacterial cells, but which are not essential to bacterial growth. Some can move from one bacterium to another. The best studied is the F plasmid. This is an episome, an element that can replicate either independently or integrate into the bacterial chromosome. In the chromosome it can cause transfer of the entire chromosome to a recipient cell. A bacterium with an integrated F factor is called an Hfr strain. During conjugation, the time at which each gene enters the recipient can be used to create a gene map. When the F plasmid is excised from the bacterial chromosome it sometimes incorporates part of the chromosome into its own structure. It is then known as an F factor. This also transfers bacterial genes but at lower frequency.

Bacteriophages and gene transfer

Bacteriophages are involved in transduction. Virulent phages replicate in the bacterium. An occasional phage particle will contain a small fragment of bacterial DNA rather than bacteriophage DNA. The bacteriophage will transfer it to a recipient cell. This is generalized transduction.

Temperate bacteriophages can integrate into the bacterial chromosome, and remain in a dormant lysogenic state. Integration occurs only at specific sites. Dormancy is lost at a low frequency and the bacteriophage genome is excised and begins to replicate. In some cases the excision is not accurate and the resulting bacteriophage particles incorporate some host DNA sequences. These can be transduced to any cell the bacteriophage subsequently infects. The bacteriophage can only incorporate host DNA that is contiguous to its site of integration, hence this is termed restricted transduction.

Recombination in bacteria	Recipient bacteria are converted to partial diploids or merozygotes by conjugation and transduction. Recombination to give new stable genotypes takes place by double crossovers between the host chromosome and the donated DNA.	
Related topics	(B2) Prokaryotic genomes	(E4) DNA cloning and
	(B6) Bacteriophages	transfection
	(C5) Linkage	

Overview

Bacteria are prokaryotes and do not show the conventional Mendelian genetics of eukaryotes. This is primarily because they are haploid and lack any cell division process that is equivalent to meiosis. However, this does not mean that they do not show genetic recombination or transfer of genes between individuals; they have a number of mechanisms that allow gene transfer, some of these are prokaryote-specific but some have an equivalent in eukaryote cells.

It is possible to transfer DNA extracted from one bacterial culture to recipient bacteria *in vitro*. This is a process known as **transformation**, or transfection. Recipient bacteria are treated so as to increase the proportion of cells that are **competent** to take up exogenous DNA. These are then mixed with purified DNA and those that take up the DNA are selected as clones on agar plates. This can be useful in constructing strains of bacteria with specific genetic markers or in moving genes between species. It can also be used to identify and map linked genes. This is achieved by comparing the frequencies of transformation for individual genes with the frequency at which they show co-transformation.

If gene A transforms recipient bacteria with a frequency of 10^{-4} and gene B with a frequency of 4×10^{-5}, the predicted frequency of bacteria co-transformed with both genes would be 4×10^{-9} (the product of the two individual frequencies). This assumes that each transformation event is due to uptake of separate fragments of DNA carrying either gene A or B. If, however, the two genes are close together on the chromosome of the bacteria used as donor for the DNA, then a proportion of the DNA molecules will carry both genes and so the frequency of co-transformation will be increased. This process only allows identification of closely linked genes, but if one gene can be shown to be linked to two other genes in separate experiments, then all three genes must lie close together on the bacterial chromosome even if the outer two are too far apart to show co-transformation. Thus a gene map can be built up in a manner similar to the linkage maps in eukaryotes where recombination frequencies are measured.

Transformation, although it may possibly occur in the natural environment of bacteria, is basically an artificial system for the mapping and manipulation of genes in the laboratory. However, natural processes in prokaryote organisms can also be utilized in mapping bacterial genes. These depend on either the action of bacterial viruses (**bacteriophages**), or chromosome-like structures known as **plasmids**.

Plasmid-mediated gene transfer

Plasmids are found in many bacterial species. They are composed of double-stranded DNA, code for proteins and are analogous to the bacterial chromosome itself. However plasmids can be removed (cured) from bacteria without harm to the host cells and so it is clear that they are nonessential structures. In many cases plasmids can move from one bacterium to another, and even between bacteria of different species. The transfer process is directed by genes carried on the plasmids. The ability of plasmids to move between bacteria can be utilized to study transfer of genes from the bacterial chromosome.

The first plasmid to be shown to move between bacteria was the F plasmid. This plasmid carries genes that direct the construction of pili, structures formed between bacteria that contain the plasmid, F$^+$, and those which lack the plasmid, F$^-$, through which the F plasmid can be transferred. The plasmid is duplicated as it is transferred. This means that the donor cell retains a copy of the plasmid, and that in a mixture of F$^+$ and F$^-$ bacteria the F plasmid will eventually spread to all bacteria in the culture.

The F plasmid is sometimes known as the F, or fertility, factor, and F$^+$ bacteria as males. This is unhelpful, as there is no parallel with the processes of sex determination in eukaryotes.

The F plasmid can exist in the bacterial cell in an autonomous state where it replicates independently of the bacterial chromosome. It can also integrate into the bacterial chromosome and replicate as part of that chromosome. We refer to plasmids that can exist in these two states as **episomes**. The process is reversible and occasionally the F plasmid is excised from the chromosome. Sometimes the excision process is aberrant resulting in variable lengths of the bacterial chromosome being incorporated into the plasmid, which is now termed an F′ (F prime). This is shown in Figure 1a. F primes are capable of moving to F$^-$ bacteria and in so doing bring their complement of host genes with them. This is known as **sexduction** (Figure 1b). Gene transfer by this process occurs at low frequencies.

Much higher frequencies of transfer are obtained when the F plasmid remains integrated within the chromosome. It still retains its ability to transfer itself to F$^-$ bacterial cells but now carries the rest of the chromosome with it. The process is known as **conjugation** (Figure 2a), and operates as follows: the bacterial chromosome is broken within the F factor and transferred into recipient F$^-$ bacterium; the portion of the F plasmid which is transferred first is called the origin of transfer; the remaining part of the F plasmid is transferred last. To move the complete chromosome into the recipient bacterium takes nearly 2 hours and this is rarely achieved because the pilus (conjugation tube) through which the chromosome moves is fragile and easily broken. However, this process results in high frequency transfer of bacterial genes, and bacteria with integrated F plasmids are referred to as **HFr** (high frequency recombination) strains.

Because transfer is initiated from a fixed point and because the chromosome is transferred as a linear structure at a more or less constant rate, the order and the time at which specific genes enter the F recipient can be utilized to construct a map of the bacterial chromosome. After mixing HFr and F$^-$ bacteria and allowing conjugation to be initiated, the process can be stopped at different times by violently agitating the culture, causing the fragile pili to rupture, and preventing further transfer of the chromosome. This is referred to as **interrupted mating**. The time at which each gene is transferred can be determined if bacteria are removed from the culture and plated out on agar containing suitable selective media at a series of time intervals during the experiment. Thus a map of bacterial genes can be constructed in map units of minutes giving the distance of each gene from the origin of transfer. An example of this process is shown in Figure 2b.

It is not technically feasible to map the full length of the chromosome in this way using a single HFr strain, but different HFr strains exist, each of which has the F plasmid integrated at a unique site. These provide overlapping maps allowing the whole chromosome

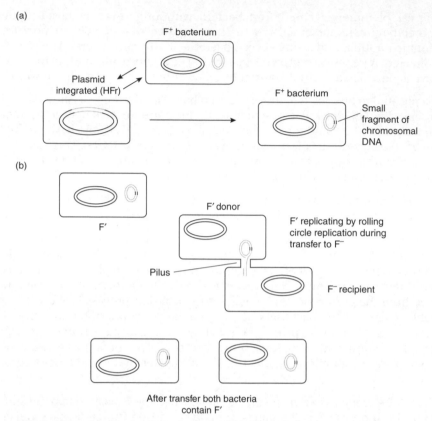

Figure 1.   (a) Relationship between F plasmid and bacterial chromosome. Plasmid DNA, shaded; chromosomal DNA, solid. (b) Transfer of F1 plasmid from F′ bacterium (donor) to F′ bacterium (recipient). This is known as sexduction.

to be mapped and showing that the genetic map of *E. coli* is circular. This is divided into 100 units each of 1 minute.

The F plasmid is not the only plasmid capable of moving between bacteria. One other class of plasmid which has serious implications in medicine is the R plasmids. These can carry multiple antibiotic resistance genes between bacteria.

Bacteriophages and gene transfer

Under certain conditions bacteriophages (usually abbreviated to 'phages'; Section B6) can facilitate the transfer of genes between bacteria. The process is known as **transduction**. It depends on errors in phage replication, and is particularly useful in fine-structure mapping of bacterial genes.

Phages can be divided into two classes, **virulent** or **temperate**, depending on how they behave after infection of a bacterium. When a virulent phage infects a bacterium it takes over the synthetic machinery of the host and uses it as a factory for the production of new phage particles. The bacterium subsequently lyses liberating a large number of phages. Temperate phages have a choice between the **lytic** life cycle described above and an alternative **lysogenic** pathway. The latter involves integration into the host's chromosome where the phage, now termed a prophage, is dormant and replicates along with the

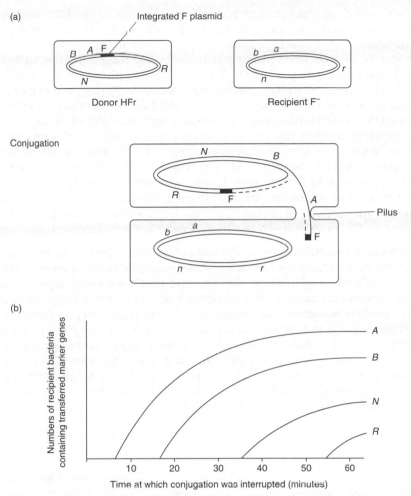

Figure 2. (a) Donor HFr and recipient F⁻ bacteria. The HFr has the F plasmid integrated into its chromosome (filled). Four selectable marker genes A, B, N, and R are marked in their respective map positions. The recipient has recessive (mutant) copies of the same genes (lowercase). During conjugation the chromosome is broken within F. Single-stranded DNA is transferred through the pilus. Note that this is replicated within the recipient, and that synthesis of a new second strand has also commenced in the donor in the region distal to F – indicated by broken lines. This is rolling-circle replication. (b) The order and relative position of the genes A, B, N, and R is determined by plotting the numbers of recipient bacteria that have gained these selectable markers at the various times when conjugation is interrupted. As shown by the graph, the approximate map position of the marker gene is: A, 8 minutes; B, 17 minutes; N, 36 minutes; R, 54 minutes from the origin of transfer.

rest of the host chromosome. An example of a temperate phage is λ. This can integrate into the bacterial chromosome only at a specific site, between the bacterial genes *gal* and *bio*. The bacterium is unaffected but is immune to further infection with the same strain of phage. Occasionally with low frequency the prophage loses its dormancy and converts to the lytic cycle, replicating phages and lysing the bacterium. The lytic phase can be induced in lysogenic bacteria by treatment with mutagens such as ultraviolet light. The mode of transfer of bacterial genes depends on the phage life cycle.

During the lytic cycle of some virulent phages errors may occur by which small fragments of the bacterial host's chromosome are randomly incorporated into new phage particles in place of phage DNA. These structures are called **transducing particles**. Figure 3a outlines the interactions between phage and bacteria and the production of transducing particles. Transducing particles may only be present in a phage stock at frequencies as low as 10^{-6}. They retain the ability to infect bacteria due to the presence of the phage coat proteins, but because they carry only bacterial DNA they simply act as vehicles for transferring DNA between bacteria. Appropriate selective media allow only bacteria that have incorporated genes in this way to grow into clones. Hence such rare events can be detected and scored. Due to its fixed size the phage can only carry a small amount of DNA and only genes very close together can be **co-transduced** in this way. Frequency of co-transduction can be used for fine-scale mapping of bacterial genes. Since any fragment of the bacterial chromosome may be transduced in this manner the process is known as **generalized transduction**. Examples of phages that can be used for generalized transduction include P22 in *Salmonella* and P1 in *E. coli*.

When a lysogenic prophage loses its dormant state and begins to replicate in the host bacterium it is excised from the host DNA. This is normally an exact process in which the phage is precisely removed from the chromosome. Rarely the excision process is inaccurate, a small amount of bacterial DNA is removed, and a reciprocal fragment of the phage is left in the host chromosome. The phages resulting from such an event lack some phage genes and are defective in that they cannot carry out their full life cycle, but like the transducing particles of the virulent phages they can transfer genes between bacteria. Because temperate phages can only integrate into bacteria at one or a few sites they can obviously only transfer genes from those regions; the process is known as **specialized transduction**. Phage λ, for example, can be used for fine-structure mapping of the *gal-bio* region. Specialized transduction is outlined in Figure 3b.

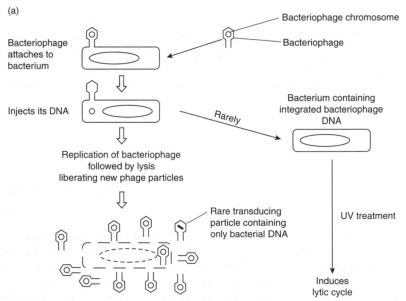

(a)

Bacteriophage attaches to bacterium

Bacteriophage chromosome

Bacteriophage

Injects its DNA

Rarely

Bacterium containing integrated bacteriophage DNA

Replication of bacteriophage followed by lysis liberating new phage particles

Rare transducing particle containing only bacterial DNA

UV treatment

Induces lytic cycle

Figure 3.   (a) Lytic infection cycle yielding rare transducing particles which allow generalized transduction.

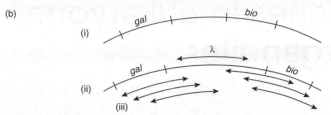

Figure 3. (b) Production of specialized transduction of the *gal-bio* region of the *E. coli* genome. (i) Position of the *gal* and *bio* genes on part of the *E. coli* chromosome. (ii) Insertion of bacteriophage λ DNA between these genes. (iii) DNA content of various defective transducing particles induced by UV irradiation that contain part of either the *gal* or *bio* genes and part of the λ genome.

Recombination in bacteria

All of the mechanisms described above relate to the moving of fragments of the bacterial chromosome between bacteria. This only allows the possibility of recombination. The recipient bacteria become partially diploid, and are termed **merozygotes**. Recombination, the production of new stable genotypes, depends on crossing-over between the host chromosome and the transferred DNA fragment. One important point must be made about this process: whereas in eukaryote recombination single crossovers are useful in mapping genes, in merozygotes there must always be an even number of crossovers. This is due to the bacterial chromosome being circular. A single crossover would result in the creation of a linear chromosome, which would render the bacterium nonviable. This is illustrated in Figure 4.

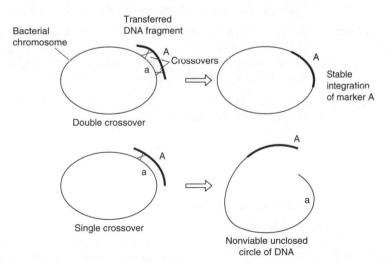

Figure 4. The need for an even number of crossovers to affect the integration of a marker from a merozygote.

C7 Genes in eukaryotic organelles

Key Notes

Mitochondrial and chloroplast genomes	The mitochondria and chloroplasts of eukaryotes contain their own DNA genomes. These vary considerably in size but are usually circular. They probably represent primitive prokaryote organisms that were incorporated into early eukaryotes and have co-evolved in a symbiotic relationship. Animal mitochondrial DNA mutates at a higher rate than nuclear sequences, but usually all mitochondria in any individual are identical. The organelles have their own ribosomes and synthesize some of their own proteins, but others are encoded by nuclear genes.
Maternal inheritance	Pollen and sperm rarely contribute cytoplasmic organelles to the zygote. Hence genes carried by mitochondria or chloroplasts can only be inherited on the maternal side. In plants the most common examples of maternal inheritance involve chloroplasts that fail to synthesize chlorophyll and hence form white regions in leaves. In yeast, mitochondrial mutants can inhibit aerobic respiration. Some inherited human diseases are caused by mitochondrial mutants, and damage to mitochondrial genomes is associated with age.
Maternal effects	In some cases the phenotype of an individual is affected by the genotype of the mother. This can mask the true genotype of the individual. This is different from maternal inheritance. A good example is shell coiling in water snails.
Related topics	(C1) Basic Mendelian genetics (D10) Human evolution (C2) More Mendelian genetics

Mitochondrial and chloroplast genomes

Eukaryote cells contain more than one genome. Most genetic studies concentrate on the nuclear genome, but the mitochondria and chloroplasts carried in the cytoplasm also contain their own DNA. It is widely accepted that these organelles represent the ancestors of prokaryote organisms that were 'captured' by primitive eukaryotes and have since lived in a symbiotic relationship. The organelle genomes differ greatly from the nuclear genome in that they resemble their prokaryote ancestors. The mitochondrial and chloroplast genomes of several species have been characterized and completely sequenced. In most plant species chloroplast genomes are circular, and range from 50 to 200 kb in size. Many copies of the genome can be found in any one chloroplast. Mitochondrial genomes are also circular but vary greatly in size between different species. Human mitochondria contain 16.6 kb of DNA, whereas the yeast mitochondrial genome is approximately five times larger. Plants have mitochondrial genomes of about 1 Mb.

In animals, mitochondrial DNA mutates at a higher rate than nuclear DNA. Despite the high mutation rate almost all mitochondria in any one individual are identical (**homoplasmy**), because only small numbers are transferred to the next generation. This allows mitochondrial DNA to be used to monitor divergence in animal populations. This is discussed in relation to the human species in Section D10. It is worth noting that plant mitochondria behave very differently and evolve slowly. Both mitochondria and chloroplasts carry their own ribosomal RNA and transfer RNA genes (Section A4), and their ribosomes are similar in size to prokaryotic ribosomes. Some of the proteins found in these organelles are synthesized on their own ribosomes from mRNA transcribed from the organelle genome, but most proteins are coded by nuclear genes and are imported from the cytoplasm. This dependence on host genes gives an indication of the large amount of co-evolution that has taken place between the organelle and the eukaryote cell.

Maternal inheritance

Although there are some exceptions, notably gymnosperms (e.g. conifers), most plant pollen does not transfer chloroplasts or mitochondria to the zygote. In animals, sperm do not contribute mitochondria. For this reason any phenotypes coded for by organelle genomes show a pattern of inheritance in which a phenotype can only be inherited from the mother; this is known as **maternal inheritance**. There are numerous examples of this.

Variegated plants, showing white patches or stripes on the leaves, are often due to the presence of a proportion of chloroplasts that have a mutation which results in loss of the ability to form chlorophyll. At each cell division the chloroplasts segregate randomly between the daughter cells, and occasionally a daughter cell will be formed that contains only mutant chloroplasts. All of its descendants will lack chlorophyll, producing a white stripe in grasses or a white patch in broad-leaved plants. The frequency of such an event will depend on the numbers of chloroplasts per cell and the ratio of mutant to normal chloroplasts. In certain species such as the four o'clock plant, *Mirabilia jalapa*, cuttings can be taken from green stems, white stems, or variegated stems. These cuttings will grow on to form flowers which produce seed. Plants derived from the green and white, cuttings breed true, but the seeds produced on variegated cuttings can give green, white or variegated plants. This is because the ova that develop into seeds after fertilization can carry all normal, all mutant, or a mixture of chloroplasts (Figure 1).

A commercially important example of cytoplasmic inheritance in plants involves male sterility in maize. Here mutant genes derived from the maternal cytoplasm prevent pollen production. This can be utilized in the production of F1 hybrid seed (Sections C1 and C8). In mixed stands of male-sterile plants of one variety and normal plants of another variety, all seed produced on male-sterile plants must have been fertilized by pollen from the other variety, and hence must be hybrid. There is an interaction between the cytoplasmic gene for male sterility and nuclear genes. Such interactions are commonly found in examples of cytoplasmic inheritance.

Cytoplasmic inheritance is also observed in fungi. Mutations in the mitochondrial DNA of two species have been associated with reduced growth rates. In the baker's yeast *S. cerevisiae* these are referred to as 'petite' mutants. Two forms of mitochondrial petite mutants are known. Neutral petites are recessive and when normal yeast are crossed with petite yeast the resulting ascospores have normal growth rates. This is because each cell contains many mitochondria, and although the hybrid contains a mixture of mutant and

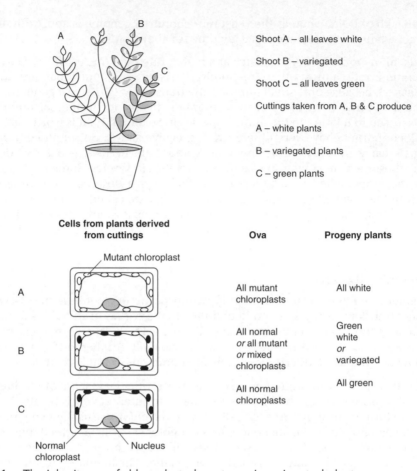

Shoot A – all leaves white

Shoot B – variegated

Shoot C – all leaves green

Cuttings taken from A, B & C produce

A – white plants

B – variegated plants

C – green plants

Cells from plants derived from cuttings

Ova

Progeny plants

Mutant chloroplast

A

All mutant chloroplasts

All white

B

All normal *or* all mutant *or* mixed chloroplasts

Green white *or* variegated

All green

C

All normal chloroplasts

Normal chloroplast

Nucleus

Figure 1.   The inheritance of chloroplast phenotypes in variegated plants.

normal mitochondria, it is highly improbable that at any cell division the distribution of mitochondria between daughter cells would result in a cell containing only mutant mito- chondria. Other petite mutants, suppressor petites, are not recessive and when crossed with normal yeast produce normal and petite progeny in non-Mendelian ratios (Sections C1 and C2). Petite mutants are incapable of supporting aerobic respiration. The precise defect differs between mutants, but frequently this involves major deletions of mitochon- drial DNA. Suppressor petites appear to be able to induce changes in the DNA of normal mitochondria in the same yeast cell.

Mutations in mitochondrial DNA can lead to antibiotic resistance in animal cells grown in culture, and certain inherited human disorders show a pattern of maternal inheritance and can be traced to mutations in mitochondrial DNA. An example of one such disease is Leber's hereditary optic atrophy. Here, as in other examples, nuclear genes interact with the expression of defective mitochondrial genotypes; there is growing evidence that an accumulation of damage to mitochondrial DNA may play a part in the aging process.

Maternal effects

Some confusion can occur between the maternal inheritance of genes encoded in organ- elle genomes and situations where the phenotype of progeny is affected by its mother, to the extent where its own genotype is masked. This can occur to a greater or lesser degree in many organisms because of maternal contribution of protein or RNA to the embryo. An extreme example of this is displayed by the pattern of shell coiling in the water snail *Limnae*. In this species a single gene with two alleles determines whether the snail shell follows a right-handed (dextral) or left-handed (sinistral) coil. There are two alleles, *D* for dextral, and *d* for sinistral coiling, dextral being dominant over sinistral. The water snail is a hermaphrodite and can either self-fertilize or mate with other snails. When a *DD* dextral snail is mated with a *dd* sinistral snail the progeny all have the same shell coiling pattern as the mother. If these F1 individuals are allowed to self-fertilize, the progeny are all dextral. A further round of self-fertilization produces an F3 generation with a ratio of 3 dextral:1 sinistral. Note that the predicted Mendelian ratios appear one generation later than would be expected in the case of conventional Mendelian genetics. Although this has some similarities with cytoplasmic maternal inheritance it is clear that the phenotype does not consistently follow that of the female parent.

Why is the true ratio delayed for a generation? The development of the snail is predetermined by the mother. Its influence directs the manner in which the shell coils. Hence the phenotype of each snail reflects the genotype of its mother (or single parent in a self-fertilization) rather than its own genotype. This is the best known example of maternal effect, but other examples are found in insects where the phenotype of the young of the next generation resembles the mother, but this is lost with age. In the case of the snail the coiling of the shell is not reversible.

C8 Quantitative inheritance

Key Notes

Quantitative traits   Most classic genetic traits are discontinuous; different genotypes produce quite different phenotypes with no overlap (e.g. brown or blue eyes). Quantitative traits such as height vary on a continuous distribution. All values of height within a range are possible. They are controlled by the cumulative effects of many genetic loci, and also by the effects of differences in the environments experienced by individuals. Quantitative traits are also called multifactorial, polygenic, or multilocus.

There is also a newly recognized category which varies discontinuously but is controlled quantitatively (continuously). A combination of genetic and environmental factors cause some individuals to cross a threshold from one phenotypic state to another. Examples are diabetes and cancer where genotype affects risk, but a combination of genotype and environment push some individuals over the threshold from health to disease.

Measuring continuous variation   Continuously variable characters are described by their mean and variance, usually assuming a normal distribution (x_i = individual height measurements of n individuals).

$$\text{mean} = \bar{x} = \frac{\sum(x_i)}{n}$$

$$\text{variance } s^2 = V = \frac{\sum(x_i - \bar{x})^2}{(n-1)}$$

Components of variation   The total phenotypic variance is equal to the sum of the variance due to all causes, namely the environment (E), gene–environment interactions (GE), additive genetic effect (GA), dominant genetic effects (GD), and epistatic interactions (GI) between different genes (loci) respectively as in the formula:

$$V_{\text{total}} = V_{\text{ph}} = V_E + V_{GE} + V_{GA} + V_{GD} + V_{GI}$$

This is often simplified by putting all the genetic components together:

$$V_{\text{ph}} = V_E + V_G$$

Broad-sense heritability

Broad-sense heritability (H^2) is the proportion of total phenotypic variance in a population sample that is due to genetic effects.

$$H^2 = \frac{V_G}{V_{ph}} = \frac{V_G}{V_E + V_G}$$

Narrow-sense heritability

Narrow-sense heritability (h^2) is predictive. It is the component of variation caused by alleles or genes with additive effects. It is passed from parent to offspring, each parent contributing 0.5 of the offspring's character:

$$h^2 = \frac{V_A}{V_{PL}}$$

Measuring heritability

(i) Controlled breeding. Environmental effects can be measured by comparing inbred (homozygous) lines in crop plants or monozygotic twins in humans. The extra variance between unrelated individuals in the same environment must be the broad-sense heritability. Controlled crosses can be used to produce genetically uniform F1 and variable F2 and backcross generations. By making the correct comparisons, the environmental and additive, dominant and epistatic genetic contributions to the range of phenotypes can be calculated. The F2 generation contains all possible genotypes, and can be used to calculate broad-sense heritability.

$$H^2 = \frac{(V_{ph.F2} - V_E)}{V_{ph.F2}}$$

Similar calculations can be performed in humans by suitable comparisons between people with different degrees of relatedness, and adoptive relatives.

(ii) Realized heritability. Realized heritability is the response of a population to selection, and can be calculated after a selective breeding program. It is the ratio of change in character to selection differential:

$$h^2 = \frac{M_O - M_T}{M_P - M_T}$$

where M_T is original mean, M_O is the mean in the next (offspring) generation after selection, and M_P is the mean of the individuals selected to be parents.

(iii) Offspring–parent regression. A graph of offspring values (y) against the average (mid-point value) of their two parents (x) for a particular trait (characteristic) will have a slope (b) equal to the narrow-sense heritability of that trait. Put simply, the correlation between a characteristic in parents and offspring reveals how much the offspring inherit that characteristic from their parents. In humans they also share their family environment, which confuses the measurement. The formula for a straight line graph is:

$y = a + bx$ where a is a constant and $b = $ slope

so

$h^2 = b$

Limitations of heritability

Heritability changes as conditions change because it is the ratio of genetic to total variance. High environmental variance causes lower heritability than low environmental variance, although genetic effects stay constant. Heritability only means something in the population and conditions where it is measured because numerical values cannot be compared meaningfully between different populations or conditions.

Quantitative trait loci

Quantitative trait loci are locatable genetic markers (typically variable number tandem repeats or SNPs) which are closely linked (physically nearby in the DNA) to the genes affecting interesting character traits. The correlation between the markers and the trait is used to find the genetic locations of the genes controlling the trait.

Human studies

Monozygotic twins are genetically identical, so all their differences should be due to their different environments. Dizygotic twins are only 50% genetically identical, and can be used as a comparison. Twins adopted separately or together at birth indicate the environmental difference between families. All the components of genetic and environmental variation in a trait can be calculated by examining the correlations between monozygotic (identical) and dizygotic twins and sibs to each other, and to natural and adoptive parents and sibs. Concordant events affect both twins of a pair (discordant events affect only one twin) and can be used to examine heritability of qualitative traits. For any particular trait (e.g. schizophrenia), the extent that concordance between monozygotic twins (80%) exceeds concordance between dizygotic twins (13%) is a measure of the effect of the 50% genetic difference between the dizygotic twins.

Human quantitative traits

Many interesting human traits are quantitative, the most intensely studied being intelligence and personality. Genetic loci that contribute to variation in these traits have been identified. The advantage of knowing the comparative effects of the environment and genetics (nature and nurture) is that resources can be directed where they are most beneficial, however there is a risk of the information being distorted and employed politically for racial or class-biased motives.

Related topics	(C1) Basic Mendelian genetics	(D4) Genetic diversity
	(C2) More Mendelian genetics	

Quantitative traits

Classic Mendelian traits that are controlled by a single gene (locus) with two alleles have two or three distinctly different phenotypes (Section C1). If one functional allele is sufficient for full activity the dominant phenotype is produced in homozygotes and heterozygotes, and the recessive phenotype is seen in homozygotes where the function is missing (Figure 1a). If one functional allele is not sufficient, the heterozygote will be intermediate in character, but may still be distinct from the homozygotes; for example when red and white homozygous plants are crossed the heterozygous offspring are pink (Figure 1b). In both cases, each allele makes a large contribution to the phenotype. The environment causes small differences between individuals with the same genotype, but there are clear differences between the phenotypes (e.g. tall vs. short), and the variation in this character is said to be **discontinuous**. Most genes that have been studied individually fall into this category. They have a large and obvious effect on the phenotype otherwise we would not notice when they were mutated.

Most genes do not have such obvious effects, instead they contribute a little to a characteristic. For example, the height of humans is not usually controlled by a single gene (although there is a dominant dwarfing allele). Most people are between 125 cm and 200 cm tall, but can be anywhere within that range. When two genes control a characteristic equally, each with two alleles and no dominance, there are five classes in an F2 ratio of 1:4:6:4:1 (Figure 1c). With some superimposed environmental variations, these classes overlap. Such variation is said to be **continuous** because any intermediate value is now possible. With more genes and more environmental variation to spread the peaks, the classes for each genotype disappear into a single broad peak. These characteristics are **quantitative** because each individual must be measured, they cannot be classified qualitatively as tall or short. The genetic control of human height is clear. Children tend to be about as tall as their parents, and some races are tall (e.g. the Masai in Africa), some

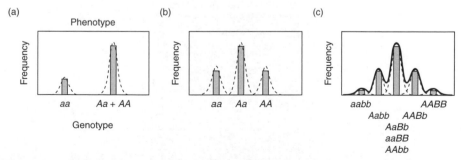

Figure 1.   (a) A single gene cross with a dominant allele gives a 3:1 ratio with discrete phenotypes (bars) broadened by environmental effects. (b) If the alleles have additive effects the heterozygotes are intermediate and the phenotypes less distinct. (c) Two genes with additive alleles give five phenotypic classes in a 1:4:6:4:1 ratio. The blurring effect of environmentally induced variation blends these into a continuous variation.

short (e.g. African pigmies). However, starved children do not grow as tall as their well-fed brothers and sisters.

The **environment** also plays a part; chronic malnutrition (protein or vitamin deficiency) or starvation in childhood limits growth to a height below the genetic capacity. Other continuous traits in humans include skin color, tendency to heart disease or diabetes, and intelligence and personality. These are said to be **multifactorial** or **polygenic** because many genes (loci, positions on the DNA) contribute to their control (hence the term **multilocus control**). Continuously variable traits include yield and quality of all agricultural products (e.g. milk, meat, fruit, and grain). It is important to know how genes and the environment contribute to the phenotype in order to maximize production by selective breeding to improve the genome, and by modifying farming practice to improve the environment.

The biochemical basis of this is easiest to see in the control of color in wheat. Some strains have white kernels, others red. Each functional allele is transcribed into mRNA which is translated into protein that contributes to the production of red pigment. The more functional alleles there are coding mRNA, the more enzyme and pigment are produced, so such alleles and genetic loci are said to be **additive** in their effect. Kernel color in wheat is controlled by three genes (loci) each with two alleles, and crossing extreme parents (red *AABBCC* × white *aabbcc*) will produce seven F2 classes with 0, 1, 2, 3, 4, 5, or 6 functional, additive alleles respectively in the ratio of 1:6:15:20:15:6:1. Only 1/64 will be the deepest red *AABBCC* class, and 1/64 white *aabbcc* like the parental generation. The Punnett square for this is shown in Figure 2. Environmental effects such as temperature (which changes enzyme-controlled reaction rates) sunlight and nutrition (which change the levels of enzymes and substrate for reactions) cause individual grains of wheat to vary within each genotype, blurring the difference between adjacent genotypic classes.

There is another category of trait where the phenotype varies discontinuously (qualitatively) but the control is quantitative under the influence of many genes as well as environmental effects. Examples are diabetes and cancer, where there are known alleles which contribute to risk, and a combination of genotype, environment, and chance events push some people over the threshold from one phenotype to another, from normal health to abnormal disease.

	ABC	ABc	AbC	aBC	Abc	aBc	abC	abc
ABC	6	5	5	5	4	4	4	3
ABc	5	4	4	4	3	3	3	2
AbC	5	4	4	4	3	3	3	2
aBC	5	4	4	4	3	3	3	2
Abc	4	3	3	3	2	2	2	1
aBc	4	3	3	3	2	2	2	1
abC	4	3	3	3	2	2	2	1
abc	3	2	2	2	1	1	1	0

Figure 2. A Punnett square showing the seven color phenotypes produced by three gene loci with additive effects.

Measuring continuous variation

The distribution of a continuously variable character such as human height is described by its **mean** and **variance**. The mean (or average) is the sum (Σ) of the heights (x_i) of all the people measured divided by the number (n) of these people. **Mean** is usually written as $\bar{x}$, pronounced exbar:

$$\text{mean} = \bar{x} = \frac{\sum(x_i)}{n}.$$

The spread of the distribution is called the **variance** (written s^2 or V). This is the sum of the squares of the difference (deviation) between each individual measurement (xi) and the mean (x), all divided by ($n-1$):

$$S^2 = V = \frac{\sum(x_i - \bar{x})^2}{(n-1)}.$$

Variance is a squared value so it is always positive, but its units are also squared (e.g. cm^2, kg^2). The square root of variance, s, called the **standard deviation**, is often used instead to restore the original units. Biological systems usually vary in a way that fits approximately to a normal distribution, an idealized mathematical distribution that gives the classic 'bell shaped curve' (Figure 3). This fit is the basis for most parametrical (number-based) statistics, but should not and must not be taken for granted. If the distribution is normal, then the expected range of results can be calculated. A range from the mean of $\pm s$ includes 68.3% of all individuals, and the mean $\pm 1.96\,s$ includes 95% of individual measurements.

Components of variation

The total amount of variance in a phenotypic character V_{ph} is the sum of the variance caused by all the separate factors. The variance due to the **environment** V_E is the

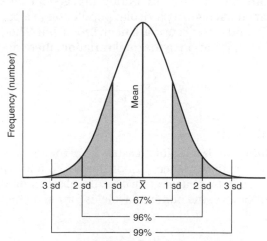

Figure 3.   The normal distribution, defined by its mean (the peak) and the breadth, or spread, measured in standard deviations (sd).

nongenetic component, and is caused by individual organisms not experiencing exactly the same conditions. Another component is the **genotype environment interaction** V_{GE}. For example, some people get fatter than others when well-fed, or starve faster when food is scarce. This is difficult to measure, but may be important (e.g. some wheat varieties increase their yield when given more nitrogen fertilizer, others do not respond to extra nitrogen, but still do well in poor soil).

There are three purely genetic components. The **additive genetic component** V_A is due to additive alleles (those with no dominance) so the *Aa* phenotype is half-way between the *AA* and *aa* phenotypes. The **dominant genetic component** V_D is due to dominant alleles. Heterozygotes express these, but are equally likely to pass the recessive allele to offspring. Expression depends upon the allele received from the other parent, so this variation does not correlate well between parents and offspring. The third genetic component is **interactions between genes** V_{GI} at different loci (epistasis, Section C2). For example *AA* may be expressed differently according to whether alleles *BB*, *Bb* or *bb* are present at another locus. These genes are likely to be inherited independently, so combinations in the parents are broken up at gametogenesis. New combinations are made in the offspring, so interactions are not passed intact from parents to offspring.

These components can be determined accurately in controlled breeding experiments between inbred lines. The total variation is therefore:

$$V_{total} = V_{ph} = V_E + V_{GE} + V_A + V_D + V_{GI} .$$

This is usually simplified by putting all the genetic components together and calling them genetic variation *V*G or broad-sense heritability:

$$V_{ph} = V_E + V_G .$$

Broad-sense heritability

Broad-sense heritability (H^2) is the contribution that genetic differences make to the variation in a particular character in a particular population in a particular environment. (The superscript 2 in H^2 and h^2 indicates variance). Broad-sense heritability is the ratio of the genetic contribution to variation to the total variation, therefore:

$$H^2 = \frac{V_G}{V_{PL}} = \frac{V_G}{V_E + V_G} .$$

Narrow-sense heritability

Narrow-sense heritability (h^2) is a useful measure for plant and animal breeders. It is the additive component of genetic variance, V_{GA}, or simple V_A, and is passed on reliably to offspring, allowing the response to a program of artificial selection to be predicted. In the example of wheat color (above) the number of alleles for red predicts the final color:

$$h^2 = \frac{V_A}{V_{PL}} .$$

The situation is quite different when dominance and genetic interactions affect the phenotype because they cannot be passed on intact.

Measuring heritability

(i) Controlled breeding
Inbred strains become homozygous (Section C12) and so should have no heritable genetic variation. The phenotypic variation V_{ph} then equals V_E. If two such strains (parental one, P1, and parental two, P2) are crossed, the F1 will again be genetically uniform and so $V_{ph} = V_E$. We now have three estimates of V_E:

$$V_E = \frac{(V_{EP1} + V_{EP2} + V_{EP1})}{3}.$$

The F2 will show the full range of genotypes and will demonstrate all the genetic components of variation (V_G). This can be calculated by subtracting the value of V_E obtained above from $V_{ph.F2}$:

$$V_G = V_{ph.F2} - V_E$$

$$H^2 = \frac{V_{ph.F2} - V_E}{V_{ph.F2}}$$

If all the genetic effects are additive, the mean value of the F1 and F2 should be midway between the means of the parent strains. Any deviation from this pattern indicates that dominance and/or epistatic interactions are occurring. The genetic variance of the F2, V_G, can be partitioned into V_{GA}, V_{GD}, and V_{GI}, although these components may not all be present in a particular case. The F1 can be crossed to the parental strains producing two backcross strains (B1 and B2). The contribution of additive and dominant effects to the variance in each (F2, B1, and B2) can be calculated and values for V_A and V_D produced from the three simultaneous equations. More complicated procedures are required to study humans.

(ii) Realized heritability
It is also possible to calculate realized heritability retrospectively from the results of selection. Suppose that a population of tomatoes has a mean fruit size M_T of 30 g, and plants with large fruits are selected, mean size $M_P = 40$ g, as parents to breed from. The selection differential ($M_T - M_P$) is 10 g. If the next generation (offspring) mean, $M_O = 34$ g, then:

$$\text{Heritability} = \frac{\text{gain}}{\text{selection differential}}$$

$$h^2 = \frac{M_O - M_T}{M_P - M_T} = \frac{(34 - 30)}{(40 - 30)} = \frac{4}{10} = 0.4$$

(iii) Offspring–parent regression
Heritability can be calculated by comparisons of relatives (in humans this is the only method) and it is possible to estimate narrow-sense heritability from such comparisons by calculating a parent–offspring regression coefficient. If all variation is additive, offspring will have a phenotype midway between their parent's phenotypes. Thus if a graph is plotted of offspring phenotype (y) against the midpoint between the parents (x) it will give a perfect straight line with a gradient (slope) of 1 if all the variance is due to additive genetic effects. A graph of one parent against offspring or brothers/sisters against each other will have a regression of 0.5 if heritability is 1, because they are related by 0.5. Environmental effects will cause scatter in the points and reduce the regression. When all

the variation is environmental there is no correspondence between parents and children ($h^2 = 0$). Points are scattered randomly across the graph and so the slope is zero. The formula for a straight line (linear) graph is $y = a + bx$ where b = slope, and so for regression of parental midpoint against offspring:

$$h^2 = b.$$

The null hypothesis is that there is no correspondence so the slope $b = 0$. The value of b is calculated by regression analysis, and its statistical significance is determined by a Student's t-test. If $b > 1$ then there must be hybrid vigor (also called heterosis, overdominance) such as might occur in a breeding program.

A simple alternative is to use a calculator to determine a correlation coefficient r between mid-parent and offspring which will indicate h^2 if offspring and mid-parent values follow the same distribution (have the same mean and variance). The principal difference between the tests is that a correlation coefficient does not assume any cause and effect relationship, whereas regression analysis assumes that the parent's genotype influences the offspring.

Limitations of heritability

Heritability is relative to the total variation. If a crop (for example, tomatoes) is raised under very variable conditions the environment will generate a high variability, and heritability will be low. If the same tomatoes are raised in a carefully controlled environment, there will be less phenotypic variation, and the heritability will be higher, although the genetic control is just the same. **Heritability is only numerically meaningful in the conditions where it is measured**. If heritability is high, breeding to improve strains will improve yield; if heritability is low, it may be more productive to change or control the environment. Within genetically uniform strains and self-pollinating species such as wheat, characters are not heritable at all because although they are still genetically controlled there is no genetic variation.

Quantitative trait loci

The many genes responsible for polygenic inheritance of particular characteristics are scattered around the genome. Their positions are known as quantitative trait loci (QTL). It is useful to know where they are for both medical and agricultural reasons. In the case of disease susceptibility, it is useful to identify the individual genes so that their normal function can be identified, and attempts made to design corrective medical treatments. In the case of animal and plant breeding it would be useful to identify young individuals with favorable alleles without waiting for their expression at maturity. Those with an unfavorable genotype could be removed earlier from selective breeding programs, while potentially high-quality types could be cloned immediately.

The procedure in breeding situations is to take inbred lines that differ in the trait of interest, and also vary for markers (typically single nucleotide polymorphisms, SNPs; Section B3) at numerous probe sites. They are crossed and both the F2 progeny and later generations are examined for the desired trait and for the variations at the probe sites. If the presence of the trait correlates with inheritance of a particular marker allele, it is likely that one or more genes affecting the trait are located on the DNA close to that marker.

The same procedure can be followed in human families, particularly for disease susceptibility loci, but is more complicated and difficult because the family sizes are smaller

(Section F1). The usual case is that one or two genes cause most of the variation, and there are increasingly more genes with smaller effects. Genes that contribute 5% or less to the variation in a trait are very difficult to find. Insulin-dependent diabetes mellitus is an autoimmune disease. The most important locus is the major histocompatability complex (tissue type), followed by the promoter of the gene coding for insulin and several loci with smaller effects.

Human studies

Monozygotic twins are genetically identical clones, so the only variation between them should be due to environmental effects. (This is not quite true for females who may differ in the extent to which they inactivated their two X chromosomes.) Comparison of the variation between monozygotic twins reared together gives the 'within twins' environmental effect, while comparison of twins separated and adopted at birth gives the 'between family' environmental effect on variance.

Dizygotic twins are simply sibs (sisters and brothers) conceived and born together. The extra variation between normal sibs compared to the variation between dizygotic twins is the variation due to being born separately. Twins adopted together or separately at birth can be correlated to each other and to both adoptive and natural parents and adoptive and natural sibs to study the environmental and genetic contribution to their phenotype. Natural children of the adoptive parents serve as controls. The correlation of adopted children to the average phenotype of their natural parents is a measure of narrow-sense heritability.

Given various means of estimating the environmental components of phenotypic variation, the remaining variation must be caused by genetic differences, that is, the broad-sense heritability. It is important to remember that families also share environments, making the separation of genetic and environmental effects difficult. Children adopted separately at birth are used as controls for this because they share genes but not environment. The environmental effects begin with the position of implantation in the womb and continue during intrauterine development. Twins and sibs must share the same mother, so they also share antenatal environmental effects as well as genes. Having obtained estimates of broad- and narrow-sense heritability the difference $(H^2 - h^2)$ may be attributed to genetic dominance effects and genetic interactions.

It is apparent that estimates of heritability for human characters are less accurate than those produced in plant breeding experiments because the samples available are smaller and they may not be fully representative of the population. They may also be compiled by combining data obtained in different groups, which is not strictly allowable.

Concordant traits occur in both twins of a pair, and can be used to estimate heritability for qualitative traits. The difference in frequency of concordance between monozygotic and dizygotic twins is taken as a measure or indication of the genetic component of a trait. A high concordance in monozygotic compared with dizygotic twins is found for manic depression (80%:20%), schizophrenia (80%:13%), and blood pressure (63%:36%), but only 95%:87% for measles. This suggests high genetic variability for schizophrenia and low genetic variability in susceptibility to catching measles.

Human quantitative traits

Most human traits are quantitative. A simple example is skin color. The data suggest that there are at least four important genes, perhaps six, with 8–12 additive alleles and 9–13 phenotypic classes. More interesting traits are intelligence and personality. It is useful to know relatively how much the environment (nurture-family and social upbringing) and genetic inheritance (nature) contribute to variation in intelligence. If heritability is high (variation is mainly genetic) then increasing expenditure on education will not make everyone a genius. The more and less able groups will require different types of education, and the weaker might benefit more from help than the geniuses who may need help less. If heritability is low (environmental effects high) it would be more productive to spend money to bring the environment of the disadvantaged up to the standards of the advantaged. Some recent studies put heritability of IQ above 0.6, often around 0.8. What does this mean? Remember that heritability is the **ratio** of genetic variation to total variation. If IQ tests are carried out on a group of children from, say, a middle-class background, where all their families have a high regard for education, and they all have similar opportunities, then the environmental differences may be low, so heritability will be high. If the whole population of a city is included, from slums to rich neighborhoods, there will be much more environmental variation, so heritability may be lower while in fact the genes involved are the same in both tests! One research group found that a particular allele of the gene for insulin-like growth factor 2 receptor is more frequent (about 30%) in 'gifted' children than in normal children (16%). Notice that nearly everyone with this allele is normal.

Personality and behavior certainly have genetic components, because they depend to a large extent upon the activity of neurotransmitters and their receptors. This is why psychoactive drugs work. There are over 100 neurotransmitters identified so far, and most have several receptors, so large genetic variation is to be expected. A serotonin receptor gene has two alleles of its promoter, a high-activity and low-activity version. People with the high-activity allele are slightly more anxious on average than people with the low-activity allele. Some comparisons of twins and siblings suggest a broad-sense heritability of 30–45% (and narrow-sense heritability of 10–25%) for scores in psychometric tests of 'agreeableness,' 'conscientiousness,' 'extrovertism,' 'neuroticism,' and 'openness.' The decisions taken as a result of genetic data are political. If people are genetically disadvantaged do you help them, prevent them from having children, or kill them? The misuse of human genetics for racial or socio-political causes has led to protests against the publication of controversial data. The scientific response is usually that knowledge is useful if it is used correctly, and cannot be worse than the ignorance which led to prejudice, witchhunts and genocide in the past.

C9 Sex determination

Key Notes

Sex determination

Primary sex determination relates to the formation of ovaries or testes. This can be regulated by genes and/or environmental factors. In alligators sex is determined by the temperature at which eggs are incubated. Many species are hermaphrodite. Secondary sexual characteristics are also under genetic regulation.

Simple genetic sex-determining systems

Unicellular eukaryotes can have sex-determining systems in which two alleles of a single gene determine sex. In Hymenoptera, heterozygosity at one gene determines femaleness.

Sex chromosome systems

Three different sex chromosome systems have been described, XX-XO, XX-XY and ZZ-ZW. The sex which has two identical sex chromosomes is called homogametic and the sex with different sex chromosomes heterogametic.

Sex determination in *Drosophila*

In this species sex is determined by the ratio of X chromosomes to sets of autosomes. If the ratio is equal or greater than 1.0 the fly is female; if 0.5 or lower the fly is male. Gynandromorphs are flies with male and female sex phenotypes in different parts of their body. This is due to changes in the number of X chromosomes in somatic cells due to nondisjunction.

Sex determination in humans

In humans the study of sex chromosome aneuploids has shown that the presence of a Y chromosome determines maleness irrespective of the presence of different numbers of X chromosomes. A gene, *SRY*, responsible for inducing maleness has been mapped to the Y chromosome.

Sex determination in plants

Most plants are hermaphrodite, but some examples of XX-XY sex chromosome systems are known in dioecious plants. In *Melandrium* sex is determined by the ratio of X chromosomes to Y chromosomes.

Secondary sexual characteristics

Mutations in genes such as the testosterone receptor gene can lead XY individuals to develop female secondary sexual characteristics. Transfer of blood cells between male and female twins in cattle can produce freemartins which have defective testes composed of cells containing two X chromosomes.

Evolution of sex chromosomes

Sex chromosomes have evolved on several different occasions. The sex chromosomes evolve from a pair of homologous autosomes. In some species the two sex chromosomes look identical, homomorphic. After divergence, a small region of homology between the two remains for meiotic alignment. It is known as the pseudoautosomal region.

Related topics	(B4) Chromosomes	(C10) Sex and inheritance

Sex determination

The sex of an individual can be determined at several levels. This topic is principally concerned with primary sex determination, which relates to whether an individual develops testes or ovaries. Secondary sexual characteristics – forms of development associated with one sex or the other – such as feather-coloring, pitch of voice, and presence of horns or manes, are also under genetic control.

Unicellular organisms can have simple systems for sex determination, but multicellular species differ greatly in the strategies they employ to generate the male and female gametes that are necessary for sexual reproduction. In many instances a single individual may have male and female reproductive organs. Such individuals are known as **hermaphrodites**. This is common in many invertebrates and in plant species. Hermaphrodites can be of both sexes simultaneously or may change from one sex to the other. The importance of genes in determining sex in such species is clearly less than in most higher organisms, however there is evidence that genes can regulate the timing and extent of different sexual phases in some hermaphrodite invertebrate species.

Hermaphroditism, in animals, is not entirely limited to invertebrates. Several fish species such as bass undergo sex-reversal, often under environmental or hormonal influence. Domestic fowl can also occasionally undergo spontaneous sex-reversal. The most clear-cut example of environmentally determined sex is found in alligators where the temperature at which eggs are incubated determines the sex of the individual.

Simple genetic sex-determining systems

Probably the simplest sex-determining mechanism is found in yeast. Two alleles of a single gene determine mating type. In *S. cerevisiae* a gene on chromosome 3 known as *MAT* has two alleles *a* and α. For most of its life cycle *S. cerevisiae* is haploid (has only one set of chromosomes), and the yeast cell will carry either the *a* or α allele. This determines mating type. Only yeast cells of opposite mating types can fuse to form diploids which undergo meiosis and release new haploid spores. Thus the *MAT* gene can be regarded as an early type of sex-determining system. A similar single-gene system is responsible for sex determination in unicellular algae species such as *Chlamydomonas*.

The Hymenoptera (ants, bees, and wasps) have an unusual method of sex determination. Male bees (drones) develop from eggs that were not fertilized and are haploid. Female bees (workers and queens) develop from fertilized eggs and are diploid. It was thought that the difference in ploidy was the sex-determining factor. However, the mechanism depends on a gene with multiple alleles. If this gene is heterozygous the bee will be female. Haploid drones cannot be heterozygous at any gene since they have only one copy of each chromosome. Intensive inbreeding results in high levels of homozygosity and in highly inbred bee stocks diploid males have been detected.

Sex chromosome systems

In many species, sex-determining genes are associated with specific chromosomes known as sex chromosomes. Several different sex chromosome systems are known.

- **XX-XO system**. This is found in many insect species. Females contain a pair of chromosomes known as X chromosomes. Males have only one X chromosome. This is the case in grasshoppers, and the bug *Protenor*, and is sometimes known as the **Protenor system**.
- **XX-XY**. This is found in mammals and also in certain insects including *Drosophila* (the fruit fly). Here females have two copies of the X chromosome and males have an X and a Y chromosome.
- **ZZ-ZW**. This is essentially the reverse of the XX-XY system, where the female is ZW and the male ZZ. It is found in birds, Lepidoptera (butterflies) and snakes.

The terms **homogametic** and **heterogametic** are used to describe these systems. Homogametic means that with respect to sex chromosomes gametes are all identical. For instance, in the XX-XY system females are the homogametic sex as all gametes will carry one X chromosome. In the ZZ-ZW system the female is the heterogametic sex as two classes of gametes containing either Z or W as the sole sex chromosome are found in equal numbers.

Although two species may share the same sex chromosome system, this does not mean the genes which determine sex operate in similar ways. To illustrate this point three different examples of the XX-XY sex chromosome system are compared below.

Sex determination in Drosophila

Sex is determined in *Drosophila* by the ratio of X chromosomes to sets of autosomes (sets of autosomes simply refers to the ploidy of the fly). When the ratio is 1.0 or greater flies are female. When it is 0.5 or less flies are male. Intermediate values give rise to intersex flies. Some typical examples are given in Table 1.

Extreme ratios such as 0.33 and 1.5 give rise to flies that are called metamales or metafemales. Although clearly of their respective sex these flies are poorly developed and have a shortened life span.

The fact that sex determination is a result of a balance of X chromosomes and autosomes suggests that genes that cause female development are clustered on the X chromosome

Table 1. Ratio of X chromosomes to sets of autosomes, and sex determination in *Drosophila*

Number of X chromosomes (X)	Number of sets of autosomes (A)	X:A ratio	Sex
3	2	1.5	Female
3	3	1.0	Female
2	2	1.0	Female
2	3	0.67	Intersex
1	2	0.5	Male
1	3	0.33	Male

and genes for maleness on the autosomes. One important point to note concerns the Y chromosome. The data above indicate that it has no role in sex determination in *Drosophila*. This is correct, but although flies that lack a Y chromosome may be male, they are infertile because a gene on the Y chromosome is essential for the development of functional sperm.

A similar genetic balance mechanism regulates sex determination in other species such as the nematode *Caenorhabditis elegans* (round worm). However this is slightly more complex as male and hermaphrodite individuals exist in this species.

One significant feature of sex determination in *Drosophila* is the presence of abnormal flies known as **gynandromorphs**. These are the result of nondisjunction (see Section B4) in the somatic cells of the flies. If this results in a change in the number of X chromosomes in a cell the X:autosome ratio will be changed and may affect the sex of the cell. This can occur because in flies sex is determined autonomously in every cell. As the cell continues to divide, its descendants will form a patch of cells (clone) which, depending on their position in the organism, may differentiate to form structures of the opposite sex to that of the rest of the fly. In the most extreme case, loss of an X chromosome in the first division after fertilization can result in a fly which develops bilaterally into two halves: one male and the other female. This type of event is not found in mammals where the production of secondary sexual characters is determined hormonally.

Sex determination in humans

Sex determination in humans is typical of the process in other mammalian species, and, although the sex chromosomes are XX and XY, the genetic basis of sex determination differs markedly from that described for *Drosophila*. Our understanding of this subject comes from the study of sex chromosome aneuploids (see Section B4). Aneuploidy of sex chromosomes arises more frequently than for autosomes because very few genes are present on the Y chromosome, and due to the phenomenon of X chromosome inactivation (see Sections B4 and A8), only one X chromosome is expressed in any cell. Hence alterations to the numbers of sex chromosomes have less effect on viability than do changes to the autosomes. The sex of several sex-chromosome aneuploids is given in Table 2. Not all of these sex chromosome configurations result in fertile individuals and individuals with the more extreme deviations from normal suffer severe mental retardation.

From these examples it is clear that at the chromosomal level the presence of a Y chromosome is the factor which determines maleness in humans. During early embryonic

Table 2. Relationship between sex chromosome numbers and sex determination in humans

Sex chromosomes	Chromosome number	Sex
X	45	Female
XXX	47	Female
XXXX	48	Female
XXXXX	49	Female
XYY	47	Male
XXY	47	Male
XXXY	48	Male

development the presence of a Y chromosome causes the undifferentiated gonad to grow more rapidly and subsequently to develop into testes. In birds, where the ZZ-ZW sex-determining system is essentially the reverse of the XX-XY, the presence of a W chromosome induces the development of ovaries from undifferentiated gonads. A specific gene, *SRY*, mapping to the Y chromosome in both humans and mice, has been isolated that is responsible for the switch from female to male development in embryos (Section C10).

Sex determination in plants

Most angiosperms are hermaphrodite, flowers contain both male and female organs. However, in some species male and female flowers are borne on separate plants (**dioecious**). One plant genus in which the chromosomal basis of sex determination has been worked out is *Melandrium* (campion). Here the ratio of X:Y chromosomes is the important factor in determining whether plants produce male or female flowers. This implies that genes on both of the sex chromosomes interact to produce the sex phenotype. Studies of plants that had radiation-induced deletions in X or Y chromosomes showed that the Y chromosome contains regions that repress female development and induce male development, whereas the X chromosome has regions that stimulate development of female flowers.

Secondary sexual characteristics

As noted at the start of this section the production of secondary sexual characteristics is also under genetic control. This is well illustrated by the gene *Tfm* in mammals. This codes for a protein that acts as the receptor for the male-specific steroid hormone, testosterone, and is expressed in both males and females. Testosterone is produced only in the testes and is responsible for secondary sexual characteristics in males. Mutant alleles of this gene are responsible for a syndrome known as androgen insensitivity. Here, the presence of a Y chromosome causes testes to form and these produce testosterone. However, the hormone has no effect on target cells because they lack functional testosterone receptors. Individuals with this syndrome develop as infertile females. Their gonads are testes but these remain internal, and their sex chromosomes are XY.

A different misalignment of primary and secondary sexual differentiation occurs in **freemartins**. These are sheep, goats, or cattle that develop as infertile females, but have defective internal testes. Cells of the testes and other organs have two X chromosomes, but the blood contains some cells that have X and Y sex chromosomes. This abnormal development is only found in females that have been a member of a pair of mixed-sex twins. The embryonic blood supplies of twins in these species are fused *in utero* and hence XY blood cells and hormones can enter the circulation of the female twin. This is sufficient to force the gonads to develop into testes, and partially to block normal female development even though the animal is genetically XX. The male twin is not affected by the presence of XX cells in its blood system.

These two examples should indicate the complex interactions that can occur between the genetic and hormonal determinants of developmental processes.

Evolution of sex chromosomes

It is generally considered that sex chromosomes evolved from a pair of homologous autosomes. This process must have taken place a number of different times during evolution. One of the best examples is found in snakes. Sex chromosomes have been studied in primitive and highly evolved snake species. In primitive species the two sex chromosomes appear identical (**homomorphic**). The two chromosomes are only differentiated by the fact that the W chromosome replicates late in S phase (Section B5). In more advanced species the W chromosome becomes reduced in size, heterochromatic and clearly different from the Z. The pair of sex chromosomes retain only a small region of homology. This is known as the **pseudoautosomal region** and is required to allow the two chromosomes to pair and segregate accurately at meiosis (Sections C3 and C10). In vertebrates the sex chromosome that is limited to the heterogametic sex (Y or W) is generally found to carry few genes and to accumulate large amounts of satellite DNA sequences and constitutive heterochromatin (Sections B3 and B4).

C10 Sex and inheritance

Key Topics	
Sex-linked inheritance	Recessive alleles of genes mapping to the X chromosome are not expressed in heterozygous female mammals but will be expressed in males because males have only one X chromosome. Males transmit the recessive allele to their daughters, where it is not expressed. They are referred to as carrier females. The daughters, in turn, transmit the allele to half of their sons, where it is re-expressed. Some genes are on the Y chromosome and are passed directly from father to son. This is known as holandric inheritance. A small region of homology exists between the X and Y chromosomes. Genes in this region, the pseudoautosomal region, do not show sex-linked inheritance.
Sex-limited and sex-influenced traits	Sex-limited traits are inherited traits caused by a single gene which are expressed only in one sex. Sex-influenced traits are those which are observed more frequently in one sex than in the other. This can be caused by dominance relationships being different in the two sexes.
Related topic	(C9) Sex determination

Sex-linked inheritance

Due to the fact that, in mammals, the X chromosome is present in only one copy in males and in two copies in females, genes which map to this chromosome show a particular pattern of inheritance that differs from the normal expectations for Mendelian inheritance (Sections C1 and C2). This is known as **sex linkage**.

Recessive alleles are not expressed in heterozygous females. The male has only one copy of the X chromosome (**hemizygous**), and hence recessive alleles present are expressed. A useful example is color blindness in humans. This is due to a recessive allele of a gene which maps to the X chromosome. The normal allele is denoted *CB* and the mutant allele responsible for color blindness *cb*. The possible genotypes are shown in Table 1.

Table 1. Possible genotypes at the color-blind locus

Normal male	*CB/Y*
Color-blind male	*cb/Y*
Color-blind female	*cb/cb*
Normal female	*CB/CB* or *CB/cb*

Y denotes the Y chromosome. The *CB/cb* female is a carrier for the syndrome.

The *cb* allele is relatively rare in human populations, with approximately one male in 40 being affected. Color-blind females must have two *cb* alleles and thus occur at a much lower frequency, approximately 1/1600 (the square of the frequency in males).

A color-blind man must have inherited the *cb* allele with his X chromosome from his mother. If she had normal vision then she must have been a heterozygote *CB/cb*. A color-blind man cannot transmit his *cb* allele to his son as, by definition, his son must inherit a Y chromosome from his father. By the same rule he must pass *cb* to all of his daughters. Any daughter that inherits the *cb* allele from her father and a normal allele from her mother will be a carrier and transmit the syndrome to, on average, half of her sons. A daughter of a carrier and a color-blind father will have a 50% chance of inheriting the syndrome (Figure 1).

Hence a phenotype present in a male disappears in the next generation, and then reappears in his grandsons. All affected males, except for new mutants, must have inherited the allele in question from their mother. These are the hallmarks of sex-linked inheritance.

Duchenne muscular dystrophy and hemophilia are other examples of sex-linked conditions in humans. In the case of hemophilia, family records show that Queen Victoria had a mutant allele for hemophilia and transmitted the disease to some of her sons and many of the European royal houses through marriage of her daughters. As there is no evidence of hemophilia in her ancestors, Queen Victoria must have inherited a mutation that arose in the germ cells of one of her parents.

Sex linkage is complicated to some extent by the process of X-inactivation (Sections B3 and A8) in female mammals. This means that approximately half of the cells of a carrier female will express the allele responsible for the syndrome. This can easily be shown, in examples such as glucose-6-phosphate deficiency or Lesch–Nyhan syndrome, by the analysis of single cells. However this is insufficient to affect the overall phenotype.

Sex linkage is not displayed by genes which map to a small segment of the X chromosome, the **pseudoautosomal region**, the part of the X chromosome which pairs with the Y chromosome at meiosis. In humans this is found at the tip of the short arm of the X chromosome. Because there is homology between the X and Y chromosomes in this region,

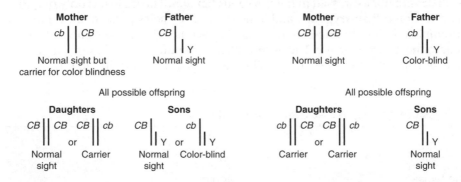

The carrier mother passes her chromosome carrying the mutant allele for color blindness on to half of her sons and half of her daughters.

The color-blind father passes his recessive *cb* allele to all of his daughters making them carriers, but with normal sight.

Figure 1. The inheritance of color blindness. X chromosomes are depicted as the long vertical lines, Y chromosomes as the short vertical lines.

crossing-over can occur and alleles can switch between X and Y. Genes mapping to this region show the same inheritance pattern as genes on autosomes.

Any genes resident on the Y chromosome are obviously passed directly from father to son. There have been some controversial examples of this, including traits such as 'porcupine man' and webbing of toes. A gene for hairs on the outer rim of the ear appears to show this pattern of **holandric inheritance**. Recently it has been shown that the sex-determining gene *SRY* and a gene for a minor histocompatibility antigen, *H-Y*, map to the Y chromosome, and hence are passed directly from father to son.

Sex-limited and sex-influenced traits

These terms relate to situations where the phenotype produced by a specific genotype is altered because of the sex of an individual. The two terms are easily confused, and care must be taken to differentiate between them.

Alleles of sex-limited genes will be expressed only in one sex. One example of this concerns mutant alleles of the breast-cancer-susceptibility gene *BRCA1*, which are dominant and cause breast cancer in females but not in males. In contrast to this, a second breast-cancer-susceptibility gene, *BRCA2*, causes breast cancer in both males and females, and is thus clearly not sex limited.

The difference between sex-limited and sex-influenced genes is subtle but important. In the former a phenotype is restricted to one sex, but in the latter the same phenotype will occur in both sexes but is more common in one. A good example of this is inherited pattern baldness. This is the form of baldness where hair loss spreads out from the crown of the head, and is controlled by a single gene with two alleles *B* and *b*. Homozygous *BB* individuals show premature pattern baldness, and *bb* homozygotes do not. The phenotype of the heterozygote, *Bb*, differs between males and females. In males the *B* allele is dominant and heterozygotes are bald, but in females it is recessive and no hair loss occurs. This is set out in Table 2. The dominance of such an allele is clearly influenced by the hormone balance of the individual. In this context it is interesting to note that this gene is also associated with polycystic ovarian disease in females.

Sex-limited and sex-influenced genes are autosomal and the genotypes follow normal Mendelian patterns of inheritance, but the phenotypes are altered by the hormonal environment. In contrast, the pattern of inheritance of sex-linked genes is due to the inheritance of genotypes caused by the genes being located on sex chromosomes.

Table 2. The expression of pattern baldness genotypes in male and female humans

Genotype	Female phenotype	Male phenotype
BB	Pattern baldness	Pattern baldness
Bb	Normal hair	Pattern baldness
Bb	Normal hair	Normal hair

C11 Genes in development

Key Notes

Three stages of development

The development of tissues in multicellular organisms requires three steps. (i) Individual cells must detect their position in the embryo by chemical cues from their surroundings, including contact with neighboring cells. (ii) The cells must respond to this by regulating their gene expression, and remember or lock the state epigenetically to control future development. (iii) The physical nature of the cells must change to fulfill their role in the organism. Understanding this process has come principally from looking at genetic mutants that do not develop properly, and looking at the expression of genes as mRNA or protein in developing embryos of insects (*Drosophila*), vertebrates (chicks, amphibians, mice), and nematodes (*Caenorhabditis elegans*).

Single cells to multicellular organisms

Single-celled bacteria (*Bacillus*) and yeasts differentiate to form dormant spores in adverse conditions. The slime molds (*Dictyostelium*) are free-living soil myxoamoebae reproducing asexually. When they reach a high density and use up the available nutrients, starved cells secrete pulses of cyclic adenosine monophosphate (cAMP) which signals surrounding amoebae to migrate into a clump which forms a multicellular grex that migrates and differentiates into stalk cells and spore cells.

Early *Drosophila* embryos

The main feature of development is that differentiation is controlled by transcription factors that regulate many other genes. *Drosophila* eggs are polarized by localization of maternal mRNAs from surrounding nurse cells localized in anterior/posterior and dorsal/ventral patterns. The relative concentrations of the proteins translated from these genes activate embryonic gap genes: *Hunchback* (anterior), *Krüppell* (middle), and *Knirps* (posterior), and others that specify cells as belonging to the main regions of the embryo. The overlaps of these gene products induces other genes in bands, and the process continues, activating pair-rule genes, until the familiar pattern of segmentation is produced, finalized by expression of segment polarity genes. The development of cells from each segment is further specified by *Hox* genes that indicate body position.

Homeobox genes (including *Hox*)

The structure of adult flies and vertebrates is controlled by homeobox genes which contain a conserved 180 bp DNA sequence coding 60 conserved amino acids (homeodomain). These are all transcription factors. One group occurs sequentially on the DNA and is expressed in a matching linear sequence, both anterior-posterior in the body and proximal-distal in the limbs. For example, mutations in *Antennapedia* cause small legs to grow from the head where the antennae should be. This type of homeotic change (one structure replacing another) gave the name to the class of genes. Gene mapping and DNA sequencing revealed a cluster (now called *Hox* genes) of these genes resulting from tandem duplications of ancestral genes, followed by diversification and specialization. The same clustered family of genes exists in all metazoans but most vertebrates have four complete sets containing 39 genes in total. Teleost fish have seven sets. Individual genes may be lost or duplicated in individual clusters.

All metazoans including vertebrates use the same developmental genes

The current evidence is that the same signaling molecules and responding transcription factors are used in the same regions of the embryo in all species. The same *Hox* genes in different tissues produce different effects. There are evolutionary differences in exact timing and duration of individual stages and in the final genes that are regulated, which give the differences in body form of different animals, including vertebrates. Examination of large, free eggs of fish and amphibians, and detection of RNA and protein in microscopy sections of mouse embryos confirm a common set of developmental genes in all metazoans.

Globins as a developmentally regulated gene

The globin genes occur as an α cluster containing ζ, $\alpha 1$, and $\alpha 2$ genes, and a β globin cluster containing ε, Gγ, Aγ, δ, and β genes. These are controlled by a large upstream locus control region (LCR) which moves along the cluster during embryo development switching genes on as it reaches them and off as it passes, regulating sequential transcription.

***Caenorhabditis elegans* development**

Caenorhabditis elegans is a nematode used extensively as a model genetic organism. Although it has a similar toolkit of developmental genes to other animals, it has a profoundly different developmental mechanism. The fate of each cell is controlled by its lineage.

Programmed cell death

Cell death by apoptosis is an essential part of development and is genetically programmed. One example is the loss of cells between the fingers and toes. Their persistence leads to webbing, as seen in ducks' feet.

Plants have similar mechanisms to animals

Plants also have developmental genes, including homeobox genes, although most plant developmental genes are MADS box genes. Plants appear to rely on diffusible paracrine inducers. In general plants are less well studied than animals.

Related topics

(A7) Regulation of gene expression
(A8) Epigenetics and chromatin modification

Three stages of development

The development of tissues in multicellular organisms is controlled by regulation of key **transcription factors** that in turn regulate a cascade of other genes, including those encoding more transcription factors (Section A7). These regulatory states are fixed for the life of the tissue by **epigenetic** modifications to chromatin structure (Section A8). The process of **differentiation** into a particular tissue requires three steps.

(i)   Individual cells must detect their position in the embryo by chemical cues from their surroundings, including contact with neighboring cells. They do this via receptors on the cell surface. These may detect diffusible **paracrine** hormones such as retinoic acid or may directly bind to signaling molecules presented externally on adjacent cells (e.g. the *notch* signaling system).

(ii) The cells must respond to this by managing their gene expression, and lock their genetic controls, a process called **determination**, to stay on their prescribed developmental track. The locking process is an epigenetic modification of the chromatin at specific genes (Section A8). This involves regulation of transcription factors by either activating chromatin (acetylating histones, demethylating DNA), or inactivating chromatin to prevent transcription. This involves specific modifications to histones. One gene which regulates development in *Drosphila* and other species is *Polycomb*. The normal *Polycomb* protein binds to modified nucleosomes where the 27th amino acid of histone H3, a lysine, is trimethylated, and joins in a complex which shuts down transcription. This state is maintained through subsequent somatic divisions. *Drosophila* males normally have a single sex comb on one segment of each front leg, but mutation in *Polycomb* causes them to form sex combs on more than one pair of legs. Once the state of expression of major developmental regulatory genes is established, the final fate of the cell lineage is **determined**. They cannot switch to a different state. Final differentiation, however, may occur much later as a restricted response to further developmental cues.

(iii) The physical nature of the cells must change to fulfill their role in the organism. This is achieved by changes in gene expression that are limited and regulated by the regulation of other genes, principally transcription factors (Section A7), earlier in development. Some of the changes control how the cell will modify its gene expression in response to further developmental cues.

These steps are repeated at successive stages of development as the cells become more and more specialized, progressing from totipotent cells individually capable of forming an embryo and extra embryonic membranes, through pluripotent stem cells capable of differentiating into any embryonic cell type, to successively committed stem cells capable of becoming a smaller range of similar cell types, until they become terminally differentiated specialized cells (e.g. cartilage or retinal neurons). Understanding this process has come principally from looking at genetic mutants that do not develop properly, and looking at the expression of genes as mRNA or protein in developing embryos of insects (*Drosophila*), mammals (chicks, amphibians, mice), and nematodes (*C. elegans*).

Single cells to multicellular organisms

Single-celled bacteria (e.g. *Bacilli*) and yeasts differentiate to form dormant spores in adverse conditions. The slime molds (*Dictyostelium*) are free-living soil myxoamoebae reproducing asexually and show all the principles of developmental control but using different genes.

When they reach a high density and use up the available nutrients, starved cells secrete pulses of cyclic adenosine monophosphate (cAMP) which signals surrounding amoebae to clump. When the cells touch they adhere using a range of proteins newly synthesized for the function, and then migrate as a multicellular **grex**, which differentiates into anterior cells that will predominantly differentiate into stalk cells and die, while the rest climb the stalk and produce spores for dispersal. This is similar to the fundamental division between somatic and reproductive germline cells that occurs in true animals.

The cAMP acts effectively as paracrine (diffusible) hormone. The cell adhesion and migration resemble the movement of cells seen in animal embryos. Thus these single-celled organisms use processes analogous to development in metazoan embryos.

Early *Drosophila* embryos

The main feature of development is that differentiation is controlled by transcription factors that regulate many other genes. The early *Drosophila* embryo is polarized by maternal nurse cells secreting mRNAs into the oocyte. Among these are *Bicoid* mRNA which is secreted at the anterior end, *nanos* mRNA bound to the posterior tip, and maternal *hunchback* and *caudal* mRNAs which are evenly distributed. Bicoid protein suppresses translation of *caudal*, while nanos represses translation of *hunchback*, producing gradients. Bicoid plus hunchback determine head, nanos plus caudal specify abdomen, and the overlap on hunchback and caudal proteins determine the thorax. The embryo continues to produce the gap-gene products, *Hunchback* (anterior) and *Krüppell* (middle), while *Giant* forms two bands, one at each end of the gradient of *Krüppell*. *Knirps* and *Giant* further divide the embryo with transverse bands. These intersecting gradients set the scene for the pair-rule genes which define segments, and the segment polarity genes which specify anterior and posterior for each segment. The homeotic genes will specify differentiation of segments, particularly in the adult.

Homeobox genes (including *Hox*)

The structure of adult flies and vertebrates is controlled by **Homeobox genes** which contain a conserved 180 bp DNA sequence coding 60 conserved amino acids (homeodomain). These are all transcription factors. One group, called *Hox* genes, was shown by gene mapping and DNA sequencing to occur sequentially on the DNA (Figure 1) and are expressed in a matching linear sequence, both anterior-posterior in the body and proximal-distal in the limbs and even along the mammalian female reproductive tract. These specify the response of cells to their position and regulate development accordingly. For example, mutations in the *Drosophila* gene *Antennapedia* cause small legs to grow from the head where the antennae should be. This type of homeotic change (one structure replacing another) gave the name to the class of genes. The cluster is a result of tandem duplications followed by divergence (Section A5). The same family of genes exists in all metazoans, but individual members may be gained or lost. The ancestral urbilatarian cluster comprised seven *Hox* genes. These correlate with development of anterior-posterior and dorsal-ventral axes, and predate evolution of a continuous gut. *Drosophila* have ten genes split into two clusters. Turtles, crocodiles, birds, and mammals have four complete paralogous clusters on different chromosomes produced by successive duplication and containing a total of 39 *Hox* genes. Teleost fishes have seven clusters, presumably resulting from genome duplication (tetraploidy) followed by chromosome loss, and a very variable complement of genes, many of which must be redundant. Homeobox genes also occur in plants but are considerably diverged from those in animals and are not the primary developmental genes.

The colinearity of the genes in the cluster and the location of their expression in embryos is remarkable (Figure 1). The genes that control head structures in *Drosophila* also do so

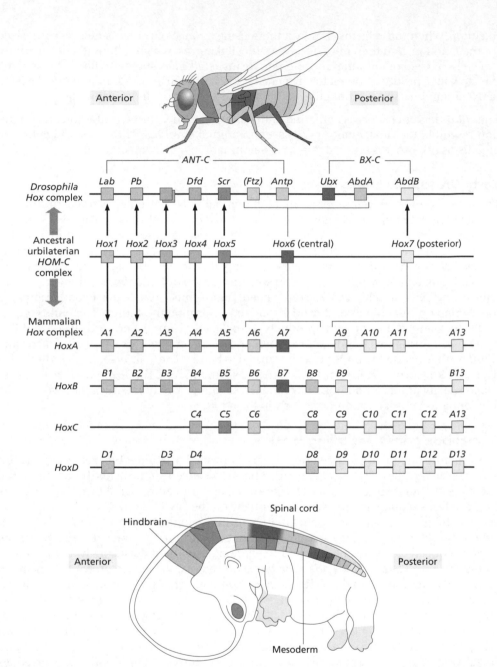

Figure 1.   The seven *Hox* genes in the oldest known arrangement, the ancestral urbilaterian *HOM-C*, have given rise to larger numbers in more complex species. The eight homeobox genes in *Drosophila* (above) are split into two clusters: *ANT-C* and *BX-C*. The 39 homeobox genes in humans (below) are split into four clusters, *Hox A*, *Hox B*, *Hox C*, and *Hox D*. In all cases, the genes are expressed in the same order as their position in the cluster. Open boxes represent genes not expressed in this spatial sequence. From Alberts *et al.* (2008) *Molecular Biology of the Cell*, 5th Edn. Garland Science, Abingdon.

in vertebrates. The extreme abdominal genes in *Drosophila* regulate the most distal parts such as toes in mammals.

All metazoans including vertebrates use the same developmental genes

It is clear from Figure 1 that the expression of *Hox* genes is spatially conserved. The evolution of body shape, clearly different between mammals, let alone between mammals and insects, seems to occur partially through changes in the timing and relative extent of expression of the main controlling genes, but also to large changes in the intracellular response to the same signals in different tissues within organisms and also between organisms. Another homeobox gene, *Pax6*, is involved in the development of eyes in all metazoan taxa, but the eyes can be quite different and even evolved independently.

Large, free vertebrate eggs (fish, amphibians) can be directly compared to *Drosphila*. They also are large enough to have differentiation in their cytoplasm. At the early stages they are similar in principle but not in detail to *Drosophila*. Mammalian eggs and embryos are much harder to study because they are very small and develop internally. Polarity is produced around the point of sperm entry which also seems to determine the first cell cleavage furrow, which in turn can produce a polarity gradient. Mutation and gene expression studies both show that the vertebrate homologs of *Drosophila* genes are functioning. Loss of the human homologue of the *Drosophila* pair-rule gene *runt* causes cleidocranial dysplasia. This causes skeletal defects including almost complete absence of collar bones (clavicles) so the shoulders can move in front of the chest. At the somite (segmentation) stage the *Hox* genes are clearly active in the same spatial order as in the DNA and in *Drosophila* (Figure 1). The multiple copies allow for greater specialization, and fish, with seven copies, show loss of individual genes. There is some redundancy between paralogous genes. Knockout of *Hoxa3* and *Hoxd3* together in mice produce much more severe abnormalities than either singly, including bones that are normal when only one of the gene functions is missing. Mutations in *HOXD13* in humans cause malformation of the feet, in keeping with the terminal position of the genes.

Globins as developmentally regulated genes

The globin genes existed before the plant/animal split. The globins that form hemoglobin are found in two clusters in humans, the alpha globin family (zeta, alpha 1, and alpha 2: ζ, $\alpha 1$ and $\alpha 2$) on chromosome 16 and the beta family (epsilon, G-gamma, A-gamma, delta, and beta genes: ε, $G\gamma$, $A\gamma$, δ, and β) on chromosome 11. The actively transcribed region moves sequentially along each cluster during embryo development. This is regulated by the locus control region (LCR), an extensive region upstream of the genes which is involved in epigenetic regulation (Section A8). The LCR binds transcription factors and proteins only found in erythroid precursor cells to form a chromatin remodeling complex called an **enhanceosome**. The complex moves along the chromosome, acetylating histones, demethylating DNA, and opening the next gene for transcription. As it passes along it opens up successive genes but it also reverses chromatin modification and shuts down the genes it has passed. This causes the embryo to make hemoglobin with an affinity for oxygen suitable for its current status in the uterus (Figure 2).

In humans the alpha group transcribes ζ mRNA for the first 5 weeks then α. The β group produces ε for the first 5 weeks then γ, then β around birth and in adulthood. A very low proportion of δ, about 1–2%, is produced with β. In a few cells the progression stops at γ. Two α and two β chains make up hemoglobin, so the succession of hemoglobin types in human embryonic stages is Gower 1 ($\zeta 2\varepsilon 2$), Gower 2 ($\alpha 2\varepsilon 2$), hemoglobin Portland ($\zeta 2\gamma 2$), and, in fetal stages, hemoglobin F ($\alpha 2\gamma 2$). In adults hemoglobin A

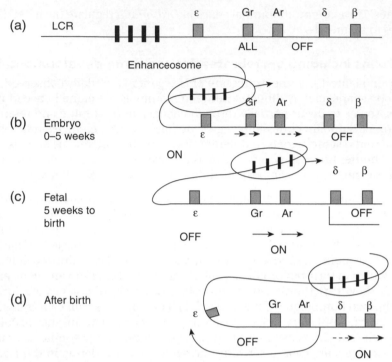

Figure 2. Progressive regulation of the human β-globin genes cluster. (a) The locus control region (LCR) attracts transcription factors, chromatin remodeling proteins, and RNA polymerase to create an enhanceosome (b) which folds over the nearest part of the cluster and opens the chromatin at the ε *globin* gene, starting transcription. It moves along slowly. By 5 weeks gestation (c) it has activated the γ genes and is inactivating ε. It remains like this until around birth (d) when it activates the β gene and inactivates the γ genes.

(α2β2) is most common, usually more than 95%, and hemoglobin A2 (α2δ2) at 1.5–3.5%. Hemoglobin F persists in some cells where the LCR does not progress completely. Persistence of hemoglobin F can ameliorate the symptoms of β-thalassemia if the β gene is deleted, because the LCR cannot continue to migrate and does not switch off the γ genes (Section F1).

Caenorhabditis elegans development

The nematode *C. elegans* is very easy to study and is a model genetic organism. It has a different sort of development from other animals and has only six *Hox* genes in three pairs. Every cell can be seen and followed through each division. Adult hermaphrodites have 959 somatic cells and around 2000 germ cells. From the very first division of the egg the fate of every cell is determined and almost completely predictable by its lineage: which cell divided to produce it, and which side of the division cleavage it came from. Removal of any cell removes all the structures that would have arisen from its clone of descendants, and prevents cellular interactions with other cells that would have induced differentiation. Gene expression seems to be programmed at each division, and when cells are moved they respond to their lineage, not their surroundings. There are cases, however, when two cells interact and arbitrarily decide which differentiates to form what (e.g. one becomes the vulva precursor cell, the other becomes the gonad anchor cell).

Programmed cell death

Genetically programmed cell death (**apoptosis**) is an essential part of development. In the vertebrate limb, cell death occurs between the digits to separate fingers and toes. This is incomplete in ducks' feet. There is also anterior and posterior cell death narrowing the wrist area and in the interior of the lower limb, separating the cartilage primordial to form two bones, the radius, and the ulna. This death is well managed; the cells shut down synthesis, chop up their DNA, and form vesicles that can be taken up by adjacent cells for recycling. Apoptosis and defects in its regulation contribute to growth of tumors (Section F3).

Plants have similar mechanisms to animals

Plants have some homeobox genes but not *Hox* genes. Most of the developmental genes in plants are from the MADS box family. Plants use diffusible paracrine inducers rather than externally cell-bound inducers and receptors. The cells cannot migrate because of their walls so they do not have to detect new cell types. One of the best understood pathways in plants is flower development. This is based on the production of three concentric rings of gene expression in the flower meristem: class A on the periphery, class C in the center, and class B between them. A give sepals, A and B together give petals, B and C together stamens, and C alone the carpel (Figure 3). A big difference between plants and animals is that plant cells do not terminally differentiate. If they are alive, individual cells can usually be induced to regenerate a whole plant.

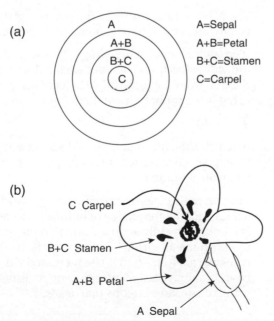

Figure 3.   Control of flower development. (a) Concentric rings of activity by three classes of gene across the meristem produce four rings of differing activity which determine flower development; (b) from the outside, sepals, petals, stamens, and the carpel.

C12 Inbreeding

Key Notes

Inbreeding

Inbreeding is mating between relatives, and the effects of this on the population. It increases the frequency of homozygotes above that expected for a random breeding population in Hardy–Weinberg equilibrium. The excess homozygosity is distinguished by calling it autozygosity to reflect the recent common origin of the alleles. Homozygosity for recessive deleterious alleles causes inbreeding depression. Crossing two inbred lines produces genetically uniform progeny with most of the deleterious loci heterozygous; this produces an F1 generation with hybrid vigor.

Effects on allele frequency

Inbreeding increases the rate at which recessive deleterious alleles become homozygous and are removed by selection, so they have a lower frequency in inbred populations. The total number of deaths due to selection is determined by the mutation rate, and is the same regardless of inbreeding, but they occur sooner in an inbred population.

Coefficient of consanguinity

The coefficient of consanguinity is the probability that two related individuals both carry a copy of a particular allele inherited from a common ancestor, for example a recessive allele causing a genetic disorder when homozygous.

The coefficient of consanguinity of two relatives is 0.5^n where n is the number of paths (gamete steps) between those relatives.

Coefficient of inbreeding

The coefficient of inbreeding (F) is the probability that an individual has both alleles of a gene (locus) identical by descent from the same allele in a common (recent) ancestor. (That is the probability of them being autozygous.)

$$F = \Sigma(1/2)^A$$

where A is the number of ancestors in each chain of descent through each common ancestor, and the values for each chain are summed.

Medical importance

Many genetic diseases are caused by recessive nonfunctional alleles, and the frequency of homozygous recessive zygotes is increased by inbreeding from q^2 in the general population to qpF in the offspring of relatives (q is the frequency of the recessive allele, $p + q = 1$). The increased risk is thus pF/q. Most cases of rare recessive human genetic disorders occur as a result of consanguineous marriages.

Related topics	(C13) Probabilities	(F1) Genetic diseases
	(D3) Genes in populations: Hardy–Weinberg equilibrium	

Inbreeding

Inbreeding is the effect of matings between relatives, or in extreme cases self-fertilization by hermaphrodites. This is quite common in plants. The effect of inbreeding is to increase homozygosity (Figure 1). Under self-fertilization conditions the proportion of the population which is heterozygous halves at each generation, because half of the combinations of gametes from heterozygotes form homozygous zygotes. (As $Aa \times Aa$ gives a *1AA:2Aa:1aa* ratio in Mendelian monohybrid crosses.) The danger in this is that deleterious recessive alleles which are present at a low frequency in the whole population will become homozygous in inbred offspring. This causes a decrease in vigor known as **inbreeding depression**. This is not a problem for plants which normally inbreed (e.g. wheat and peas) because selection has nearly eliminated deleterious alleles. In fact, inbreeding helps farmers because, when a cultivar is homozygous, offspring are all genetically identical to their parent. The crop will be predictable and consistent from year to year, and all the plants will have the same characteristics (e.g. size, ripening time, quality). In organisms that normally outbreed, inbreeding depression is a serious problem. It is necessary to cross two inbred lines to make genetically uniform but fully heterozygous F1 plants. The restored quality is known as **hybrid vigor** but it is not passed on to the F2 generation, which shows the full range of phenotypes including the phenotypes of both parents. Inbred maize is very weak, and four inbred lines are used to make two F1 strains which are then crossed to produce sufficient double hybrid seed for commercial production.

Inbreeding produces **excess homozygosity**. Many loci are homozygous in most organisms, simply because there is a limited amount of genetic variation. Alleles with a frequency of q will be homozygous with a frequency of q^2 in a randomly breeding population (Section D3). Most specific mutations only arose once, and to some extent everyone is descended from a common ancestor if their family trees are traced back far enough. Inbreeding is specifically about mating between comparatively close relatives. This puts the frequency of homozygotes in the population above the level expected from random mating at the Hardy–Weinberg equilibrium. We distinguish this by saying that the

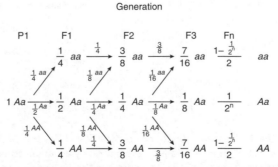

Figure 1.   Increase in homozygosity with successive rounds of self-fertilization (extreme inbreeding).

inbreeding coefficient is the probability of **autozygosity**, a special case of homozygosity, where both alleles of a locus have a recent common ancestor.

Effects on allele frequency

In an inbred population, deleterious recessive alleles become homozygous and are subject to selection more quickly than they would be in a large outbreeding population. They are removed more quickly after they are formed by mutation, and do not survive so long in the population. For this reason deleterious recessive alleles are present at a lower frequency in inbred populations than in outbred populations. The total number of deaths is not changed by inbreeding because deaths are due to mutations, not inbreeding. Every new deleterious mutation must be removed by a death. Mutations just become homozygous and are removed sooner in an inbred population.

Coefficient of consanguinity

Consanguineous marriages are marriages between relatives. The coefficient of consanguinity is the probability that two relatives will both have alleles which are identical by descent from a specific allele in a common ancestor. The use of this is that their consanguinity equals the inbreeding coefficient of their offspring, but it only works for a specific allele. This is medically useful if an identified common ancestor (say a grandparent) is known to have carried a particular recessive deleterious allele. It then predicts the chance of descendants (e.g. grandchildren, who are cousins) producing children who will be affected by the genetic disorder because they are homozygous for that allele.

Coefficient of consanguinity is calculated as follows. Each gamete carries half the alleles from the parent that produced it, so children have half their alleles in common with their father and half in common with their mother. What is the probability that a son and daughter (brother and sister) each carry a copy of the same allele? We can calculate this using Figure 2. The probability of son C getting allele z from his father A is 1/2 (0.5) and daughter D has the same probability of getting z, so the probability of them both getting allele z is $(1/2)^2 = 1/4$. This is their coefficient of consanguinity. (Note that they also have a probability (1/4) of both inheriting the other allele Z, so the probabilities of them both having copies of the same unspecified allele are actually 1/2.) If we continue down the family tree, there is a 1/2 probability of C carrying z, and if he does, a probability of 1/2 of passing it (z) to A's grandson E, so E has $1/2 \times 1/2 = 1/4$ chance of getting z.

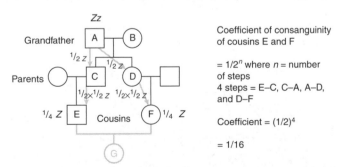

Figure 2.   The coefficient of consanguinity of cousins E and F is the probability of them both inheriting a copy of the same allele (say z) from one particular common parent (in this example their grandfather, A) and is 1/16, the same as the inbreeding coefficient of their child G. (Circles, females; squares, males.)

The consanguinity of nephew E and aunt D is $1/4 \times 1/2 = 1/8$ (=0.125). (It is the same for uncle C and niece F.) Similarly there is a $1/2 \times 1/2 = 1/4$ probability of D having and passing on a copy of z to granddaughter F. The consanguinity of first cousins E and F therefore $= 1/4 \times 1/4 = (\frac{1}{2})^4 = 1/16$ or 0.0625. This calculation can be summarized as **path analysis**. The formula is

$$\text{coefficient of consanguinity} = 0.5^n$$

where n is the number of paths (gamete steps) between them. For E and F (Figure 2) these paths are E–C, C–A, A–D, and D–F, which equals four paths or steps. (They could be written E–C–A–D–F.)

Coefficient of inbreeding

The coefficient of inbreeding (F) is the probability that an individual has both alleles of a gene identical by descent from the same allele in a common (recent) ancestor. The inbreeding coefficient can be calculated by following the paths of descent and calculating the probability of the same allele being inherited from both parents. Figure 3a shows the paths for the product of a brother–sister mating. Consider a particular allele z_1 in the grandfather A. There is a probability of $1/2$ that it will be in the gamete contributed to the father C, and a probability of $1/2$ that it will be passed on again to the child I (sib), so the child has a probability of $(1/2)^2$ of getting z_1 from his father. On the other side of the pedigree he also has the same probability of getting z_1 from his mother D. If he gets z_1 from both parents he will be homozygous. The probability of getting z_1 from both parents and so being homozygous is

$$(1/2)^2_{(\text{father})} \times (1/2)^2_{(\text{mother})} = (1/2)^4 = 1/16 .$$

However, there are four alleles in the grandparents (z_1, z_2, z_3, z_4) and I (sib) is equally likely to be homozygous for any of them, so the probability of him being homozygous for any allele at this locus (his coefficient of inbreeding) is $F = 4 \times (\frac{1}{2})^4 = 4 \times 1/16 = 1/4$. (This is

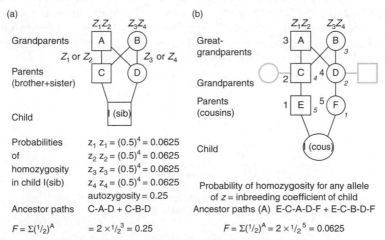

Figure 3.   (a) Inbreeding coefficient F of progeny (I) of mating between sister D and brother C (sibs). (b) Inbreeding coefficient F of a mating between cousins. Normal numbers to left show chain of ancestors via A, small italics to right show ancestors via B. (Circles, females; squares, males.)

numerically the same as the coefficient of consanguinity of the parents C and D.) Following this to the offspring of a cousin marriage ($I_{(cous)}$, Figure 3b) we can see that here there is another step in each side of her ancestry, so the chances of her getting any particular allele are reduced by $(\frac{1}{2})^2$ and the inbreeding coefficient is $4 \times (1/2)^6 = (1/2)^4 = 1/16$.

A formula for calculating F is

$$\text{inbreeding coefficient} = F = \Sigma(1/2)^A$$

where A is the number of ancestors in each chain of descent through each common ancestor, and the values are summed for each chain. For the example of a brother–sister marriage ($I_{(sib)}$, Figure 3a) the paths are C–A–D and C–B–D, so $A = 3$ for two paths and $F_{(Isib)} = (1/2)^3 + (1/2)^3 = 1/4$. In the case of a cousin marriage ($I_{(cous)}$, Figure 3b) there are five ancestors via A: E–C–A–D–F, and there are another five via B: E–C–B–D–F. Thus $A = 5$ for each chain and

$$F_{(cous)} = (1/2)^5 + (1/2)^5 = 1/32 + 1/32 = 1/16.$$

A more complicated example that might arise in a breeding program is shown in Figure 4. Each ancestor in each chain must be used once only in each chain. There are three chains, each with four ancestors: L–J–M–O, L–K–M–O and L–K–N–O:

$$F_{(I)} = 3(1/2)^4 = 3/16.$$

Medical importance

The frequency of homozygosity for rare recessive alleles in a random-breeding population is given by the Hardy–Weinberg equilibrium (Section D3) and is q^2 where q is the frequency of the recessive allele in the population. Many genetic diseases are caused by homozygosity for rare recessive alleles. If we consider an allele with frequency $q = 0.0001$ (one in 10 000) then $q^2 = 1 \times 10^{-8}$ (1 in 100 000 000); however, for the child of a cousin marriage it is $qpF = 0.0001 \times 0.9999 \times 1/16 = $ (approximately) 6.25×10^{-6} (1 in 160 000). The frequency of homozygotes is increased 625 times compared with the general population. The increase equals pF/q. The majority of cases of the rarest human recessive genetic diseases occur as a result of consanguineous marriages. Studies suggest that marriages between cousins have at least two times, and up to 10 times the rates of miscarriage and neonatal mortality as marriages between unrelated parents.

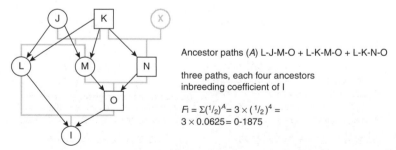

Ancestor paths (A) L-J-M-O + L-K-M-O + L-K-N-O

three paths, each four ancestors
inbreeding coefficient of I

$Fi = \Sigma(1/2)^A = 3 \times (1/2)^4 =$
$3 \times 0.0625 = 0\text{-}1875$

Figure 4. A more complicated pattern of inheritance from a breeding program. J and K produce L and M; K also sires N. Half sibs (N × M) produce O, which is mated to L, producing I, with an inbreeding coefficient of 3/16. Female X does not contribute to inbreeding. (Gray lines are family pedigree, black arrows are paths of alleles. Circles, females; squares, males.)

C13 Probabilities

Key Notes

Probabilities

When we toss a coin we cannot say which way up it will land because each landing is a random or stochastic event. We can say that it has a probability of 0.5 of being 'heads' and 0.5 of being 'tails'; both outcomes are equally likely. Probability theory tells us how often we can expect something to turn out in a particular way when we cannot predict the outcome of an individual event. In science, our hypothesis (about the mechanism of inheritance) predicts the probabilities of particular results, and allows us to test that hypothesis.

The sums rule (OR)

The sums rule (addition) is applied to combine probabilities of events which are mutually exclusive. A die has six numbered sides and each side has a probability of 1/6 of being on top. The probability of getting a 5 or a 6 (getting one excludes getting the other) is

$$1/6 + 1/6 = 2/6 = 1/3 \qquad P_{(5 \text{ or } 6)} = p_{(5)} + p_{(6)}.$$

The products rule (AND)

The products rule (multiplication) is applied to independent events. The probability of a child being a boy is 0.5 and of being a girl is 0.5 ($p_{(B)} = 0.5$ and $p_{(G)} = 0.5$). These can be multiplied to give the probability of two consecutive babies being boys, which is

$$0.5 \times 0.5 = 0.5^2 = 0.25 \text{ (i.e. } 1/4).$$

Calculating probabilities

Independent events are not affected by previous events. If the first three children in a family are boys the fourth still has a probability of 1/2 of being a girl. The product rule is often combined with the sums rule, for example to calculate the probability of two consecutive babies being both boys or both girls:

$$0.5^2_{\text{(both boys)}} + 0.5^2_{\text{(both girls)}} = 0.5 \text{ (i.e. } 1/2).$$

Punnett squares can be used to calculate probabilities, but probability paths are less error prone. In a two-factor cross ($AABB \times aabb$), the probability of F2 individuals being homozygous recessive (aa) at the first locus is 1/4 and of being heterozygous (Bb) at the second locus is 1/2, so the probability of being ($aaBb$) is $1/4 \times 1/2 = 1/8$. The sum of the end-points of all paths must be 1.

Permutations

Permutations are different ways (arrangements, combinations) which give the same result. There are two permutations by which two babies can be one boy and one girl; the boy can be the first one or the last one. The probability of a boy (0.5) followed by a girl (0.5) is $0.5^2 = 0.25$. This must be multiplied by two to include the probability of a girl followed by a boy, so the probability of a boy and girl in any order is $0.5^2 \times 2 = 0.5$. The probabilities of all possible outcomes add up to 1 (unity), 0.25 of two boys, 0.5 of a girl and a boy, and 0.25 of two girls.

The number of permutations is the coefficient of the terms of a binomial expansion:

$$P = (n! / s! t!)$$

where $n =$ total number of trials (e.g. babies), $s =$ number of one outcome (e.g. a boy) and $t =$ number of other outcome (e.g. a girl) and $s + t = n$.

The '!' means factorial, the product of all integers down to 1, so $3! = 3 \times 2 \times 1 = 6$. If the probability of the first outcome is p and of the second is q (where $p + q = 1$) then the probability of s first events and t second events in n trials (where $n = s + t$) is

$$P_{(s \text{ and } t)} = (n! / s! t!) p^s q^t.$$

The coefficient ($n!/s!t!$) is the number or permutation, and $p^s qt$ is the probability of a specific sequence of events. With n events, each with two possibilities, there are ($n+1$) numerical categories: if there are three children, zero, one, two, or three could be boys. The coefficients (number of ways of getting) for each category can also be read from the ($n+1$) row of Pascal's triangle. When there are more than two alternative outcomes for each event (e.g. six numbers on a die) a multinomial expansion must be used.

Related topics	(C1) Basic Mendelian genetics	(D3) Genes in populations: Hardy–Weinberg equilibrium
	(C2) More Mendelian genetics	(C14) Tests for goodness of fit: chi-square and exact tests

Probabilities

In genetics it is often necessary to calculate the expected frequency and numbers of particular genotypes, for example in predicting the outcome of Mendelian crosses (Sections C1 and C2) or predicting the frequency of particular genotypes in a population for comparison with Hardy–Weinberg equilibrium (Section D3). The scientific method requires that predictions are made so that they can be tested (Section C14).

We know that the probability of a tossed coin falling heads up is 0.5, but we cannot predict what the result of tossing a coin once will be, because each toss is unique and is effectively a **random** or **stochastic** event. Similarly we may know the proportion of flowers that will be pink from a particular cross, but we cannot predict which color each individual plant will produce, we must examine it to see. **Probability theory** allows us to calculate what we expect to get.

The probability (*P*) that something will happen is the number of times it *does* happen (*a*) divided by the total number of times it could *possibly* have happened (*n*):

$$P=a/n.$$

P can be measured directly or deduced from the nature of the event. We can assume that the probability of a coin falling heads up is 0.5, or of getting a 6 when we throw a die is 1/6 (because a die is equally likely to fall with 1, 2, 3, 4, 5, or 6 uppermost). **The sum of all the probabilities must add up to 1** because **something** must happen; one of the six sides *must* be uppermost. If something is certain to happen it has a probability of 1 (often called unity), if it can never happen it has a probability of 0 (nil or zero).

The sums rule (OR)

The sums rule is applied to mutually exclusive events. The die can give *either* a 5 *or* a 6, but not both. What is the probability of getting a 5 or a 6? Each has a probability of 1/6. These are added to give the probability of getting both:

$$P_{(5 \text{ or } 6)} = 1/6 + 1/6 = 1/3.$$

A genetic example would be a cross where we expected to get 1/4 red, 2/4 pink, 1/4 white flowers. The probability of any particular flower being red or pink is 1/4+2/4=3/4.

The products rule (AND)

The products rule is applied to independent events and events happening in a specific order. What is the probability of getting two sixes if we throw two dice? The fall of one die will not affect the fall of the other. The probability of the first being 6 is 1/6, and that will not affect how the second one falls, it also has a probability of 1/6, so in 1/6 of our throws the first die gives a 6, and in 1/6 of these the second die also gives a 6, so

$$P_{(6 \text{ and } 6)} = 1/6 \times 1/6 = 1/36.$$

A genetic example is a marriage between two people heterozygous for a recessive genetic disease, say phenylketonuria (PKU). If they have three children, what is the probability (chance) of all three being affected? The probability of a child being homozygous for the recessive allele is 1/4, therefore

$$P_{(\text{all 3 PKU})} = 1/4 \times 1/4 \times 1/4 = (1/4)^3 = 1/64.$$

Calculating probabilities

Independent events are not affected by the outcome of previous events. Consider the PKU family above. $P_{(\text{all 3 PKU})} = 1/64$ is only true while they have no children. If they already have two children, and both children have PKU, the probability of the third child having PKU is not changed, it is still 1/4 because the third child is an independent event. Similarly if you throw 10 heads consecutively with a coin, the next toss is still equally likely to be heads or tails.

The sums rule and products rule are often combined. The probability of two children being a boy and a girl is the probability of the first being a boy multiplied by the probability of the second being a girl ($1/2 \times 1/2 = 1/4$) summed with the probability of a girl first followed by a boy (also 1/4), so the boy–girl combination in any order is $1/4 + 1/4 = 1/2$. The segregation of each different genetic locus in a cross is an independent event (Mendel's second law). The easiest way to calculate the frequency of combinations of independent events (genotype or phenotype) is to make a probability path. They are simpler than Punnett squares, and less likely to lead to mistakes. For example, suppose we consider a three-factor cross with Mendel's peas. If we cross a strain with the dominant characters round, yellow, gray-coated seeds (genotype *RRYYGG*) with a strain carrying the recessive characters wrinkled, green, white coat (genotype *rryygg*), the F1 (genotype *RrYyGg*) will produce eight different gamete genotypes, and a Punnett square will have 64 cells in it. It is easier to work out the F2 probabilities directly. What is the probability (i.e. what is the proportion) of an F2 seed being round, green, and gray-coated? Of the total, 3/4 will be round (round is dominant), 1/4 will be green, (green is recessive), and 3/4 will be gray-coated (dominant) so we can construct a probability path (Figure 1). The answer is 3/4 round×1/4 green×3/4 gray=9/64 round, green, gray-coated. The probabilities of all the possible phenotypes add up to 64/64=1, as they must if our math is correct. The same principle can be applied to genotypes. What will be the frequency of *RryyGg* genotypes in the F2? The frequency of *Rr* will be 1/2, of *yy* will be 1/4 and of *Gg* will be 1/2, so the frequency of *RryyGg* will be $1/2 \times 1/4 \times 1/2 = 1/16$ (or 4/64 in a Punnett square).

Permutations

Permutations are different ways of doing the same thing. There is only one way that tossing two dice can give two sixes: both dice must fall 6-up. The same is true if we specify the order of events (e.g. getting a 3 then a 6 from two throws: $P_{(3 \text{ then } 6)} = 1/6 \times 1/6 = 1/36$). However, if we wanted a 3 and a 6, this could happen in two ways: the first die 3 and the second die 6, or the first die 6 and the second die 3. These are two permutations or arrangements:

$$P_{(3 \text{ and } 6)} = 1/6 \times 1/6 \times 2 = 1/18 .$$

Boys and girls are equally frequent ($P=1/2$), so the probability of two children both being boys is $(1/2)^2 = 1/4$; the probability of both being girls is similarly 1/4, but the probability of one being a boy and the other a girl is $1/2 \times 1/2 \times 2 = 1/2$ because they could occur in the order boy–girl or girl–boy These are two permutations (also called **combinations** or

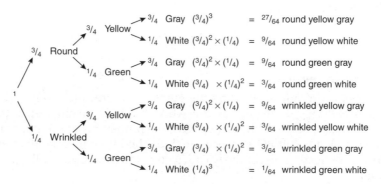

Figure 1.   Probability paths for phenotype frequencies in the F2 of a Mendelian trihybrid cross.

arrangements). In the case of two parents heterozygous for PKU who have three children, what is the probability that they will have one affected child and two normal children? The affected child could be the first, second or third born, so there are three permutations.

Probability of one PKU affected child (A) and two normal (N)
$= P(\text{A then N then N}) + P(\text{N then A then N}) + P(\text{N then N then A})$
$= (1/4 \times 3/4 \times 3/4) + (3/4 \times 1/4 \times 3/4) + (3/4 \times 3/4 \times 1/4)$
$= 3[(3/4)^2 \times 1/4] = 27/64.$

The number of permutations (combinations) is given by the formula for the terms of the **binomial expansion**: $(p+q)^n$ where p is the probability of one outcome (say a normal child, 3/4) and q is the probability of the other outcome (a child being affected by PKU, 1/4) and n is the number of trials (total number of children). Note that $(p+q) = 1$. If there are n children, the probability of s children being normal and t children being affected is

$$P = (n!/s!t!)p^s q^t.$$

The number of permutations is given by the part $n!/s!t!$ of the equation and is called the **coefficient**, whereas $p^s q^t$ is the probability for each particular ordered occurrence. Note that $(p+q) = 1$ and $(s+t) = n$. The symbol ! as in $n!$ is called **factorial**, (pronounced n-factorial) and means the product of all the integers from n down to 1 (e.g. $5! = 5 \times 4 \times 3 \times 2 \times 1 = 120$). $0! = 1$, the same way that $n0 = 1$. If we recalculate the probability of one child in three having PKU ($t = 1$) and two being normal ($s = 2$) we get:

$$P = (n!/s!t!)p^s q^t$$
$$P = (3!/2!1!) \times (3/4)^2 \times (1/4)^1$$
$$= [(3 \times 2 \times 1)/((2 \times 1) \times 1)] \times 9/16 \times 1/4$$
$$= 3 \times 9/64 = 27/64$$

This gives the same result as the direct calculation above, but has the advantage that it is much easier for complicated combinations, such as the chance of having three affected children in a family of seven ($n = 7$, $s = 4$, $t = 3$).

Pascal's triangle has all the coefficients (number of permutations) of the terms of the binomial expansion (Figure 2) and can be used by those who find the formula difficult. Each row starts and ends with 1, and each coefficient is the sum of the two adjacent terms in the row above. In the three-children example above ($n = 3$) we use the fourth ($n+1$) row of Pascal's triangle. We could have 0, 1, 2 or 3 children affected. The fourth row of Pascal's triangle shows that there is one way of getting zero, three ways of getting one affected/two normal, three ways of getting two affected/one normal, and one way of getting three

Pascal's triangle

Row		Total
1	1	1
2	1 1	2
3	1 2 1	4
4	1 3 3 1	8
5	1 4 6 4 1	16
6	1 5 10 10 5 1	32
7	1 6 15 20 15 6 1	64
8	1 7 21 35 35 21 7 1	128
9	1 8 28 56 70 56 28 8 1	256

Figure 2.   Pascal's triangle shows the number of permutations (coefficients) of binomial expansions.

affected (1:3:3:1). It is easy to generalize upwards to three or more alternatives using the general **multinomial expansion.** If the probabilities of flowers being red, pink or white are p, q and r respectively, the probability of getting s red, t pink and u white flowers is

$$P = (n!/\, s!t!u!)p^s q^t r^u$$

where the total number of flowers $= n = s + t + u$. In fact this can be extended to any number of classes.

C14 Tests for goodness of fit: chi-square and exact tests

Key Notes

Goodness of fit	Goodness-of-fit tests calculate the probability of observed data being obtained by chance if that data is a sample from a hypothesized population or distribution (i.e. if we count 673 round peas and 117 wrinkled peas, what is the probability of getting these results from a Mendelian 3:1 distribution?). The null hypothesis is that the observed data does not deviate significantly from the hypothesized ratio. The usual test is to calculate chi-square.
The chi-square test	Chi-square = χ^2 = sum [(observed−expected)^2/expected] usually written $\chi^2 = \Sigma(O-E)^2/E$
	Chi-square is always calculated on original data, never on percentages or frequencies.
Degrees of freedom	Degrees of freedom are one less than the number of independent classes measured. In a Mendelian cross with three phenotype classes there are two degrees of freedom. In a test of Hardy-Weinberg equilibrium with three genotypes (one locus, two alleles) there is only one degree of freedom because the expected values are calculated from the observed values of allele frequency; they are not independent.
Using χ^2 probability tables	The probability of getting a result by chance is looked up in tables of χ^2. Looking along the row for the appropriate degrees of freedom, find the values greater and smaller than that calculated, and read the probability at the top. A one-tailed test is used because we are interested only in detecting large deviations, not good fits. Large values of χ^2 have low probabilities of occurring by chance, and indicate a large deviation from expectation.
Level of significance	The level of significance is the probability below which we assume the data observed does not fit expectations, and so reject the null hypothesis. This is conventionally 0.05 (1/20). When there is a probability of less than 0.05 (5%) of getting a particular set of data by chance if the null hypothesis is true, then it is conventional to reject the null hypothesis as probably being false and say the data does not fit. This

	decision will be correct 19 times out of 20. One time in 20 a real fit will be rejected.
Yates' correction	If the expected number in a class is five or less, the numerical value of the difference between observed and expected is reduced (moved closer to zero) by 0.5 before it is squared. This is to allow for the fact that only integers (whole numbers) can be counted, whereas we may expect fractions.
Fisher's exact test	Fisher's exact test is simply to work out the exact probability of obtaining the observed result, plus the probability of obtaining an even more extreme result. For example, a man is known to be heterozygous for phenylketonuria but his wife's genotype is unknown. The null hypothesis is that she is heterozygous, in which case each child will have a 3/4 chance of being normal. If they have 11 children all normal, with a probability of $(3/4)^{11} = 0.04223$ then we can reject the null hypothesis with a significance level of 4.22%. That is the exact chance of us being wrong, and her having so many normal children if she is heterozygous. (If they have one affected child, she must be heterozygous.)
Related topics	(C1) Basic Mendelian genetics (D3) Genes in populations: (C2) More Mendelian genetics Hardy–Weinberg equilibrium (C13) Probabilities (G1) Genetics in forensic science

Goodness of fit

It is often necessary to discover whether observed numbers fit an expected hypothesis. For example, Mendelian genetics suggest a 3:1 ratio in the F2 generation of a monohybrid cross (Section C1). Suppose we cross a red flower with a white flower, interbreed the F1 (which are all red) and observe a ratio of 63 red:37 white in the F2. Is 63:37 a good fit for a 3:1 ratio allowing for chance error in sampling? Chi-square ($\chi 2$) is a statistical test which tells us the probability of obtaining the observed result by chance if the null hypothesis is true. It does not tell us whether the null hypothesis is actually true. The null hypothesis is that the data fits the ratio we suggest, and this is what is tested. We must have a null hypothesis before we can do any meaningful test. Our null hypothesis is that the results (63:37) do fit a 3:1 ratio.

The chi-square test

Chi-square (χ^2)=sum[(observed–expected)^2/expected]

$$\text{usually written } \chi^2 = \sum \frac{(O-E)^2}{E}.$$

χ^2 is always calculated from original data, never from percentages, frequencies, or proportions.

The χ^2 for comparing the 63:37 ratio above is calculated in Table 1. The expected numbers are calculated by multiplying the total by the predicted frequency (e.g. $100 \times 1/4$ white). The deviation, or difference, between the observed and expected number is squared to remove negative numbers, and divided by the expected value to undo the squaring and standardize the numbers. This is repeated for each class (the red class and the white class in this example), and the values are added (summed) to give the overall value of χ^2. Notice that the value of χ^2 increases when the deviations from expected are large, so large values of χ^2 lead us to reject the null hypothesis. The data do not fit if χ^2 is large. A perfect fit gives a χ^2 of zero. We are not interested in identifying unusually close fits, so we use a **one-tailed test** which tells us the probability of getting a large deviation from the expected. The example (Table 1) gives a χ^2 value of 7.68. What does this mean? We need another piece of information before we can look this up in statistical tables. We need to know the degrees of freedom.

Degrees of freedom

Degrees of freedom are one less than the number of classes. They tell us something about the number of independent numbers we have, and this relates to the usefulness of our data. In this case we have two classes: red and white. All the plants that are not red must be white. When we have counted the red ones, the number of white ones is fixed so we have only one degree of freedom. If we had red, pink, and white flowers – three classes – there would be two degrees of freedom; when two classes had been counted, the third would be fixed. When testing Hardy–Weinberg equilibrium with one locus and two alleles (three genotypes) there is only one degree of freedom. This is because the expected numbers are calculated from the observed allele frequencies, they are not independent (whereas a 3:1 ratio prediction is independent of the observed results). The only independent variable is the frequency of one allele (e.g. p): this fixes q because $(p+q=1)$ and this fixes $2pq$.

The number of classes has a practical effect on the value of χ^2 because an extra value is added into the calculation for each extra class. Therefore χ^2 gets bigger as the degrees of freedom get bigger, and we must look at a different line in the table of probabilities.

Using χ^2 probability tables

Now we can look up the level of significance of our result. Go to Table 2, and follow the line for 1 degree of freedom (d.f., top line) to find the nearest values of χ^2 above and below our value. We can see that the value of 6.6349 is exceeded with a probability of 0.01 (that is 1%, or 1/100) and 7.87944 is exceeded with a probability of 0.005 (i.e. 0.5%, or 1/200). Our value, 7.68, is between these. This means that if we repeated the experiment 100 times we would not expect to get this big a deviation by chance once. This suggests that our data do not fit the expectation of a 3:1 ratio.

Level of significance

The conventionally accepted significance level is 5% (0.05); if the deviation is so large that the probability of it occurring by chance is less than 5%, we reject the null hypothesis because the data differ **significantly** from the predictions of the null hypothesis. So if χ^2 exceeds the value which has a 5% (i.e. 1/20) probability of occurring by chance, we will be correct to reject the null hypothesis 19 times out of 20. Note that one time in 20 we will reject the null hypothesis when it really is correct. Statistics do not 'prove' anything; they just tell us the probability of certain things happening if our particular assumptions are correct. In this case we reject the null hypothesis that the results fit a 3:1 ratio because a 63:37 ratio will only occur by chance on less than one occasion in 100.

Table 1. χ^2 **calculations**

Null hypothesis (predicted ratio)

3:1	(100 total)	63 Red	37 White
Observed		63 Red	37 White
Expected		$100 \times 3/4$	$100 \times 1/4$
		75	25
Observed−expected		−12	12
(Observed−expected)^2		144	144
(Observed−expected)^2/expected		144/75 = 1.92	144/25 = 5.76

$\chi^2 = 1.92 + 5.76 = 7.68$
Degrees of freedom = 2 − 1 = 1 d.f.
From Table 2 probability (p)(1 d.f.) $0.01 > p > 0.005$

9:7	(100 total)	63 Red	37 White
Observed		63 Red	37 White
Expected		$100 \times 9/16$	$100 \times 7/16$
		56.25	43.75
Observed−expected		6.75	−6.75
(Observed−expected)^2		45.5625	45.5625
(Observed−expected)^2/expected		45.5625/56.25 = 0.81	45.5625/43.75 = 1.04

$\chi^2 = 0.81 + 1.04 = 1.85$
Degrees of freedom = 2 − 1 = 1 d.f.
Probability (p)(1 d.f.) $0.20 > p > 0.05$

Table 2. Values of the χ^2 distribution exceeded with probability P

d.f.	0.2	0.1	0.050	0.025	0.010	0.005	0.001
1	1.642	2.706	3.84146	5.02389	6.63490	7.87944	10.828
2	3.219	4.605	5.99146	7.37776	9.21034	10.5966	13.816
3	4.642	6.251	7.81473	9.34840	11.3449	12.8382	16.266
4	5.989	7.779	9.48773	11.1433	13.2767	14.8603	18.467
5	7.289	9.236	11.0705	12.8325	15.0863	16.7496	20.515
6	8.558	10.645	12.5916	14.4494	16.8119	18.5476	22.458
7	9.803	12.017	14.0671	16.0128	18.4753	20.2777	24.322
8	11.03	13.362	15.5073	17.5345	20.0902	21.9550	26.125
9	12.242	14.684	16.9190	19.0228	21.6660	23.5894	27.877

The highest applicable value of χ^2 is normally quoted to emphasize the quality of a conclusion. For example, if the probability of obtaining a particular value of χ^2 by chance is less than 0.01 it is normal to quote this value (1%) rather than the conventional 5%. The 1% significance means we will only be wrong in rejecting the null hypothesis once in 100 experiments.

There is another possible explanation for the 63:37 ratio which is that there are two genes that differ between the original red- and white-flowered strains. If these code for enzymes that work in the same biochemical pathway they will both be needed to produce red pigment. The new null hypothesis is that the ratio will be 9:7 (Section C2). The χ^2 based on this null hypothesis is also calculated in Table 1, and is 1.85. Table 2 tells us that the value of χ^2 will be less than 1.64 in 0.20 (one in five) of our experiments, and greater than 3.84 in 0.05 (5% or one in 20) of our experiments. Our result, 1.85, has a probability of occurring by chance between these values. Since a deviation this great or greater will occur by chance in more than one experiment in 20, we will accept the null hypothesis that the data fit a 9:7 ratio and assume that two genes are involved in producing the red color. There is a small chance that we are wrong, but we have no better hypothesis to test at present.

Yates' correction

Yates' correction should be applied in χ^2 calculations in any class where the value is less than five. χ^2 is a continuous distribution (it expects fractional values) but we count in integers (whole numbers) – you cannot grow half a pea. So if we expect 3.5 white flowers, the nearest we can get is three or four, a minimum error of 0.5. Yates' correction is applied by reducing the numerical value of (observed–expected) by 0.5 if the value expected is five or less. If observed minus expected $=-1.6$ then it is adjusted to -1.1, if it is 2.7 it is adjusted to 2.2, before it is squared.

Fisher's exact test

Fisher's exact test can be applied when there are only a few events, or when some or all of the classes are rare, and only small numbers are expected, because it gives an exact probability for any observed result. The probability of the particular outcome, and all more deviant outcomes, can be calculated from the probabilities of each individual event. These can be summed to give the probability of getting a deviation as great as, or greater than, the observed deviation by chance. If this probability is less than 0.05 we reject the null hypothesis. For example if the null hypothesis is that babies are equally likely to be boys or girls (each has a probability of 0.5) the probability of two consecutive births both producing girls is exactly $0.5^2 = 0.25$. Thus, if two consecutive children are both girls, this is not significantly different from the expected 1:1 ratio. The probability of getting six girls consecutively is $0.5^6 = 0.015\,625$ so (if we had no other information) a family with six children, all girls, would lead us to reject the null hypothesis that boys and girls are equally probable with a significance level of 1.5625%. Consider a more complex example: a pair of short-haired dogs are mated and produce four puppies, three long-haired and one short-haired. The parents must be heterozygous, and short hair must be dominant, so our null hypothesis is that the data do not deviate from the expected ratio of three short-haired to one long-haired. What is the probability of getting this particular reversed result by chance? What is the probability of getting a deviation this great or greater by chance? The numbers are much too small to use χ^2. The greatest deviation would be four long-haired puppies, so we start there. Table 3 shows the individual and cumulative probabilities of all the possible combinations of long-haired and short-haired puppies in a litter of four. The probability of getting our family of three long-haired and one short-haired or the more extreme case of four long-haired puppies is $0.050\,781\,25$, slightly greater than one in 20, so we have no reason to reject the null hypothesis. Even if all the puppies were long-haired (probability 0.0039, one in 256) we really should repeat the experiment because a sample of four is not big enough to justify profound conclusions. Sometimes there are too many permutations to calculate (e.g. in possible human genetic profiles for forensics, some genotypes never occur, see Section G1). A computer simulation can be used, effectively picking thousands of genotypes from a hat, to approximate Fisher's exact probability.

Table 3. Exact test for hair length in puppies

Hair length		Individual probability × permutations	Probabilities of specific family	Cumulative probabilities
Long (n)	Short (n)			
4	0	$1/4 \times 1/4 \times 1/4 \times 1/4$	0.00390625	0.00390625
3	1	$1/4 \times 1/4 \times 1/4 \times 3/4 \times 4$	0.046875	0.05078125
2	2	$1/4 \times 1/4 \times 3/4 \times 3/4 \times 6$	0.2109375	0.26171875
1	3	$1/4 \times 3/4 \times 3/4 \times 3/4 \times 4$	0.421875	0.68359375
0	4	$3/4 \times 3/4 \times 3/4 \times 3/4$	0.31640625	1

D1 Introduction

Key Notes

Overview	Population genetics is the study of the frequencies of alleles in populations, and evolution is the change in allele frequencies over time. The topics in this section are highly interrelated.
Natural selection	Darwin observed that only a few progeny in any species survive and reproduce. Nature selects these, and they pass their genetic attributes on to their offspring.
Hardy–Weinberg equilibrium	It is possible to calculate genotype frequencies from allele frequencies on the assumption that they are not being changed. This allows quantifiable analysis of changes, and the equations can be extended to predict the changes in allele frequency resulting from continuing those changes. Hence evolutionary outcomes of particular scenarios can be predicted.
Genetic diversity	There is a lot of genetic variation in most populations. It originates from mutations and is the essential raw material for selection; however, it is difficult to account for its quantity. Too much death from natural selection would be required to maintain all the variation by selection. Much of the variation may be selectively neutral, and subject to drift (chance changes). There is debate about the relative importance of selection and drift in maintaining variation.
Neo-Darwinian evolution	The New Synthesis combines natural selection with a knowledge of genetic inheritance and population genetics. Selection is seen as acting on phenotypes to change the frequency of genetic units, alleles, which persist through generations.
Chromosome evolution	The chromosomes in each species have a distinctive size, shape, and number. The complement of chromosomes is called the karyotype of the species. Chromosomes, the carriers of genetic information, change in shape and number through evolutionary time. This contributes to diversity within species, and contributes to speciation events.
Species and speciation	Members of a single species can potentially share in a common gene pool. The essential event in speciation is when members of two populations of a species stop interbreeding and stop interchanging genetic material. Genetic differences can then accumulate, causing divergent ecological and behavioral changes until each population is a new species.

Polyploidy	Occasionally individuals occur with extra complete sets of chromosomes, usually as products of cross-species hybridization. These can create new species because they cannot interbreed with diploids, and about half of all plant species are polyploid. The extra genomes are also free to mutate and some new gene functions may emerge.	
Evolution and populations	Comparison of DNA sequence and allele frequency allow retrospective study of the separation and divergence of populations and species.	
Related topics	(D2) Evolution by natural selection (D3) Genes in populations: Hardy–Weinberg equilibrium (D4) Genetic diversity (D5) Neo-Darwinian evolution: selection acting on alleles	(D6) Chromosome change in evolution (D7) Species and speciation (D8) Polyploidy (D9) Phylogeography, molecular clocks and phylogenies

Overview

Population genetics and evolution are inextricably linked. Population genetics is the quantitative study of the frequencies of alleles and genotypes in populations, whereas evolution is the change in those frequencies over time. Factors which change allele frequencies are the factors which cause evolution. The topics in this section all interrelate, and the approach taken here is to concentrate on key concepts as a focus for learning, but the interactions are clearly indicated. The order of presentation is also problematic, because it is often necessary to understand two things together, but something must come first. It is important to remember that many topics in this section are the same subject examined from different perspectives. The topics, and their place in this section, are outlined below

Natural selection

Natural selection (Section D2) examines Darwin's original observations and deductions. These are the foundation for our understanding of factors which cause adaptive genetic changes in populations in response to their environment, and these changes over time are evolution. Any habitat can only support a limited population of any species, far less than the potential for the population to reproduce. As individuals compete to reproduce, the most successful pass on to their offspring any genetic variant that made them successful. Thus evolution proceeds.

Hardy–Weinberg equilibrium

Hardy and Weinberg independently developed this calculation (Section D3) to show that an equilibrium in genotype frequencies will occur after one generation of random mating, and those genotype frequencies can be calculated from allele frequencies. The equilibrium frequencies will be maintained from generation to generation unless some force changes the allele frequencies. Hardy and Weinberg disproved the erroneous notion that dominant characteristics become more common. The introduction of simple algebra started the science of population genetics, and allows us to quantify the evolutionary effects of different levels of selection on homozygotes and heterozygotes, or on dominant or recessive alleles. This topic is fundamental to neo-Darwinian evolution (Section D5) and also to understanding the distribution of genetic disease in human populations.

Genetic diversity

Most populations of most species show extensive genetic variation between individuals (Section D4). The source of this variation is mutations, but why is there so much variation persisting in the population? There are two alternative explanations: natural selection and chance. Natural selection could be continually replacing alleles, so we see a snapshot of the changes, or there could be frequency-dependent selection favoring rare alleles, keeping them in the population. Both explanations require a large number of deaths by natural selection in each generation. There is so much variation that the required level of selection would probably be intolerable. The alternative view is that most of the variation does not affect fitness, so it is selectively neutral, and the frequency of the alleles involved fluctuates by chance (drift). The real explanation is some combination of these two, but arguments continue about their relative importance.

Neo-Darwinian evolution

The combination of population genetics and natural selection (the New Synthesis) allows us to examine the effect of selection on genotypes and predict the outcome (Section D5); it turns the study of genetic evolution into a quantitative science. We can calculate the number of generations needed for a particular level of selection on a dominant or recessive allele to lead to replacement of one allele by another. We find that recessive alleles are less affected when rare because they are mainly hidden in heterozygotes, and selection for heterozygotes preserves both alleles, whereas selection against heterozygotes removes rare alleles, even if they are fit as homozygotes. The important outcome is that it is the frequency of alleles (alleles are regions of DNA) that changes over time. This leads us to consider selection, on average, as acting on alleles rather than on phenotypic characters, and so establishes the concept of 'selfish genes' (Section D2) which secure their own transmission, not the fitness of the individual.

Chromosome evolution

Chromosomes carry the genetic information. Their shape and number is a characteristic of each species, its **karyotype**, and can change over time. Gametogenesis (Section C3) requires the accurate division of genetic material in meiosis, and this is difficult to achieve if the chromosomes do not match. There are species where heterozygosity for chromosome shape is common, parts of chromosomes being inverted or translocated to other chromosomes. Some special genetic effects result from this. Meiotic recombination (crossing-over; Section C3) in a heterozygous rearrangement often produces inviable products, so recombination is prevented in the rearranged chromosome segment, and all the alleles there can co-evolve. In other cases, chromosomal differences cause sterility in hybrids, and can act as an agent towards speciation. Comparison of chromosomes between species may indicate evolutionary change.

Species and speciation

The formation of new species is the key occurrence in evolution. Members of different species do not interbreed and cannot share the same gene pool. This block on genetic exchange is the only essential requirement for speciation, and must ultimately be genetically controlled when two species meet and have an opportunity to interbreed. If two populations are separated physically by either time or space when mating, they cannot exchange genetic material. As neutral and selected changes arise, they remain within one population. Eventually the differences between populations become so large that members of the two populations do not interbreed successfully when they do meet, and speciation has occurred.

Polyploidy

Sometimes individuals occur with three or more sets of chromosomes. They are polyploid as opposed to the normal diploid with two sets (Section D8). Organisms with three sets are triploid, with four sets are tetraploid, and so on: pentaploid (five), hexaploid (six), octaploid (eight). These usually arise as hybrids between two related species. They have a set of chromosomes from each parent, but they do not match. Odd number polyploids are sterile (unless they are parthenogenetic) because they cannot segregate chromosomes evenly into gametes in meiosis; odd numbers are not divisible by two. However, they may accidentally double the number of chromosomes in mitosis and each chromosome can then pair with its sister; thus fertility can be restored. This is a common event in plants where 47% of angiosperms (flowering plants) are polyploid. Because their gametes only form balanced zygotes when they fuse with similar ploidy gametes, the original polyploid individual can start a new species with a functional species isolation mechanism.

Evolution and populations

Comparisons between species show us what has happened in the past. Species which share a recent common ancestor are more alike than species which diverged longer ago. Comparison of the amino acid sequence of particular proteins shows us that they change (diverge) at different rates. Proteins which interact with many other molecules must maintain their structure to keep their function, and therefore evolve very slowly. Other proteins, which are soluble and mainly interact with water, are less constrained and change much more quickly. Examination of the DNA shows that sequences which do not code for anything evolve fastest of all. In this way we can develop **molecular clocks**, based on the basic intrinsic mutation rate, filtered (in coding or regulatory sequences) by selection against deleterious mutations. This reinforces the view that most surviving mutations are neutral. It also allows us to deduce the evolutionary distances between species by comparing their DNA sequences. A similar process can be applied to populations. Particular attention is currently being given to the spread of ancient and modern humans from Africa during the last few hundred-thousand years (Section D10). On a finer scale, the spread of tribes across Europe and Asia, and into Oceania and the Americas during and after the last ice age, can also be examined, with much more emotive prospects.

D2 Evolution by natural selection

Key Notes

Evolution by natural selection

Natural selection is the constraint that natural conditions put on the size of populations, forcing individuals of the same species to compete for limited resources. Those types which use the resources most successfully in order to reproduce, pass on their genetic material to their offspring, and are selected for. Inefficient or unsuccessful types fail to reproduce as successfully and are selected against. The process is analogous to selection, by farmers, of the best types of stock and plants for breeding. This causes a gradual change in the genetic makeup of the population. The phenotypic characters which are selected for will become more frequent providing that they are genetically controlled. Relative fitness is the differential average reproductive success of individuals with different genotypes (combinations of alleles). Natural selection was first proposed as the cause of evolution by Charles Darwin, but he did not know how inheritance worked.

Darwin's observations and deductions

Darwin's observations were: (i) the potential number of descendants from any species is infinite – many more offspring are produced than can survive; (ii) this potential population growth is prevented by limited resources – populations are relatively constant in size; (iii) there are many differences between individuals which affect their ability to survive and reproduce.

Darwin deduced that, therefore: (i) there is a struggle for the limited resources; (ii) those individuals which survive and reproduce successfully will pass on to their offspring the characteristics which helped them to survive and reproduce; (iii) gradually, the inherited characteristics which help in the competition to reproduce will become more frequent in the population. This gradual change is evolution.

Modes of selection

(i) Stabilizing selection removes phenotypes which deviate too far from the optimum (norm). In a stable environment, change or deviation usually produces less fit phenotypes. (ii) Directional selection selects for a new optimum, and against one end of the range of phenotypes, causing the mean for the character to change towards the new optimum. This can occur after a change in environment, including competition with another species. Character displacement occurs when two species compete. Selection is greatest where their niches overlap, so each species is selected to be different from the other (e.g. one species becomes larger, the other smaller, reducing competition). (iii) Disruptive selection selects against the mean phenotype and favors two different optimum phenotypes. It can develop as a sexual difference (e.g. female sparrow hawks are larger than males and catch larger prey). Disruptive selection can work when there are two qualitatively different types, but there are theoretical problems with continuously variable characters because of the continued production of unfit intermediates.

Sexual selection

In polygamous species, individuals of one sex, usually the males, compete for mates. Intra-sexual selection is competitive interaction between individuals of the same sex for access to mates. Typically it is competition between males and selects for increased body size and weapons such as antlers. Inter-sexual competition is an interaction between members of opposite sexes, when one sex displays to advertise their fitness and attract mates. The other sex (typically 'choosy' females) select their mate on the basis of the display. This selection produces extravagant display ornamentation (e.g. the peacock's tail). These may be described as 'handicaps' but the terms 'advertisements' or 'status symbols' would reflect their role more accurately.

Incipient speciation

Races are incipient species. Genetic differences between separated populations can increase until they first become different races then eventually gain the status of separate species.

Selfish DNA/genes

Only copies of DNA or genes can survive through generations (phenotypes die with individuals, genotypes are disrupted at meiosis). Transposable elements are short regions of DNA (many related to retroviruses) which are copied and inserted into all the chromosomes in a cell, thereby increasing their frequency without increasing the fitness of the individual they are in. DNA sequences are selected to enhance transmission of copies of themselves, even to the detriment of other genes in the genome, or the host individual. In extreme cases DNA or a whole chromosome may show meiotic drive, distorting segregation at cell division to increase their frequency in gametes and zygotes, or causing destruction of gametes which do not carry a copy of themselves (e.g. segregation distorter in *Drosophila melanogaster*), reducing competition. Such selfish or parasitic DNA can reduce the phenotypic fitness of the individual carrying it, providing meiotic drive increasing its transmission exceeds fitness reduction.

Historical perspective	Charles Darwin recognized natural selection as the mechanism causing evolution in 1838, but did not publish the *Origin of Species* until 1859; Alfred Russel Wallace reached the same conclusion independently in 1858. Previous theories of evolution supposed that organisms improve their status by their own efforts (inheritance of acquired characteristics), a theory now known as 'Lamarkian' after its last great proponent. It was supposed that there was a progression from slime up to humans.	
Related topics	(D4) Genetic diversity (D5) Neo-Darwinian evolution: selection acting on alleles	(D7) Species and speciation (D9) Phylogeography, molecular clocks, and phylogenies

Evolution by natural selection

Natural selection is the process by which the environment limits the size of populations and forces individuals to compete for resources. Some individuals inherit characteristics which help them to survive and reproduce better than individuals with other characteristics. As a result, copies of the survivors' genetic determinants (which we know as alleles of genes) for favorable characteristics become more frequent (common) in the population in future generations. Darwin made a direct analogy with selection of plants and animals by human breeders, saving individuals with favored characteristics for breeding, and destroying the rest. He used fancy pigeons as an example. Darwin did not know how inheritance worked, and like others at the time he incorrectly assumed a blending of characteristics from both parents, rather than discrete Mendelian genes. Blending should lead to a uniform type, reducing variation. Darwin's main obstacle was not understanding the mechanism of inheritance, although he recognized new variant types (our mutations).

Natural selection acts on the variation already existing in populations; where there is no variation, there can be no change (Section D7). However, species do not become extinct because they fail to adapt. Species become extinct when their habitat is removed or changed to a state where not enough individuals with an adapted genotype remain to maintain the species. There have been many large animals which are now extinct because of predation by humans. They were well adapted and carried well-adapted fleas. When the animals became extinct, so did their fleas. Both had lost their environmental niche, and had nothing else to adapt to.

Darwin's observations and deductions

Darwin made observations and deductions which may be presented simply.

Observations

- *Observation 1.* The number of individuals in any population can increase exponentially when conditions allow them all to survive and reproduce. Any species could soon cover the entire Earth if allowed to do so.

- *Observation 2.* This potential reproduction is always limited eventually by lack of resources. At all stages in the life cycle, different factors will limit survival or reproduction, but their relative importance may change with time, both in one life cycle, or historically over generations. Examples of limiting factors include availability of resources: food, shelter, territory, nest-sites, mates, sunlight (essential for plants), and the effects of predators, diseases, and parasites.
- *Observation 3.* Individuals are not all the same. There are differences (variations) between individuals in every species, which may, however slightly, affect their ability to compete, survive, and reproduce. Some of these variable characters are inherited.

Deductions

- *Deduction 1.* From observations 1 and 2, there must be a 'struggle' or competition for survival and reproduction in which some individuals are less successful than others.
- *Deduction 2.* From observation 3, some of the differences in characteristics between individuals will affect their success in the struggle for survival and reproduction. Therefore the elimination will be selective. Individuals with favorable characteristics will be more successful and produce more offspring than less successful individuals.
- *Deduction 3.* Those characters which are genetically determined, and which assist individuals to survive and reproduce (characters which are selected for) will be more frequent in the next generation and so will produce a gradual change in the genetic makeup of the population over time.

Evolution only acts on the portion of variation which is controlled by genetic factors. Differences caused by environmental factors, for example differences in size due to being born where food was plentiful or scarce, will affect an individual's ability to leave offspring, but these differences will not be passed through successive generations of offspring, so will not contribute to evolution.

Darwin thought that Herbert Spencer's phrase 'survival of the fittest' was not the best way to describe this process because of its misleading suggestion of physical fitness. The important aspect of fitness is the individual's relative success in producing successful progeny, in comparison to other individuals of the same species. Individuals may adapt to changes in the environment (e.g. grow thicker hair in the cold) but this is not evolution because it does not cause a change in the genetic makeup of the population and is not passed on to the next generation. The ability to adapt is passed on, however.

Modes of selection

There are three main modes of natural selection (Figure 1).

Stabilizing selection

Stabilizing selection (Figure 1a) removes individuals deviating too far from the norm, and maintains the population near the optimum. In a stable environmental niche most individuals are well adapted and new types or deviants are likely to be less fit. For example, house sparrows (*Passer domesticus*) killed in a severe snowstorm in New York were found to have wings markedly longer or shorter than the mean. Human babies with birth weights deviating furthest from 3.6 kg are more likely to die or suffer damage at or soon after birth because they are either too big to pass through the birth canal, or too small to survive. This selection is now very much reduced in countries where medical support is available.

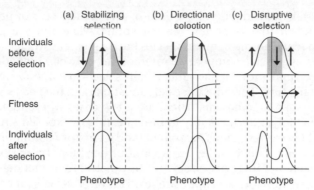

Figure 1.   Modes of selection. The shaded region is selected against. Arrows show changes in phenotype frequency expected over time. Vertical solid lines show means, dashed lines allow comparisons.

Directional selection

Directional selection (Figure 1b) favors individuals with characteristics at one end of the range found in the population. An example which changes direction with the climate relates to Darwin's finches, *Geospiza fortis*. During drought in the Galapagos Islands, grasses and small plants do not grow well, but bigger deep-rooted plants with larger seeds still produce fruit. There is then selection which favors individual finches with larger beaks capable of eating large seeds, and selection against those with smaller beaks best-suited for eating grass seeds. This is reversed after wet years when small seeds are plentiful. A common form of directional selection causes **character displacement** when two species compete. If there are two similar species of finch on one island (sympatry), one species will specialize in large seeds and evolve a large beak, the other will specialize in small seeds or insects and evolve a suitably small beak. Either species alone (allopatry) has an intermediate size beak. This effect is very common among related species.

Disruptive selection

Disruptive selection (Figure 1c) selects for two distinctively different phenotypes in a population. Examples are size differences (e.g. female sparrow hawks, *Accipiter nisus*, are twice as large as males and catch larger prey), and behavior (e.g. some male salmon mature precociously in rivers while small without going to sea; others go to sea and get much larger, return to breed, and can aggressively displace small males and fertilize more eggs, but they have the disadvantage of a longer life cycle and greater predation). Disruptive selection in a single population will work if the variation is discontinuous, with two clearly different alternative types, as in these two examples. There are, however, reasons to believe that two phenotypes cannot coexist for a continuously variable trait (Section C7) if interbreeding between them produces hybrid offspring with unfit intermediate phenotypes. The less common of the two fit phenotypes may then be lost (Section D7). Disruptive selection might cause one species to evolve into two, each adapted to one of the optima, but this again requires the existence of premating isolation mechanisms to prevent continual interbreeding.

Sexual selection

Selection can affect the two sexes differently, mainly because microgametes (sperm and pollen) are small and cheap to produce, so males can father many offspring from many females, while female reproduction is limited by the greater cost of producing macro-gametes (eggs) and in many species, rearing young. Males have a greater potential reproduction than females, and in many species have a greater variance in success. **Intra-sexual selection** is aggressive competitive interactions between individuals of the same sex (usually between males for females) and usually selects for large size and weapons (e.g. antlers) useful for fighting. The differences are most marked in species such as elephant seals where males hold large harems. There are a few cases where females aggressively compete for males (polyandry), notably in birds nesting in exposed situations. Females of some species of lily trotter (genus *Jacana*) defend a territory containing several males, each male incubating a clutch of her eggs without her help. **Inter-sexual selection** is competition between members of one sex (usually males) to attract mates, and involves interactions between individuals of opposite sex. This often selects for ornaments for dis-play such as peacocks' tails and bower birds' bowers, which act as status symbols adver-tising the surplus resources of the fittest males to the females. The females are said to be **choosy**. These costly male adornments are often called **handicaps**, but this is misleading. The human equivalent would be owning a luxury yacht, an advantage in attracting mates, and not a handicap if you can afford it. The terms **status symbol** and **advertisement** are more accurate. The males must have good health and fitness to produce the advertise-ment because it does cost resources. In group displays it is often not clear whether males are threatening other males or courting females, or doing both simultaneously.

Incipient speciation

Darwin considered varieties to be incipient species. There is a continuous range between slightly different races and fully diverged species. Divergence occurs when two popula-tions undergo different genetic change with time. This may be due to directional selec-tion for different phenotypes in the two populations. However, even if the selection on the phenotype (e.g. bill size, optimum temperature, etc.) is in exactly the same direc-tion in both populations, which is unlikely, a different combination of alleles may finally be selected in each population. For example, suppose two genes *A* and *B* both code for similar enzymes, and there is selection to produce less enzyme. Now *AABB* has too much enzyme, and either *A* or *B* must be lost. *AAbb* may become fixed in one population while *aaBB* becomes fixed in another (*aabb* is lethally deficient because it has no enzyme at all). There are also differences which arise slowly by chance (drift or mutation). When the two populations are sufficiently different in appearance, behavior, ecology, or genetics, they may be considered to be separate species (Sections D7 and D9).

Selfish DNA/genes

The genome of a sexually reproducing organism is disrupted at gametogenesis by recom-bination and reassortment of chromosomes. Individuals and their genomes are tempo-rary; only copies of regions of the DNA pass unchanged through the generations. This leads to the concept of 'selfish DNA' and 'selfish genes' because any piece of DNA or gene which can enhance transmission of copies of itself to the next generation has a selec-tive advantage over other allelic DNA sequences. One example of selfish DNA is a gene for sacrificial behavior in one individual to enhance reproduction by a relative who is likely to pass on copies of the same DNA. This is called **kin selection**. It is demonstrated by worker bees, ants, and wasps who help a sister to become a queen and reproduce. Males of this group (*Hymenoptera*) are haploid, so full sisters all inherit the same genes from their father and are related to each other by 0.75 of their genome, rather than the 0.5 of their genome they pass to each of their offspring. Thus genes are transmitted more successfully through sisters than through offspring. This is probably why the group has evolved so many social species. Somatic cells in all multicellular animals similarly abandon long-term reproduction to facilitate sexual reproduction of specialized germ cells (Section C3), which are usually genetically identical to the soma (body tissue). **Transposable elements** (Sections B3, B4, and B7) are short DNA sequences which replicate and are inserted around the genome into all uninfected chromosomes. They are presumably 'selfish DNA.' Their transposition may be performed by proteins coded by the transposable element itself (e.g. some are similar to retroviruses), or by proteins coded by genes elsewhere in the genome. If any piece of DNA enhances its own transmission into gametes and zygotes at greater frequencies than expected by Mendelian segregation then it has a selective advantage. This is known as **meiotic drive**. In some cases this may be accompanied by a reduction in the total reproduction of the individual. For example the allele *SD* (*segregation distorter*) in *Drosophila* actually does this by destroying sperm not carrying copies of itself. It is then truly parasitic selfish DNA. There are also parasitic chromosomes which persist in germline cells, but are lost from the soma where they might reduce fitness. Other parasitic chromosomes get into the egg nucleus and avoid the polar bodies.

Historical perspective

Evolution by gradual change through successive generations was already an accepted idea when Darwin proposed natural selection as a mechanism for evolution in his book *Origin of Species* in 1859. Darwin had formed the idea in September 1838, but only went public and published it in 1858 after Alfred Russel Wallace came to the same conclusion independently. Previous theories were based on an ancient idea of a **Great Chain of Being** with slime at the bottom and humans, angels and God at the top, all organisms striving to become more like angels. **Lamark** tried to write this formally as a scientific hypothesis in which improvements made by the efforts of parents were passed on to their offspring (**inheritance of acquired characteristics**). Evidence shows this to be incorrect; amputating the tails of mice at birth for several generations does not produce a tailless race of mice.

D3 Genes in populations: Hardy–Weinberg equilibrium

Key Notes

Introduction

Population genetics is the study of alleles of genes in populations (often called demes), and the forces which maintain or change the frequencies of particular alleles and genotypes in populations. The total genetic stock of the population is its gene pool. Individuals have a selection of alleles from that gene pool, possibly taken randomly. The Hardy–Weinberg equilibrium is a means of calculating expected genotype frequencies from allele frequencies determined in the same population (and vice versa) assuming random mating, equal reproductive success, no mutations and no effects of selection or migration affecting particular genotypes. If a population does not fit Hardy–Weinberg predictions then that is evidence of some real effect (e.g. natural selection) operating to disturb it.

Measuring allele frequency

Allele frequency is found by adding up the number of copies of each allele in a population and expressing it as a frequency. A population of N diploid individuals has $2N$ alleles. Each Aa heterozygote has one A allele, each AA homozygote has two A alleles, so

frequency of allele A is $\dfrac{(n_{Aa} + 2n_{AA})}{2N}$

where n is the number of individuals with the respective genotype. The frequency of A is usually called p, the frequency of other alleles (a or non A) is denoted by q. As a check, $p + q = 1$, but this only checks your arithmetic, not whether p and q are correct.

Equilibrium genotype frequencies

Equilibrium genotype frequencies will be reached in zygotes after one round of random mating. The expected genotype frequencies are predicted by the Hardy–Weinberg equation, a simple binomial:

$p^2 + 2pq + q^2 = 1$.

If there are two alleles, A and a, with frequencies p and q respectively, then the expected frequency of AA homozygotes is p^2, of Aa heterozygotes is $2pq$, and of aa homozygotes is q^2. As a check of your arithmetic $p^2 + 2pq + q^2 = 1$. When the frequency q of an allele is low, the frequency of occurrence of homozygotes, q^2, is very low (e.g. recessive genetic diseases). Uncommon alleles (small value of q) are usually present in heterozygotes, so cannot be identified if they are recessive and are not expressed, so selection cannot act against them.

Distorting effects

A range of natural phenomena can distort genotype frequencies from those predicted by the Hardy–Weinberg equation.

(i) Selection reduces the fertility or survival of certain genotypes.

(ii) Migration replaces local individuals with immigrants from a population where the allele frequencies, and hence expected genotype frequencies, are different. If some genotypes are more likely to emigrate than others this will also change their frequency locally.

(iii) Assortative mating of either similar or dissimilar genotypes produces an excess of homozygotes or heterozygotes respectively.

(iv) Subpopulations exist. These are locally mating groups in a larger, possibly continuous population. This may be common in humans due to ethnic or class groupings, and also in organisms with low mobility, and always increases the frequency of homozygotes. In an extreme case this becomes inbreeding.

(v) Mutations will produce new alleles, but the rate is too low to be noticeable practically, and selection will exactly counteract this if the mutants are deleterious.

(vi) Drift (chance) will also cause small deviations from the predicted frequencies, especially in small samples, and accumulate large effects over long periods. These deviations are expected and allowed for in statistical tests.

Testing the fit

A chi-square (χ^2) test (Section C14) is used to compare observed genotype frequencies to expected frequencies estimated assuming that the population conforms to Hardy–Weinberg equilibrium:

$$\chi^2 = \mathrm{sum}\left[\frac{(\mathrm{observed} - \mathrm{expected})^2}{\mathrm{expected}} \right].$$

For n allele, there are $(n^2 - n)/2$ degrees of freedom because once the frequency (p) of one allele (of two alleles) is known, the frequency of the other is fixed ($p + q = 1$) and so are the expected genotype frequencies. Small values of χ^2 indicate a good fit to expectation, large values of χ^2 indicate a large deviation.

Estimating allele frequency

If heterozygotes cannot be identified, the frequency of recessive alleles can still be estimated from the frequency of homozygotes by assuming Hardy–Weinberg equilibrium:

the frequency of homozygous recessives is q^2, so
the frequency of the recessive allele is q, and
the frequency of the dominant allele is $p = (1 - q)$.

There is a large error associated with this method. It is difficult to measure q^2 accurately because: (i) it is often very small so a very large population must be sampled; and (ii) Hardy–Weinberg equilibrium may not apply, usually because of population structure and local inbreeding subpopulations.

Related topics   (D4) Genetic diversity   (D7) Species and speciation
(D5) Neo-Darwinian
evolution: selection acting
on alleles

Introduction

Population genetics is the study of alleles of genes in populations, and the forces which maintain or change the frequencies of particular alleles and genotypes in populations. The local interbreeding population is called a **deme** to distinguish it from a geographical population. A deme may be difficult to define, for example humans living in one area may actually marry partners chosen on criteria of religion, race, wealth, or social class. Similarly, insects on trees in a forest may mate with others on the same tree, and rarely travel to the next tree to mate. The insect population may fill the forest, or each tree may have its own semi-isolated population. Natural populations are composed of many interbreeding individuals, each with a unique combination of genes and alleles, but the population shares a **gene pool**. Evolution is the change of frequencies of alleles in the gene pool, so population genetics is of fundamental importance. Some genotypes, such as those associated with human genetic diseases, are rare, and population genetics is important for understanding why this is so, and for predicting changes in the incidence of such diseases. Most individuals have phenotypes close to the average, but some carry rare combinations of alleles which cause their phenotype to deviate markedly from a typical member of the species.

There was a fear, expressed by the **eugenics movement**, that if 'defective' humans were allowed to reproduce, their defects would become more common, and the human race would degenerate. Eugenics refers to the principle of selective breeding of humans by encouraging desirable types to have more children (positive eugenics) and discouraging undesirable types from having any children (negative eugenics). This has been applied in varying degrees, 'desirable' generally being interpreted to mean the ruling class or race. It is becoming technically possible to pursue this by genetic manipulation of embryos (designer babies).

Hardy and Weinberg independently published calculations showing that undesirable characteristics would not become more common, and that they were due to rare combinations of relatively common alleles. Hardy and Weinberg showed that the frequency of particular genotypes in a population of a sexually reproducing diploid species reaches equilibrium after one generation of random mating and fertilization, and will stay constant unless something changes the frequencies of alleles in the population. Hardy–Weinberg equilibrium requires random mating, random fertilization, and an absence of differential selection or migration affecting particular alleles or genotypes. (A population can be in equilibrium in the sense of being constant from generation to generation without being in Hardy-Weinberg equilibrium.)

The goodness of fit of observed genotype frequencies to frequencies expected from the Hardy–Weinberg equilibrium can be tested statistically using a chi-square (χ^2) test (Section C14). Deviations may reveal evidence of natural selection, assortative (like with like) or disassortative (choose unlike type) mating, migration from an unsampled population, population substructure (increasing inbreeding raising the frequency of homozygotes), or selective fertilization. (Selective fertilization can occur, for example, when pollen of the wrong compatibility type is rejected by the stigma of a female flower.) The observed frequencies will also deviate from the expected frequencies due to chance (also known as drift, sampling, stochastic, random, or statistical error). This is greater in small samples and is allowed for in the statistical tests.

Measuring allele frequency

The frequency of an allele (e.g. A) in a population is the number of A alleles divided by the total number of alleles of that gene locus. In a diploid species, a population of N individuals will be produced from $2N$ gametes and contain $2N$ alleles for each genetic locus. Suppose that there are two alleles, A and a, of a particular gene in this population. The number of A alleles is twice the number of AA homozygotes plus the number of Aa heterozygotes. (Each homozygote has two A alleles, each heterozygote has only one.) The frequency of A is the number of A alleles divided by the total, $2N$, and similarly for a. If we use n to represent number, and the suffixes A and a for alleles, and AA, Aa and aa for genotype, then

$$n_A = 2n_{AA} + n_{Aa}.$$

If p is the frequency of allele A then

$$p = \frac{n_A}{2N} = \frac{2n_{AA} + n_{Aa}}{2N}.$$

Similarly the frequency of a is q:

$$q = \frac{n_a}{2N} = \frac{2n_{aa} + n_{Aa}}{2N}.$$

Notice that all alleles must be accounted for:

$$n_a + n_A = 2N, \text{and } p + q = 1.$$

If there are more than two alleles of a gene, they can all be counted in a similar way. The frequency of the third allele is usually denoted by r.

The frequency of alleles can only be determined accurately by counting homozygotes and heterozygotes in a sufficiently large population. It was often impossible to identify heterozygotes, but direct DNA-based techniques involving polymerase chain reaction techniques (Section E2) are now helping to solve this problem.

Worked example

There are two alleles of the L human blood group gene on chromosome 2, L^M, and L^N (usually called M and N respectively) (Table 1). These alleles are codominant (both are expressed in heterozygotes) so they can be identified in heterozygotes by antibody tests. In a population of 1000 white Americans, 357 were MM, 485 were MN, and 158 were NN. The 357 MM individuals have 714 M alleles, and the 485 MN individuals have another 485 M alleles, so the frequency of M ($=p$) is:

$$p = \frac{(2 \times 357) + 485}{2000} = 0.5995.$$

Table 1. Measuring the frequencies of N and M in a population

	Genotype (=phenotype)			
	MM	MN	NN	Total
Number of individuals	357	485	158	1000
Number of M allele	714	485	0	1199
Number of N alleles	0	485	316	801
Total number of alleles	714	970	316	2000

Frequency of M in population: 1199/2000 = 0.5995; frequency of N in population: 801/2000 = 0.4005.

Similarly the frequency of $N=$

$$q = \frac{(2 \times 158) + 485}{2000} = 0.4005.$$

Equilibrium genotype frequencies

Equilibrium will be reached in a single generation. We have already assumed random mating and fertilization. The frequency with which a male gamete carrying allele A (frequency p) fuses with a female gamete carrying A (also frequency p) will be $p \times p = p^2$. Similarly the frequency with which a male gamete carrying allele a fuses with a female gamete also carrying a will be $q \times q = q^2$.

Heterozygotes (Aa) are produced by the fusion of gametes carrying different alleles. The frequency with which a male gamete carrying allele A fuses with a female gamete carrying a will be $p \times q = pq$, and the frequency of a male gamete carrying allele a fusing with a female gamete carrying A will be $q \times p = qp = pq$, so the total frequency of Aa heterozygotes will be $2pq$.

This is all derived from the binomial expression (see Section C13):

$$(p+q)(p+q) = p^2 + 2pq + p^2.$$

This is shown graphically for the frequency of MM, MN, and NN blood groups in white Americans as calculated above (Figure 1a) and for Australian aborigines (Figure 1b). Notice that the allele and genotype frequencies are quite different for the two populations.

The frequency of homozygotes is the square of the appropriate allele frequency, the frequency of heterozygotes is twice the product of the two allele frequencies. This can be applied to any number of alleles. Figure 2 shows the frequencies of three alleles and six genotypes for the human ABO blood-group gene.

Points to notice

(i) When the frequency (q) of an allele is low, the frequency of occurrence of homozygotes, q^2, is very low. Notice the rarity of MM homozygotes among Australian aborigines (Figure 1b) and of BB genotypes (Figure 2).

(ii) Uncommon alleles (small value of q) are usually present in heterozygotes (this is the case for M alleles in Australian aborigines). If they are recessive they are not expressed and so cannot be identified, so neither natural selection nor eugenics programs can act against them.

(a)

Sperm pool

	Frequency of M=p=0.6	Frequency of N=q=0.4
Frequency of M=p=0.6	MN p²=0.36	MN pq=0.24
Frequency of N=q=0.4	MN pq=0.24	NN q²=0.16

(Egg pool)

(b)

Sperm pool

	Frequency of M=p=0.178	Frequency of N=q=0.822
Frequency of M=p=0.178	MM p²= 0.0317	MN pq=0.146
Frequency of N=q=0.822	MM pq =0.146	NN q²=0.676

(Egg pool)

Figure 1.   Frequencies of *MM*, *MN*, and *NN* blood groups in (a) white Americans and (b) Australian aborigines. Allele L^M, frequency p, codes for production of *M* antigen, and L^M, frequency q, code for N antigen. These are codominant, so genotype $L^M L^M$ is phenotype M, $L^M L^N$ is phenotype *MN*, and $L^N L^N$ is type *N*.

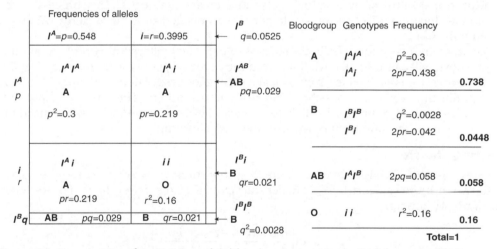

Figure 2.   Frequencies of A, B, AB, and O blood groups at Hardy–Weinberg equilibrium. Frequencies of alleles are p (I^A), q (I^B) and r (i). I^A and I^B encode enzymes which produce A and B blood-group antigens respectively, and are codominant, whereas i is a recessive null mutation, of no effect.

Distorting effects

If a population is found not to be in Hardy–Weinberg equilibrium, there are several possible causes:

*   **Selection** which affects the viability or fertility of individuals or gametes of particular genotypes in the parental generation will distort the genotype frequencies in the zygotes. Differential survival of genotypes will distort the genotype frequencies in adults, in which case the deviation from Hardy–Weinberg equilibrium may be seen to change (increase) as progressively older individuals are sampled.

- **Migration** will bring in individuals with genotypes from a gene pool different from the one sampled, and obviously introduce errors. Mixing populations always gives too many homozygotes. If different genotypes show different tendencies to migrate in or out, the deviation will be in a particular direction.
- **Assortative mating** whereby individuals choose mates with similar genotypes to themselves will produce an excess of homozygotes ($AA \times AA$ gives AA) while if they choose opposites there will be an excess of heterozygotes ($AA \times aa$ gives Aa).
- **Subpopulations** cause inbreeding within the sample and will increase the frequency of homozygotes (Section C12). This can arise if the organism's range of movement and interbreeding is much smaller than the area sampled, so individuals only have a small gene pool available for mating, and they cannot mate randomly. Because of this effect, pooling genotypes from different populations always increases the frequency of homozygotes. Social structures in human societies have the same effect. The extreme arrangement is inbreeding between members of the same family (relatives). This is demonstrated by rare human genetic disorders. For example 33% of cases of alkaptonuria and 54% of cases of microcephaly sufferers have parents who are cousins. These cases show that, in humans, the frequency of rare homozygotes is considerably increased by assortative mating with relatives. The frequency of homozygotes goes from q^2 for random mating to $pq/16$ for cousin marriages (Section C12).
- **Mutations** will destroy existing alleles and create new ones. This process is too slow to have an effect in the single generation needed to achieve Hardy–Weinberg equilibrium.
- **Drift**, or chance differences in the success of particular alleles from one generation to the next, will cause small deviations from the expected genotype frequencies, but this random error is allowed for in statistical tests of goodness of fit to the Hardy–Weinberg equilibrium. There is a possible source of confusion here. The term 'drift' properly applies to chance changes in allele frequencies between generations, which need not involve any deviation from Hardy–Weinberg equilibrium.

Testing the fit

A χ^2 test (Section C14) is used to compare the proportions of two or more samples falling into two or more categories. The smaller the value of χ^2, the better the fit. For testing fit to Hardy–Weinberg:

$$\chi^2 = \mathrm{sum}\left[\frac{(\mathrm{observed} - \mathrm{expected})^2}{\mathrm{expected}}\right].$$

There are two alleles, so there is one degree of freedom. For n alleles of one locus there are $(n^2 - n)/2$ degrees of freedom. The fit of our sample of M, MN and N blood groups in white Americans (above) is tested in Table 2. The null hypothesis is that the sample frequencies are not different from those predicted by assuming Hardy–Weinberg equilibrium. The value of χ^2 obtained is 0.4236, very small, meaning that there is very little deviation between the observed and expected values. Using χ^2 tables we can see that we are very likely to get a deviation larger than this by chance, so we accept the null hypothesis: the frequencies fit the expected values. A value of χ^2 (with 1 degree of freedom) greater than 7.879 would be required to indicate that there was only a 5% probability of getting the fit by chance, and thus a 95% probability that there was a genuine deviation from Hardy–Weinberg equilibrium. Note that a large deviation from the expected values gives a large value for χ^2, a small deviation (good fit) gives a small χ^2 value.

Table 2. Testing the fit of observed M, MN, and N blood groups to Hardy–Weinberg using χ^2

	Genotype			
	MM	MN	NN	Total
Observed	357	485	158	1000
Expected	$1000p^2$	$1000 \times 2pq$	$1000q^2$	
	359.4	480.2	160.4	1000
Observed–expected	–2.4	4.8	–2.4	
(Observed–expected)^2	5.76	23.3	5.76	
(Observed–expected)^2/expected	0.016	0.0485	0.3591	

Frequency of M alleles=p=0.5995; frequency of N alleles=q=0.4005; $\chi^2_{(1df.)}$=0.016+0.0485+0.3591=0.4236.

Estimating allele frequency

In many cases (e.g. recessive alleles causing diseases in humans), it is easy to identify individuals homozygous for the recessive allele (who have the disease) but impossible to distinguish between heterozygotes and homozygous normal individuals. The frequency of the recessive allele can be estimated roughly by assuming Hardy–Weinberg equilibrium. The frequency of homozygous recessive individuals should then equal q^2, so q=square root of the frequency of homozygous recessives:

$$q = \sqrt{q^2} = \sqrt{\frac{\text{homogygous recessives}}{\text{total population}}} \; .$$

About 1/2500 Caucasian humans is born with cystic fibrosis. They are homozygous for nonfunctional alleles of a gene for a chloride transport protein involved in mucus secretion. Without this function, mucus is very viscous and accumulates causing damage to the pancreas, intestines, and most acutely the lungs. Because the frequency of homozygotes=q^2=1/2500=0.0004, then

$$q = \sqrt{q^2} = 0.02 \quad (1/50).$$

This method is not very accurate for several reasons. (i) Most populations are inbred. People marry those they meet, such as relatives (cousins), and also choose mates by neighborhood, race, religion, education, even height, increasing the frequency of homozygotes. (ii) The frequency of homozygous affected individuals is low, and there are large stochastic errors when counting such rare events. For example, if the true frequency was four affected homozygotes in a population of 100 000 people, any one study might find two more or less, that is two or six homozygotes, giving a twofold range of estimates of allele frequency. (iii) Some genetic diseases have low penetrance, which means that not all homozygous individuals show the disease. This is because other conditions (genetic or environmental) are needed to trigger the disease, and so its frequency is underestimated. Some serious conditions have high prenatal mortality so many affected fetuses are never counted.

We can calculate the frequency of heterozygotes (carriers) for cystic fibrosis as:

$$2pq = 2 \times (1 - 0.02) \times 0.02 = 2 \times 0.98 \times 0.02 = 0.0392 \, (\text{approximately} \, 1/25).$$

In other words, one Caucasian in 25 is a carrier of this lethal recessive allele. This is a remarkably high frequency for such an allele. The explanation seems to be that carriers of cystic fibrosis have an increased tolerance of cholera toxin, which causes death by loss of fluid in the gut. (They may also have increased tolerance to some *Salmonella* species.) Heterozygotes for the cystic fibrosis allele show approximately one-third the rate of fluid loss of their homozygous normal relatives when exposed to cholera toxin. Mice homozygous for cystic fibrosis showed no fluid loss in response to the toxin. This is an example of a selective equilibrium due to heterozygote advantage, similar to sickle cell anemia and malaria (Section D4).

D4 Genetic diversity

Key Notes

Introduction	There is a large amount of genetic variation in the amino acid sequence of proteins in most populations. Molecular techniques, especially DNA sequencing, reveal further variation in DNA which has no apparent effect. Synonymous changes (e.g. AAA to AAG; both code for lysine) are silent, and affect only mRNA secondary structure and the efficiency of protein synthesis. There is a continuing debate about the relative importance of mutation (introducing new alleles), drift (chance changes in the frequency of alleles which may be selectively neutral), and natural selection (either directional, changing alleles, or frequency-dependent selection maintaining an equilibrium) in maintaining this variation.
Source of variation	All new variation starts as a mutation, a change in the sequence of bases in the DNA. The mutated sequence may code for a protein which does not adequately fulfill its function, or has a reduced activity (rarely increased activity), or is phenotypically normal (a neutral mutation), or, rarely, may code for an advantageous variant. Null mutations have lost all function (i.e. a deletion of the gene) but loss of activity can follow a single amino acid change. Many, perhaps most, surviving mutations have an insignificant effect on function, so are neutral to selection, and susceptible only to drift.
Mutation vs. selection	Defective alleles are removed by selection, and an equilibrium frequency (q) will exist where rate of removal of defective alleles by selection equals rate of formation of new mutations. For fully recessive alleles, $q^2 = \mu/s$ (where q = frequency, s = selection against homozygotes and μ = mutation rate). For recessive lethal alleles, $s = 1$ and $q = \sqrt{\mu}$.
Drift	Drift is the chance difference in transmission of alleles between generations, which causes fluctuations in allele frequency. This can cause differences in allele frequency between separate populations of one species. It has most effect in small populations, and on rare alleles, and its effects accumulate with time. Drift is important for establishing (or eliminating) new favorable mutations until they reach a frequency where selection becomes more important than drift. This is especially the case for recessive alleles which are hidden in heterozygotes. It is the main factor (with mutation) affecting the frequency of neutral alleles. The founder effect occurs when a small group is isolated and founds a new

	population that has a gene pool with allele frequencies different from those in the parent population. Bottle-necking is similar to the founder effect, but refers to a population or species reduced to a very small number of individuals by adverse circumstances.
Frequency-dependent selection	In some circumstances (e.g. a heterogeneous habitat) individuals which are different from each other can exploit the range of habitats and reduce competition. Common phenotypes which exceed their ideal habitat will face more competition than rare phenotypes which have surplus habitat. A range of genotypes will persist. Disease strains adapted to common genotypes will be less harmful to rare genotypes, again making rare genotypes fitter.
Balanced selection – heterozygote advantage	Selection which favors heterozygotes (heterozygote advantage, overdominance) will maintain two alleles and favor rare alleles which are more frequently found in heterozygotes than are common alleles.
	An example is sickle cell disease caused by a mutated allele of hemoglobin A. Normal homozygotes Hb^AHb^A suffer more from malaria than heterozygotes Hb^AHb^S who have a tolerance for malaria. Homozygous Hb^SHb^S individuals suffer from sickle cell disease. In the presence of malaria both the normal allele Hb^A and the sickle cell allele Hb^S are maintained in the population.
Related topics	(D2) Evolution by natural selection (D5) Neo-Darwinian evolution: selection acting on alleles

Introduction

There are several alleles for most genes in most populations, in fact the limit to genetic variation seems to be the technical ability of researchers to discover it. DNA sequencing reveals differences which have no detectable phenotypic effect. This **synonymous variation** does not change the coded amino acid (e.g. codons AAA and AAG both code for lysine) so do not affect the polypeptide or protein product, and are called **silent mutations**. However, the alternative codons are often not used in equal numbers, which suggest that one codon may be translated more efficiently than another on the ribosomes. Sequence variation also exists at noncoding sites, which may or may not have regulatory roles. Only regions where the sequence is important will be subject to selection. There is thought to be too much variation in most populations for it all to be due to selection, because too many individual zygotes would have to die for sufficient selection to act to maintain the variation. The mechanisms which generate and maintain variability can be considered in turn to attempt to decide how much variation is due to selection, and how much is due to other forces like mutation and chance (drift).

Source of variation

All genetic variation originates from **mutations** (Section A5), which are changes in the sequence of bases in the DNA (or RNA of RNA viruses). These usually destroy the allele's ability to function, and so are usually detrimental to the fitness of the genotype. **Null mutations** are due to complete destruction of the gene's function, typically by deletion. Stop codons (nonsense mutations) or frameshifts early in the coding sequence, or point mutations changing essential amino acids (e.g. in enzyme active sites) have similar effects. All **loss-of-function** alleles can be considered together. They are recessive if a single copy of a functional allele is sufficient to produce a normal phenotype. They are usually deleterious (often lethal) when homozygous, and cause the typical human genetic diseases. Some null mutations may be advantageous (e.g. loss of the I^A allele prevents expression of the A blood group antigen, which is not essential, and its loss confers some resistance to smallpox).

Point mutations which change a single amino acid in the encoded polypeptide may have no significant phenotypic effect. These may be very slightly deleterious, but not enough to matter. Such mutations are termed **neutral** because selection does not affect their frequency. An allele can be defined as neutral if its frequency is controlled by chance (drift, see below) more than by selection. Between these two extremes are mutations which vary in their deleterious effects. Very rarely, mutations will improve the fitness of the genotype. Such advantageous mutations will be selected for, and may replace the pre-existing alleles.

Mutation vs. selection

Dysfunctional alleles are continuously being created by mutation. Selection acts against them, but most are recessive, so selection only acts against homozygotes. An increase in frequency of mutant alleles causes an increase in homozygotes relative to heterozygotes, and increased selection against these homozygotes prevents further increase in the frequency of the mutant allele. At equilibrium the number of new dysfunctional mutations produced equals the number lost by selection. For simplicity we will assume that dysfunctional alleles are recessive, so selection only acts against homozygotes, which occur at a frequency of q^2. In a diploid population of N individuals there are $2N$ alleles. Assuming the frequency of mutants (q) is low, there will be $2pN\mu$ new mutations (where μ is the mutation rate per generation and p is the frequency of functional alleles), and $2Nsq^2$ mutant alleles will be lost by selection, where s is the coefficient of selection against homozygotes. ($s = 1$ for recessive lethal alleles.)

$$\text{At equilibrium,}\ 2pN\mu = 2Nsq^2$$

p is usually assumed to be 1, in which case, dividing through by $2N$ gives:

$$\mu = sq^2 \text{ and } q^2 = \frac{\mu}{s} \text{ and } q = \sqrt{\frac{\mu}{s}}\ .$$

If there is no dominance, and heterozygotes are also subject to selection of $0.5s$, then $q = 2\mu/s$ (approximately).

Drift

The genotype of any gamete or zygote depends on which of the two homologous segments of chromosome it inherited by chance from its diploid parent(s). Each normal (Mendelian) allele has a probability of $1/2$ of being transmitted to any one offspring, just as a tossed coin has an equal chance of coming down heads or tails. Two diploid parents will produce on average two offspring, so on average each allele will be transmitted once, each copy in the parental population will be represented once in the next generation.

However, individual alleles will be more or less frequent, just as tossing many coins does not give exactly equal numbers of heads and tails. The gene pool will be changed in the next generation, and the whole process will be repeated, so the frequencies of particular alleles will drift (change) through time. The variance in frequency derives from the binomial distribution and can be calculated (but do not try to remember how). Variance S^2 of the frequency q of allele a is $S_q^2 = p_0 q_0 / 2N$ (where p_0 and q_0 are the initial frequencies of the two types of allele). Over a period of t generations this becomes:

$$S_q^2 = p_0 q_0 \left\{ 1 - \left[1 - \frac{1}{2N} \right]^t \right\}.$$

There are three important points to note:

(i)   The proportional change depends on initial allele frequencies. Rare alleles are more susceptible to drift. If an allele only occurs in a single codfish, it might be included in half a million eggs, of which 1000 might survive (a huge increase) or it might be caught and eaten before it breeds, and the allele will be extinct. The more individuals an allele is found in, the nearer its transmission frequency will be to the average value of 1.

(ii) The process is cumulative over time, but the frequency can go up or down in each generation (Figure 1).

(iii) The change is inversely related to population size (for mathematicians, it is inversely proportional to the harmonic mean). Larger samples, like tossing more coins, more often give results near the expected mean value (Figure 1).

Drift is important in three cases. (i) It is important for removing or promoting very rare alleles, especially new favorable mutations before they become established, because drift has a greater effect than selection on the transmission of rare alleles. It is important for increasing the frequency of new recessive mutations to a frequency where homozygotes occur in sufficient frequency for selection to become effective. (ii) Drift is responsible for

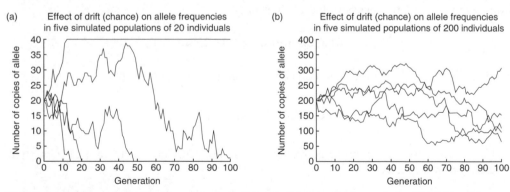

Figure 1.   Change in frequency of neutral alleles in five populations each of (a) 20 or (b) 200 diploid individuals produced by a computer simulation. The y axis shows the number of copies of a specific allele starting from a frequency of 0.5. Each trace shows the frequency (number of copies) of that allele in one population, changing by chance over successive generations. In four of the five populations of 20 individuals, the allele was fixed or lost (p=1 or 0) within 50 generations.

changing the frequency of neutral mutations, (caused by recurrent mutation) which by definition are affected more by drift than by selection. The probability of any new neutral mutation eventually becoming fixed in the population is $1/(2N_e)$ where N_e is the effective (breeding) population size. (Assuming they are neutral, each of the $2N_e$ copies which exist for each genetic locus in the population has an equal chance of becoming fixed.) (iii) Drift in small populations can produce unrepresentative allele frequencies which would be very unlikely to occur in a large population. This is called **founder effect** (see Section D7) when a small and unrepresentative group founds a new colony. A human example is the Amish, a religiously united group in the USA, established by a small number of immigrant families. It is also termed **bottle-necking** when a population is reduced to a small number (e.g. by disease or famine) who become parents for a large later population. Drift may cause small isolated populations to be very different from the species norm, and can be important in the production of new races or species. Inbreeding can be a special case of founder effect where the effective population is a small group of related individuals, possibly a single hermaphrodite individual.

Frequency-dependent selection

An allele may have an advantage when it is rare, especially in a heterogeneous environment. If a species of finch lives on an island where there are several plant species producing seeds of different sizes, birds with larger beaks can exploit the larger seeds, those with smaller beaks exploit the small seeds, and there is less competition between them. If either large or small beak size becomes too frequent for the current seed crop, it will face more competition, and selection will decrease its frequency. For example, in *Geospiza fortis*, a species of Darwin's finches on Daphne Major in the Galapagos Islands, the mean beak size was found to be decreased after wet years when small seeds were numerous, and increased after drought years when only large-fruited plants were successful.

Disease is another selective agent which maintains variation in its host population. Strains of disease organism are adapted to particular genotypes of host, and the most successful disease strain is the one which can spread fastest through the most common host genotypes, reducing that genotype's fitness and frequency. This is a serious problem for agriculture, where large areas of a single variety of a single species are grown (monocultures). Merely mixing three varieties of barley seed, each variety susceptible to a different strain of fungal pathogen, can remove the need for fungicide treatment, because no single strain of fungus can spread through the entire crop. In a genetically variable population, no single disease strain spreads well, and this is the normal condition in natural populations.

Balanced selection – heterozygote advantage

If both homozygotes are less fit than the heterozygote, then both alleles will be maintained in the population (e.g. normal and **sickle cell disease** allele). An allele of beta hemoglobin, Hb^S, codes for a polypeptide with valine as the third amino acid, instead of the normal glutamic acid. When homozygous this allele causes sickle cell anemia, which is serious and often fatal, but the allele is common where malaria is prevalent because heterozygotes $Hb^A Hb^S$ have some tolerance to malaria. In parts of Africa the frequency of Hb^S reaches 0.2. At this frequency, the frequency of homozygous recessive genotypes is $0.2^2 = 0.04 = 4\%$ of the population, four out of five of whom might die of sickle cell anemia. This is compensated by an increased susceptibility to malaria in normal homozygotes ($Hb^A Hb^A$), to which heterozygotes $Hb^A Hb^S$ have some resistance. If the frequency of $Hb^S = q = 0.2$, and the fitness of sickle cell sufferers is 0.2 (four out of five do not reproduce), the selection due to malaria on the $Hb^A Hb^A$ genotype can be calculated. The equilibrium frequency of $Hb^S = q = s/(s+t)$, where $s =$ selection coefficient on $Hb^A Hb^A$ and $t =$ selection on $Hb^S Hb^S$. The ratio of allele frequencies is the inverse of the ratio of selection coefficients:

Genotype	$Hb^A Hb^A$	$Hb^A Hb^S$	$Hb^S Hb^S$
Fitness	0.8	1	0.2

The actual fitnesses depend on the local incidence of malaria and availability of healthcare.

D5 Neo-Darwinian evolution: selection acting on alleles

Key Notes

Introduction

Neo-Darwinian evolution (the New Synthesis) considers evolution in terms of changes in allele or gene frequencies over time, and the average action of selection on genotypes (particular alleles), rather than on individuals. This combines an understanding of the mechanism of inheritance with the action of Darwinian natural selection.

Units of selection

Selection operates on individuals, because it is individuals which either survive or die, and which reproduce more or less successfully than others. In diploid species, however, the genetic combination which made that individual is disrupted at gametogenesis. The only continuity is in the continuation of copies of alleles, and the New Synthesis considers selection to act on particular alleles in relation to their average contribution to their own transmission through all the individuals that carry copies of them.

Fitness

Fitness (W) of a particular genotype is usually defined as the survival and reproductive success of that genotype relative to the optimum genotype. The fitness of a gene or allele is its relative transmission from generation to generation compared with that of other alleles. Those alleles with the highest transmission rates are fittest and increase in frequency. Selection (s) against a particular genotype is the proportional reduction of reproductive success of individual organisms with that genotype. If zygotes with the genotype aa die as embryos, or are sterile, then $W = 0$ and $s = 1$.

$$\text{Fitness} = W = (1 - s) \text{ where } s = \text{selection} .$$

If aa individuals have 0.99 the fitness of AA or Aa individuals, then

$$W_{(aa)} = 0.99, \ s_{(aa)} = 0.01.$$

Industrial melanism

The peppered moth, *Biston betularia*, is typically gray and camouflaged on lichen-covered tree trunks where it hides by day. Pollution during the industrial revolution killed the lichen and exposed dark bark. A new dark carbonaria mutant was then better camouflaged than the typical phenotype, and increased in frequency. Birds eat a higher proportion of typical forms in polluted cities but eat more carbonaria in unpolluted rural areas, maintaining a polymorphism related to habitat. In a mark-release-recapture experiment in a rural habitat, $30/473 = 0.063$ carbonaria and $62/496 = 0.125$ typical moths were recaptured. The relative fitness of carbonaria over that part of the life cycle in those conditions was $0.063/0.125 = 0.504$.

Selection against recessive alleles

Recessive alleles are only expressed when homozygous, but as they become rare they are mostly present in heterozygotes (Section D3). Selection therefore has little effect on rare recessive alleles, which tend to persist.

Selection against dominant alleles

Dominant alleles are always expressed and exposed to selection. When they are frequent, their removal also causes loss of most favorable recessive alleles present as heterozygotes, so the response to selection is slow initially. The response becomes faster as the frequency of recessive alleles increases towards 0.5. Dominant deleterious alleles can be completely removed from a population by selection (barring recurrent mutation).

Selection against additive alleles

If there is incomplete dominance the heterozygote will have a phenotype (=fitness) somewhere between that of the two homozygotes. When there is no dominance, the effect of two alleles is additive and the phenotype of heterozygotes is halfway between the two homozygotes. Selection will act as it would on dominant alleles but with less severity because they are only incompletely expressed in heterozygotes.

Selection for heterozygotes

Heterozygote advantage or overdominance maintains a polymorphism in the population. The most frequent allele has most low-fitness homozygotes, the rare allele occurs mainly in high-fitness heterozygotes, so the alleles have a fitness inversely related to their frequency. Equilibrium frequencies depend on the relative fitness of the two homozygotes.

Selection against heterozygotes

Selection against heterozygotes removes the rarest allele because it is most commonly present in heterozygotes. This will continue until it is entirely lost, and the frequent allele is fixed in the population, irrespective of their relative fitnesses in homozygotes. The situation is only likely to arise where two separated and divergent populations come into secondary contact between adjacent geographical ranges. A hybrid zone is then formed.

Related topics

Introduction

When the principles of genetic inheritance and mutation were discovered it was realized that this filled in the detail about inheritance needed to complement Darwin's theory of evolution. Evolution can be considered as a change in the frequency of alleles and genes in a population over time, and selection can be considered as acting on particular alleles or genes through the average effect they have on all the individuals in a population which carry them. This approach was termed the New Synthesis, or neo-Darwinism.

Units of selection

It is usually considered that the unit of selection is the individual organism, because it is individuals which either survive or die, and which reproduce more or less successfully than others. This is true within any one generation, but there is no continuity in individuals in a sexually reproducing species. Individuals grow old and die, and their genome is split up at gametogenesis. However, copies of their alleles are passed on into new combinations. The only continuity is in the continuation of copies of alleles, and the New Synthesis considers selection to act on particular alleles in relation to their average contribution to all the individuals that carry copies of them.

Fitness

Fitness (W) has two components: viability (survival) and fecundity (reproductive success), and operates on individuals, each with a particular genotype. For simplicity, it is necessary to consider genes individually. The fitness of a genotype is the ratio of the average fitness (success in genetic transmission) of individuals with that genotype compared with the fitness of individuals which have the optimum genotype. For example, consider an allele C which codes for a metabolic enzyme, and is dominant. Homozygous CC individuals and heterozygous Cc individuals have a fitness of 1. If the enzyme is essential then all cc homozygotes will die, their fitness $W_{cc}=0$, and the selection coefficient $s_{cc}=1-W_{cc}=1$.

If the enzyme is not essential, but 1 in 100 cc individuals lacking it fail to reproduce successfully because they lack it, then the fitness of cc genotype individuals (W_{cc}) is 0.99, and the selection coefficient (s_{cc}) is 0.01. The rate at which the frequency of allele c changes in the population will be much faster with the higher selection coefficient (Figure 1). Note that selection against a dominant allele is the same as selection for a recessive allele, and vice versa.

Industrial melanism

Biston betularia, the peppered moth, provides a classic example of natural selection. They are nocturnal, and the 'typical' form is gray and speckled, which is camouflaged very well on lichen-covered trees where they rest during the daytime. In 1849 the first dark melanic specimen was collected in Manchester, England. Collectors are eager to obtain rarities, so we can assume they were very rare before this. The melanic form is known as 'carbonaria' and is caused by a dominant allele of a single gene (C) so typical moths are homozygous cc. Carbonaria turned from a collectors' rarity to the common form and by 1895 about 98% of that population of moths was dark. Pollution, principally acid rain from the industrial revolution, had killed the lichens on the trees, exposing the dark bark, which may have been darkened further by soot. The gray typical moths were now not camouflaged as well as the dark carbonaria. Direct observations and mark–release–recapture experiments supported the hypothesis that predators, in this case birds, were catching a higher

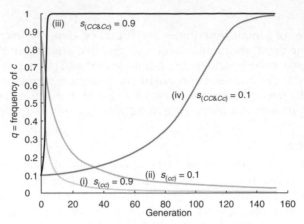

Figure 1.   Changes in frequency of the recessive allele under different types of selection (Table 1). Each trace starts with the less fit allele at a frequency of 0.9. Data was calculated by repeatedly applying selection to genotypes at the Hardy-Weinberg equilibrium. The four conditions shown are selection against the recessive allele c at $s_{(cc)} = 0.9$ (i) and $s_{(cc)} = 0.1$ (ii); and selection against the dominant allele C at $s_{(CC\&Cc)} = 0.9$ (iii) and $s_{(CC\&Cc)} = 0.1$ (iv). $s_{(cc)} = 0.9$ means that 90% of individuals with the cc genotype are removed each generation.

proportion of typical than carbonaria in polluted industrial regions. When both types were released in unpolluted rural woods, the situation was reversed and carbonaria were eaten proportionately faster than typical.

In one experiment in Birmingham, UK, 123 out of 447 (0.275) released carbonaria were recaptured, and 18 out of 137 (0.131) typical moths. The carbonaria are given a fitness of 1 (=0.275/0.275), while the typical form had a fitness of 0.131/0.275=0.476. (This assumes equal migration and equal efficiency of re-trapping of both forms.) A similar experiment in rural Dorset, UK, recaptured 30 out of 473 carbonaria (0.063) and 62 out of 496 typical (0.125). Here typical forms are designated a fitness of 1 (=0.125/0.125) and carbonaria a fitness of 0.063/0.125=0.504. The relative fitness reversed between the polluted and unpolluted habitats.

The importance of these experiments has led to criticism by opponents of evolution through natural selection. The moths tend to rest on twigs, not trunks where Kettlewell studied predation; the moths might already have laid their eggs before being eaten, and the color affects temperature and metabolism in caterpillars. No experiment is perfect, and these critics would be extremely perverse to suggest that changes to the moths' coloration to match the habitat are not related to effective camouflage from predators. The trend is correct. In another experiment, Bishop calculated a fitness of typical, relative to carbonaria, of about 0.8 in Liverpool and 1.8 in rural Wales, 50 km away, based on egg-laying rates and survival of adults. Genotype frequencies at all stages of the life cycle must be determined for more than one generation before really accurate numerical conclusions can be drawn. The effects of selection on allele frequency over a long period of time can be calculated if the selection coefficients are known (Figure 1).

Selection against recessive alleles

Recessive alleles are only exposed to selection in homozygous *cc* genotypes. As the frequency (q) of *c* decreases, q^2 becomes very small (Hardy–Weinberg equilibrium,

Section D3) and most c alleles are in dark heterozygotes ($2pq$) where they are not expressed and so cannot be acted on by selection. For this reason, selection is very ineffective when acting on rare recessive alleles, whether they are of high or low fitness in homozygotes (Table 1a, and Figure 1). The typical allele will tend to persist at low frequency even in highly polluted areas.

Selection against dominant alleles

The carbonaria allele is dominant, and therefore subject to selection in both homozygous (CC) and heterozygous (Cc) individuals. Selection against carbonaria in an unpolluted wood will tend to remove all the C alleles because they are all expressed in the phenotype, even when rare (Table 1b and Figure 1c, d). Notice that selection against dominant alleles does not lead to an immediate replacement by rare recessive alleles because these are mostly in the heterozygotes being selected against because of their one dominant allele. It is only when the recessive alleles reach a frequency (q) around 0.3, and the frequency of homozygotes (q^2) reaches about 0.09 that they start to increase in frequency rapidly (Figure 1). Even then, more than two-thirds of the recessive alleles are in heterozygotes ($2pq = 0.42$).

Selection against additive alleles

In cases where there is no dominance the effects of the two alleles adds up, so the heterozygotes have a fitness halfway between those of the homozygotes. If there is incomplete dominance, the relative selection against heterozygotes (denoted by h) will be somewhere between that of the homozygotes. In these cases the rate of change of allele frequency due to selection will be intermediate between the effects of selection against wholly dominant or wholly recessive alleles (Table 1c).

Selection for heterozygotes

Overdominance, or heterozygote advantage, occurs when the heterozygote is fitter than either homozygote. A classic example are carriers (heterozygotes) of sickle cell anemia

Table 1. Change in allele frequency per generation produced by selection against different genotypes

Type of selection	Genotype fitness			dq
	CC	Cc	cc	
a Complete dominance selection against c (in cc)	1	1	$1-s$	$-sq^2\dfrac{(1-q)}{1-sq^2}$
b Complete dominance selection against C (in CC and Cc)	$1-s$	$1-s$	1	$\dfrac{sq^2(1-q)}{1-s(1-q^2)}$
c Incomplete dominance selection against c (in cc and Cc)	1	$1-hs$	$1-s$	$\dfrac{hsq(1-q)}{1-sq(2q(2hp+q))}$

q is the frequency of the recessive allele, dq is the change in its frequency, s is the fraction of that genotype lost to selection, and h is the extent of dominance, between 0 in a and 1 in b. These values were obtained by multiplying the frequencies of zygotes of each genotype by the appropriate selection factor to obtain the new allele frequencies, and subtracting algebraically to find the difference. The derivation of this can be found in specialized population genetics textbooks.

trait (Section D4), who are more tolerant of malaria than normal homozygotes, and who do not get the sickle cell disease of homozygous recessive individuals. The rarest allele is found most frequently in heterozygotes and so is favored. This gives a frequency-dependent advantage to the rare allele and maintains a polymorphism. Both alleles persist in the population at an equilibrium frequency determined by the relative fitness of the two homozygous genotypes. Selection for heterozygosity may be an important factor in maintaining genetic variation in a population.

Selection against heterozygotes

Heterozygote (or hybrid) disadvantage occurs when heterozygotes are less fit than homozygotes. The rare allele is most commonly found in heterozygotes, so suffers the greatest rate of loss. It is lost faster as it becomes rarer and there are fewer homozygotes. The rare allele will therefore become extinct once a trend is established. This extinction depends on frequency rather than on the fitness of the two homozygous genotypes. This situation is only likely to arise when two adjacent populations are isolated and become homozygous for different alleles, and then come into secondary contact at the borders of their ranges. If some members of the two homozygous groups do not interbreed (i.e. they have a premating isolation mechanism) then they may become two species (Section D7) but this outcome is uncertain.

D6 Chromosome change in evolution

Key Notes	
Karyotypes	Karyotype describes the number and shape (physical morphology) of chromosomes. Diploids have *2n* chromosomes, where *n* is the haploid chromosome number found in gametes. Those chromosomes found equally in both sexes are called autosomes, and there is also a specific number of sex chromosomes. Each species has a characteristic karyotype, the set of chromosome pairs with particular shape and size. Chromosome shape is described by centromere position, from metacentric (centromere near center), through submetacentric (centromere offset), acrocentric (centromere near end), to telocentric (centromere adjacent to telomere). There may be variation in chromosome size and shape (polymorphism) within a species, and it may change over evolutionary time scales. Changes are dramatic in some organisms, and very slight in others.
Polymorphism and fertility	Heterozygosity for chromosomal rearrangements can prevent correct segregation of genes in meiosis. Recombination (chiasmata, crossovers) between the normal and rearranged segments tends to produce unbalanced gametes with two copies or no copies (duplications or deletions) of some chromosome segments. These reduce fertility by failing to produce viable zygotes. This makes it difficult for new rearrangements to become established, and can act as a postmating isolation mechanism between species by causing hybrid sterility. When chromosome polymorphisms persist a special mechanism is usually found that reduces the deleterious effects and maintains the polymorphism.
Mechanisms of rearrangement	Rearrangements require at least two breaks and one rejoin in DNA/chromosomes, but almost anything is possible. The probable causes are illegitimate recombination between dispersed repeated sequences, particularly transposable elements, and incorrect repair of double-strand breaks in DNA caused by radiation or other mutagens. Rearrangements may occur within a chromosome, or involve exchange or transfer between chromosomes.

Deletions	Deletions involve loss of genetic material, which is more or less deleterious, depending on the number and role of the genes which are lost. Acentric chromosome segments (that is, detached from a centromere) are lost.
Duplications	Duplications give an extra copy of a block of genes. They are generally less harmful than deletions, and one copy of the duplicated genes can become mutated and evolve a new role. This is the origin of all new genes throughout most of evolution. Strong selection for multiple copies of a gene (e.g. multidrug resistance) usually detects cells with the necessary multiple duplications (gene amplification), often visible as a heterogeneous staining region on the chromosome.
Centromeric fusions and fissions	Two telocentric or acrocentric chromosomes (with negligible short arms) may fuse at the centromere to produce a single metacentric chromosome (centromere near the middle) or submetacentric (two arms clearly unequal). This is called a centric fusion or Robertsonian translocation. The reverse is centric fission, which is splitting of the centromere in a metacentric to give two telocentrics. In this way chromosome number can change. Correct meiotic segregation in a heterozygous cell requires that the two telocentrics segregate to one pole, and the metacentric (with both arms) goes to the other pole.
Translocations	Translocations involve exchange of distal regions of nonhomologous (genetically different) chromosomes. It is necessary that either both rearranged or both original chromosomes occur in the same gamete to ensure genetic balance. In meiosis, normal chromosomes and chromosomes with translocations pair up alternately to form a ring or chain of several chromosomes. If two adjacent chromosomes segregate to the same pole, genetically unbalanced gametes will be produced, but if alternate chromosomes go together the gametes will contain completely balanced genomes. Some species are heterozygous for many translocations, and form long chains or rings of chromosomes in meiosis.
Inversions	Inversions turn part of the chromosome around, reversing its polarity. A pericentric inversion includes the centromere, a paracentric inversion is contained in one arm of the chromosome and does not move the centromere. A chromosome with an inversion can pair with a normal chromosome in meiosis in a heterozygote if they form an inversion loop. Meiotic recombination within an inversion loop duplicates one end of each chromatid and deletes

the other end in a reciprocal manner. These recombined chromatids are therefore not viable, and do not reach the next generation. Inversions therefore act as recombination suppressors, preventing recombination between the inverted segment and its normally oriented counterpart. Some species have polymorphic chromosomal inversions and have mechanisms to prevent loss of fertility (see below).

Paracentric inversions

These do not include the centromere. Meiotic recombination in a paracentric inversion loop generates a dicentric chromatid (two centromeres) and an acentric (no centromere) fragment. Most dipterans (true two-winged flies) can tolerate this because they have no recombination in males so sperm all contain complete chromosomes, and female dipterans have mechanisms to ensure that only unrecombined chromatids reach the egg cell, so zygotes are viable.

Pericentric inversions

These invert the centromeric region, and the effects of meiotic recombination in the loop duplicating and deleting the ends of the chromatid cannot be avoided because all chromatids have one centromere. Many zygotes will have unbalanced genomes. Pericentric inversions only persist as polymorphisms in species where they do not recombine at meiosis.

Changes in sex chromosomes

Y chromosomes have few genes and are lost in some species (XX females, XO males). Fusion of an X chromosome with an autosome causes a copy of the free autosome to become a neo-Y chromosome, only found in males which have another copy attached to their single X. Fusion of a Y chromosome with an autosome causes the unattached autosome to become a neo-X chromosome, one free copy being in males (to balance the copy attached to the Y), two copies in females as before. When two or more different X or Y chromosomes exist they are numbered X_1, X_2, Y_1, Y_2, Y_3, etc.

Evolutionary effects

Polymorphisms for paracentric inversions are common in dipterans where they may contain sets of alleles coadapted to control sex, or adapted to particular ecological environments. The suppression of recombination protects the cluster of genes and facilitates their co-evolution. Chromosomal change is common in evolution, and can cause hybrid sterility between closely related species, forming a postmating isolation mechanism.

Related topics

(B4) Chromosomes   (D7) Species and speciation
(C3) Meiosis and gametogenesis

Karyotypes

The term **karyotype** describes the number, size, and centromere position/arm length ratio of the chromosomes in a cell (Section B4). **Autosomes** are the chromosomes found in all individuals, irrespective of sex. **Metacentric** chromosomes have a centromere near the center, and roughly equal arm lengths. **Submetacentric chromosomes** have noticeably unequal arm lengths. **Acrocentric chromosomes** have one long arm and one very short arm. **Telocentric chromosomes** have their centromere adjacent to one telomere. Telomeres are specialized sequences and associated proteins which protect the end of the DNA from degradation, and interact with the nuclear envelope and the cytoskeleton during meiosis.

The number of chromosomes in a single set, as found in a gamete, is called the **haploid number** and is denoted by n, so **diploids** (which are zygotes formed by fusion of two haploid gametes) have $2n$ chromosomes. Put another way, diploids have n pairs of chromosomes. In some species the two sexes have different numbers of sex chromosomes, so males and females can have different diploid numbers. The **sex chromosomes** may be noted separately (e.g. in a species where $n=8$, females have two X chromosomes, males have one X, and there is no Y chromosome, then in males $2n=16$XO, and females are 16XX). The karyotype is usually a constant feature within a species, but variant types of chromosome carrying homologous information may exist in a species. These are known as **chromosomal polymorphisms** (many shapes).

Karyotypes may be very stable through evolution. Most dragonflies have $n=13$; snakes and birds both have about 18 large chromosomes and several microchromosomes. These data suggest karyotype conservation over hundreds of millions of years. On the other hand, closely related species may be very different suggesting rapid change. The Indian muntjac, *Muntiacus muntjac vaginalis*, has $2n=7$ for males and $2n=6$ for females, whereas the Chinese muntjac, *Muntiacus muntjac reevesi*, has $2n=46$. Different populations (or races) of the European mole cricket, *Gryllotalpa gryllotalpa*, have diploid numbers from 12 to 23. This variation may relate to speciation (Section D7).

Polymorphism and fertility

Sexual reproduction requires meiosis to segregate genetic information in the DNA of chromosomes so that each gamete gets one copy of each gene (and each chromosome), and all zygotes get two copies of each (except the sex chromosomes of course). If the cell is heterozygous for a chromosomal rearrangement then the homologous genes are not in the same structural locations, and segregation may be distorted, producing genetically unbalanced gametes with missing or duplicated information. The problem is complicated further if there is a chiasma (also called meiotic recombination or crossover; Sections C3 and B4) in the heteromorphic (different shaped) chromosomes. This is only a problem in heterozygotes, and therefore has its worst effects on rare chromosome types (morphs), making it difficult for new chromosomal rearrangements to become established in a population, because rare genetic elements are usually heterozygous (Section D3). Where chromosomal polymorphisms are found in populations, they are either transient states during replacement of one chromosome type by another, or there is likely to be some special genetic mechanism operating to achieve balanced gametes or to minimize fertility losses. Chromosomal differences between races can act as an effective postmating isolation mechanism by causing hybrid sterility (Section D7).

Mechanisms of rearrangement

Rearrangement of chromosomes requires at least two breaks (four ends) in DNA (chromosomes), followed by efficient rejoining and/or healing. These break/rejoins may occur by recombination between DNA sequences repeated and dispersed throughout the genome. Transposable elements (Section B3) dispersed throughout the genome may promote such rearrangements directly by their transposition mechanism. Incorrect repair of ends of double-strand breaks caused by ionizing radiation (or other mutagens) is another source of rearrangement which can be demonstrated in cultured cells (Section B3). All linear chromosomes must have a telomere, special repeated sequences with associated proteins which protect the end of the chromosome from degradation (Section B4). Chromosome breaks which do not rejoin to a normal chromosome end or acquire a telomere of their own remain unstable and reactive and promote further cycles of chromosome breakage and possible rearrangement. **Terminal** rearrangements involve the end of a chromosome, **interstitial** rearrangements are internal. The rearrangement may only affect one chromosome, or may involve transfer or exchange between two nonhomologous (genetically different) chromosomes.

Deletions

Terminal deletions involve a single break, then the fragment with a centromere obtaining a new telomere. The **acentric fragment** (no centromere) will be lost at cell division. Interstitial deletions involve the loss of a region within the chromosome, followed by rejoining of the ends of the flanking pieces. This may occur by recombination between repeated sequences in similar orientation in the same chromosome to delete the intervening region as a closed circle (Figure 1a–c). Alternatively, misaligned recombination between repeats on homologous chromosomes can create a duplication-deletion pair (Figure 1d and e, see 'Duplications' below). Deletions are also produced by meiotic recombination

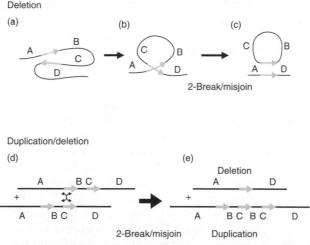

Figure 1.   Deletion of region BC as a closed circle (a–c) or duplication/deletion of region BC (d–e) by misaligned breakage-rejoining or recombination, possibly between repeated sequences or transposable elements (gray arrows) on the same chromosome (a–c, deletion) or on homologous chromosomes or sister chromatids (d–e).

in chromosome regions heterozygous for other rearrangements (e.g. translocations or inversions, see below).

Heterozygous zygotes carrying a deletion will be partially haploid, unless the region is duplicated elsewhere. In the fruitfly, *Drosophila melanogaster*, there are two loci which are dominant lethals if present in a single copy (haplolethals), 11 are deleterious, and 41 give small flies (haplo-minutes) if haploid. Single copies of genes may often be inadequate for full fitness, and combined loss of about 3% of the genome is lethal, probably because of the combined effects of loss of one copy of many genes. Deletions are probably less harmful in polyploid species. Heterozygous deletions, especially if large, may inhibit meiotic pairing/synapsis, and so cause nondisjunction and lower fertility. Thalassemias are blood disorders caused by a loss of hemoglobin genes by deletion, or more rarely imbalance by duplication, of alpha and/or beta hemoglobin. They are relatively common in malarial areas outside Africa because subclinical imbalance gives some protection.

Duplications

Since cellular life arose, and possibly before, duplications have been the main source of new genetic material, followed by divergence and separate evolution of the two copies. Sequencing entire eukaryote genomes consistently reveals large numbers of duplicate chromosome segments, both within and between chromosomes. In some cases where numerous copies of a gene are required (e.g. ribosomal RNA genes) the tandem copies evolve in concert in one species, all with the same sequence. Misaligned recombination between dispersed repeated sequences on homologous chromosomes is the most likely origin of tandem duplications on one chromosome (Figure 1d and e). Occasionally a small translocation will survive in its new location as a duplication. When selection acts for a large increase in copy number (**amplification**) of a particular gene, duplications usually appear, often as microscopically visible heterogeneous staining regions containing many tandem repeats. Examples are selection by insecticide on aphids to amplify the gene for an esterase which breaks down the insecticide, thus becoming insecticide resistant, and by chemotherapy on cancer cells to amplify genes for a protein which pumps toxic drugs out of the cell. Duplications of the centromere do not survive because the two centromeres may orientate to opposite poles at division, breaking the chromosome. Duplications are generally less harmful than deletions.

Centromeric fusions and fissions

Two telocentric chromosomes, or acrocentric chromosomes with no significant genetic material on the short arm, may fuse at the centromeres to give one metacentric or submetacentric chromosome, reducing the number of chromosomes (Figure 2a). This is also called a **Robertsonian fusion** or **Robertsonian** translocation. The short arms are lost. The reverse of fusion is **centric fission** where the centromere of a two-armed metacentric chromosome splits into two and heals, producing two one-armed telocentric chromosomes, increasing the chromosome number (Figure 2a). When heterozygous, the two telocentric chromosomes may pair with the metacentric homolog to make a meiotic trivalent (three chromosomes paired up). If the two telocentrics orientate to the same pole opposite the metacentric (Figure 2b) segregation is normal and the gametes will be balanced, containing either the two telocentric or the single metacentric chromosome. If a telocentric enters the same gamete as the metacentric chromosome (**missegregation** or **nondisjunction**), that gamete carries two copies of the genes on the telocentric, and will produce a trisomic (one chromosome too many) zygote. The corresponding gametes will have no copies of that telocentric chromosome, and will produce monosomic zygotes

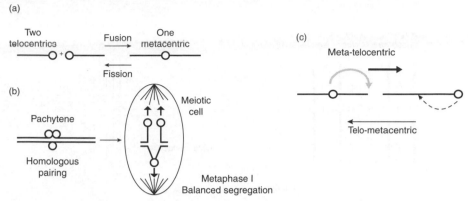

Figure 2. (a) Interconversion between two telocentric chromosomes and a single metacentric chromosome by centromeric fusion and fission respectively (circles represent centromeres). (b) Meiosis in a heterozygote produces balanced gametes if the two telocentric chromosomes go to the pole opposite the metacentric. (c) Movement of the centromere by pericentric inversion can change the chromosome's shape between telocentric and metacentric.

(one chromosome missing). In mice and humans, heterozygous Robertsonian fusions are associated with much more sterility than expected from observed nondisjunction, but the reason for this is unknown. Inversion (see below) with a breakpoint near a centromere can convert between metacentric and acrocentric chromosome shapes (Figure 2c). Further centromeric fusion can follow, reducing the chromosome number still further.

Translocations

A translocation is the transfer of material between two nonhomologous chromosomes. This usually involves a reciprocal exchange of the ends of chromosome arms (Figure 3a). Human chromosome 2 was formed by translocation close to the ends of two chromosomes, followed by the inactivation of one centromere. All apes still have two separate chromosomes, so the event happened after the split between humans and chimpanzees. Heterozygotes will only produce genetically balanced gametes when both original chromosomes or both rearranged chromosomes pass into the same gamete. Recombination complicates this segregation, but there is one arrangement, **alternate disjunction**, which ensures balanced gametes. This requires metacentric chromosomes, and no chiasmata (i.e. no recombination) between the translocation breakpoint and the centromere. This is favored if the translocation breakpoints were near the centromere, and chiasmata are near the telomeres (Figure 3b). (These conditions may arise by centric fusion as easily as by translocation.) Translocation heterozygotes can form **multivalent rings** or **chains** in meiosis. These are closed rings or open chains composed of several chromosomes joined by chiasmata. The translocated chromosomes alternate with their normal homologs, and alternate centromeres must align towards opposite poles for correct segregation (Figure 3d), hence the name alternate disjunction. **Adjacent segregation** (Figure 3c) where any two chromosomes which are side-by-side (adjacent) go to the same pole after meiosis produces gametes with duplications and deletions.

In the evening primroses, genus *Oenothera*, many or all of the chromosomes are involved in translocations, forming multivalents in meiosis. In one species, *O. lamarkia*, six out

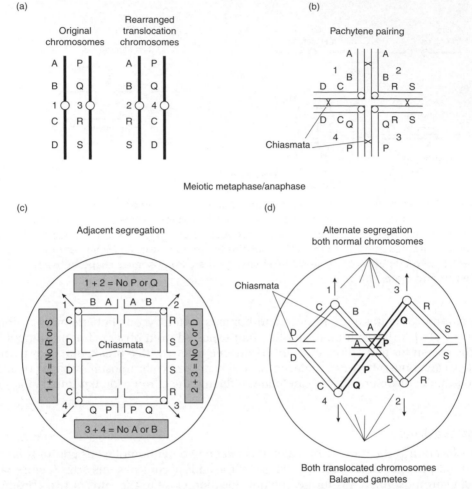

Figure 3. (a) A translocation involves transfer or exchange between two nonhomologous chromosomes. (b) Two normal and two rearranged chromosomes provide a diploid genome and can pair to form a quadrivalent ring or chain in meiosis. The centromeres are numbered in their sequence in this ring. (c) If any two adjacent chromosomes segregate to the same pole they produce unbalanced gametes with a duplication and a deletion. (d) Segregation of alternate chromosomes to the same pole gives gametes with a copy of all genetic segments ABCD and PQRS.

of the seven pairs of chromosomes are always heterozygous for translocations. All of the rearranged chromosomes go to one pole, the original type to the other (original and recombined are arbitrary), so each set of chromosomes always segregates together, linking most of their genes, which cannot be reassorted by recombination. The two sets of chromosomes may carry alleles for quite different phenotypes, one adapted to a desert habitat, the other to a humid habitat. Each set of chromosomes in *O. lamarkia* also has gametic lethal alleles, so one set dies in egg cells and only survives in pollen, the other only survives in egg cells, and all zygotes (and so all individual plants) are fully heterozygous. In primitive termites (e.g. genus *Kalotermes*), the X and Y chromosomes are translocated into a chain with most of the autosomes, effectively sex-linking a large part of the genome. This makes genetic relatedness brother-to-brother and sister-to-sister high, greater than between parent and offspring, and encourages the evolution of sociality in termites through kin selection (Section D2).

Inversions

An inversion occurs when part of a chromosome is turned around, possibly by recombination across an omega-loop (named after its shape, see Figure 4). Heterozygous inversions can pair in meiosis to form an **inversion loop**. Recombination in an inversion loop produces a duplication of one end of the chromosome and loss of the other, so recombined chromatids do not survive. The exact outcome depends on whether the centromere is inside (**pericentric**) or outside (**paracentric**) the inversion. As with other heterozygosities which may reduce fertility, rare inversions are likely to be selected against. Some species avoid the adverse affects by not pairing the inverted segment. Inversions are **recombination suppressors** because only unrecombined parental chromosome arrangements of the inverted region can be transmitted, allowing genes linked in the inversion to evolve together, differently from homologous genes on the standard chromosome.

Paracentric inversions

Paracentric inversions have breakpoints on the same side of the centromere (Figure 5a). Meiotic recombination (a chiasma) in an inversion loop (at pachytene, Figure 5b) produces one recombined chromatid with two centromeres (**dicentric**) and another chromatid with no centromeres (**acentric**; Figure 5c–e). The acentric fragment lags at the first anaphase and is lost, and the dicentric bridge breaks to give incomplete chromosomes. Dipterans are often polymorphic for paracentric inversions, which are important in their ecology and evolution. In females the bridge holds the dicentric in the center of the spindle (Figure 5d) so that the genetically correct unrecombined chromatids are at the outer poles. In female meiosis it is one of these outer daughter nuclei which becomes the egg nucleus (gamete; Figure 5e), the other three become polar bodies. This mechanism ensures that the egg is viable with unrecombined chromosomes, all nonviable products are excluded in polar bodies, and fertility of the females is not impaired. Male dipterans generally have no recombination, so they do not generate acentric/dicentric chromatids, their sperm are normal, and so they also escape the ill effects of heterozygous paracentric inversions.

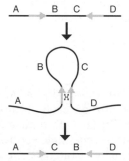

Figure 4.   Production of an inversion by recombination between inverted repeats (gray arrows) in an omega loop configuration.

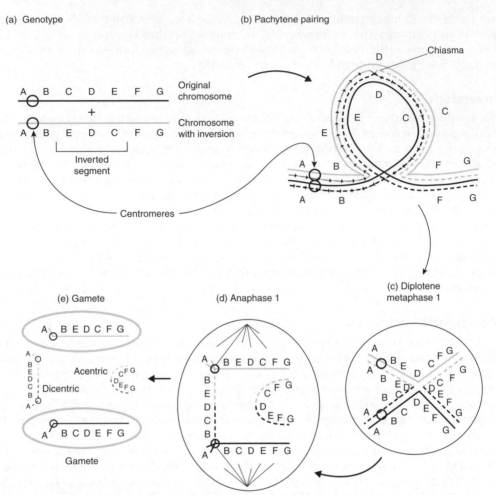

(a) Genotype

(b) Pachytene pairing

(e) Gamete

(d) Anaphase 1

(c) Diplotene metaphase 1

Figure 5.   Effects of recombination in a heterozygous paracentric inversion in a female *Drosophila*. (a) The polymorphic chromosome pair; the gray chromosome has the inversion. (b) At pachytene the inverted region pairs to form an inversion loop. There are two chromatids in each chromosome. The chiasma creates a dicentric chromatid ABEDCBA, traced by small arrows, and an acentric fragment GFEDCFG. (c) At diplotene the homologous pairing ceases and a chiasma is visible in the inverted region. (d) At anaphase the dicentric chromatid forms a bridge between the poles and breaks, and the acentric fragment is lost. (e) One of the unrecombined chromatids at the outside of the cell enters the gamete nucleus with a full complement of genes.

Pericentric inversions

Pericentric inversions span the centromere (Figure 6a). A crossover within the inversion loop swaps the ends of one chromatid on each chromosome (Figure 6b and c), duplicating one end and deleting the other (as with paracentric inversions). With pericentric inversions, however, all chromatids have a single centromere (Figure 6d, compare Figure 5e), so they can all segregate normally at meiosis and pass into gametes which can be genetically unbalanced in either sex, and so produce inviable zygotes. Polymorphisms for pericentric inversions are not found in *Drosophila* but are found in other species,

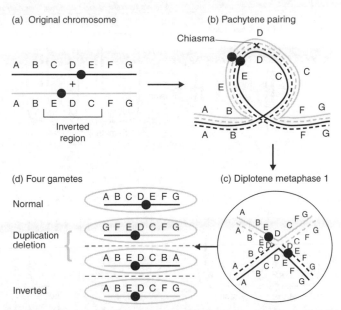

Figure 6.   Effects of recombination in a heterozygous pericentric inversion (a). This is similar to the paracentric inversion (Figure 5) but the centromeres are now within the inversion loop at pachytene (b). A crossover in the inversion gives duplications and deletions (c) but all chromatids now have a single centromere, so the duplication and deletion chromatids can be incorporated into gametes (d).

for example where the inverted region pairs nonhomologously in meiosis, preventing crossing-over (e.g. in the grasshopper *Keyacris scurra*).

Changes in sex chromosomes

The chromosome only found in the heterogametic sex (e.g. the Y chromosome in humans and *Drosophila*) has few genes on it, and in some species it is lost, giving XO males and XX females a reduction in the number of sex chromosomes. Fusions or translocations between sex chromosomes and autosomes can also increase the number of sex chromosomes by converting autosomes into neo-sex chromosomes. These are forced to continue to exist in the free state to balance segregation of the homologous copy attached to a sex chromosome and to maintain diploidy. For example, when an X chromosome fuses with an autosome, that autosome becomes part of the X. This is balanced in females with two X chromosomes plus attached autosome, but in males one copy of the autosome will be attached to the X, the second will be free, only found in males, and behave just like a Y chromosome (Figure 7a). It is known as a **neo-Y chromosome**, and is found after an X-autosome fusion. If a Y already exists they will be numbered Y1 and Y2. If a Y chromosome fuses to an autosome (Figure 7b), the male will still have one free autosome, and the female will have two free autosomes. These are called **neo-X chromosomes** and form as a result of a Y-autosome fusion. In this way, multiple X and Y chromosomes can evolve from autosomes. Multiple sex chromosomes are numbered X1, X2, X3, and Y1, Y2 and so on.

Evolutionary effects

Dipterans (two-winged flies) have no recombination in males, an adaptation which avoids fertility loss from paracentric inversions. The females also avoid passing recombined

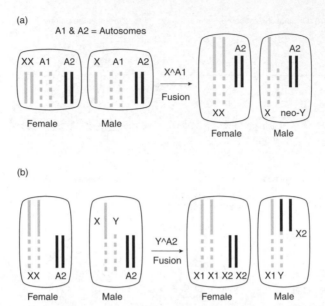

Figure 7.　(a) Fusion of an X chromosome to an autosome A1 (creating A1^X) causes the free copy of the autosome to behave as a neo-Y chromosome only found in males, as a second copy to the one attached to the X. (b) Fusion of a Y chromosome to an autosome A2 (creating A2^Y) causes the free copy of A2 to behave like an X-chromosome (called X2), with two copies in females, but only one in males to balance the single copy attached to the Y.

chromatids with duplications and deletions into eggs, so heterozygous paracentric inversions can persist in dipterans, where they may be common. Suppression of recombination between the inverted and normally orientated chromosome can allow co-evolution of the genes locked together in the inversion. They may become adapted to a different habitat (e.g. hotter or colder, higher or lower altitude) than the alleles on the other chromosome. Particular paracentric inversions in *Drosophila* species are associated with sex determination, taking the role of sex chromosomes. Many populations have stable polymorphisms for inversions, which are said to be 'floating' (as opposed to 'fixed' when only one chromosome type is found in a population or species). There are about 2000 species of *Drosophila*, and their evolution is estimated to have seen the evolution of some 350 million paracentric inversions, of which 20000–60000 have become fixed, and 18000–28000 are still floating. Different types of chromosomal rearrangements are common in different evolutionary lines. In primate lineages, great apes (including humans) have mostly pericentric inversions, lemurs mainly have centric fusions, and *Cercopithidae* have many fissions.

D7 Species and speciation

Key Notes

Species

The genetic (biological) definition of a species is a set of organisms actually or potentially interbreeding, and sharing a common gene pool. Most species are clearly different (domestic cats and dogs) but two closely related species may appear identical to humans (they are hidden or cryptic), and members of a single species may be quite different. Taxonomy is the scheme or science used to classify species. Species which share a more recent common ancestor should be classified closer together than distantly related species. A morphological species may be defined on the basis of appearance or shape as a group of organisms sharing a unique set of morphological characters (this is the only technique that can be applied to fossils). An ecological species can be defined as a group of (interbreeding) organisms utilizing a unique resource/habitat combination (a niche). These definitions are complementary, and should not conflict.

Species isolation mechanisms

Isolation mechanisms prevent species interbreeding or exchanging genetic material (gene flow). This is the most important requirement for speciation, because populations can only diverge into species if they stop exchanging genes. Gene flow dilutes the differences. Premating isolation mechanisms are most efficient because they avoid wasting resources. These are: (i) behavioral isolation (mate choice, sexual, ethological); (ii) spatial isolation (mating in different places, ecological or habitat isolation); (iii) temporal isolation (mating at different times); (iv) mechanical isolation (mating physically impossible); (v) pollinator isolation (some plants); or (vi) gametic isolation (restricted attraction and/or fusion between gametes).

Postmating isolation mechanisms are inefficient because mating produces unfit or inviable offspring, wasting resources and reducing successful reproduction. They are not compatible with extended cohabitation (sympatry). They are: (i) hybrid inviability; (ii) hybrid sterility; and (iii) hybrid breakdown (where second generation hybrids and backcross progeny are unfit).

Speciation

Speciation is the process by which one species splits into two or more. Study of this splitting is called cladistics, and the tree of descent of related species is called a phylogeny. (The science is phylogenetics.) Speciation requires that populations are isolated so that they do not interbreed to exchange genetic information, and their genomes can then diverge. This isolation can happen either because they mate (or live) in different places (geographic isolation or allopatry), or they mate (or live) in different habitats in one geographic area (habitat isolation in sympatry). The physical separation required depends on the motility of the organism, and may be small for sedentary species. Divergence occurs by accumulated genetic differences which may prevent successful interbreeding when these populations establish secondary contact. It will be faster if the populations experience different selection pressure. When they develop premating isolation mechanisms they are separate species.

Secondary contact

The test of speciation occurs when two divergent populations meet and mix following range expansion because of habitat change. The outcome depends on the extent of development of pre- and postmating isolation and ecological (niche) divergence. If hybrids are fit the two populations merge into a hybrid swarm. If there is some hybrid unfitness for genetic or ecological reasons then there will be selection to reinforce premating isolation. If, however, limited interbreeding and low postmating isolation transmits some genes between them (gene flow) then divergence may be reduced or prevented and a permanent hybrid zone can be established between parapatric (side-by-side) populations. If there is little or no gene flow because hybrids are either inviable or sterile (strong postmating isolation), divergence can continue until the populations become isolated species. If premating isolation exists or arises after contact between the populations, then individuals of the two populations can intermingle as different species, but if they still utilize the same ecological niche, the more efficient species will replace the other.

Phylogenetic patterns

There is a complete gradation between races (no isolation), subspecies (postmating isolation), semispecies (some premating isolation) to sibling species which look alike but do not interbreed. A single species may change through time. The succession appears in the fossil record as a series of chronospecies. Phylogenetic algorithms typically produce a dichotomous pattern of species splitting, but in reality environmental change may divide a widespread species into many isolated refugia, each population a potential new species. Expansion of a few species after an environmental catastrophe may also produce an enormous adaptive radiation with populations diverging as they enter new habitats free from competition from adapted species (e.g. mammalian radiation after the extinction of the dinosaurs).

Related topics

(D9) Phylogeography, molecular clocks, and phylogenies

(E5) Bioinformatics

Species

The **genetic** or **biological definition of a species** is a set of actually or potentially inter-breeding organisms. Members of a species share a common **gene pool** with all other members of that species. Members of different species do not exchange genetic material. There are clear differences in most cases (e.g. between cats and dogs), however in many species there are geographic races which look different but which can interbreed if they come into contact. This makes it difficult to define species, and causes conservation problems, e.g. Mallard drakes from Europe mate with Canadian Black Duck females, and Ruddy Duck from America mate with European White-Headed Duck, in both cases destroying the native species by hybridization. Conversely, similar organisms which do not appear so noticeably different to humans, may be classified as single species. One named species may contain many hidden (**cryptic**) **species** which we simply do not notice. The commonest British bat, *Pipistrellus pipistrellus*, has recently been discovered to include two species, first differentiated by the use of electronic bat detectors which indicated two types of ultrasonic call, 45 kHz for the common species and 55 kHz for the 'soprano' sister species.

Taxonomy is a scheme of **classification** for defining all species and grouping related species together. This should ideally group together species which share a common ancestral species. The first step is usually based on a **morphological species** concept. All members of a species share a set of physical characteristics which can distinguish them from members of other species. This is the only criterion which can be applied to fossils. It has problems with very variable species; in severe cases even males and females may be described as different species. The reverse problem arises with very similar species such as sheep and goats whose skeletons cannot be distinguished unambiguously. There is also an **ecological species** concept. Each species uses a particular set of resources, a **niche**, from their environment, and they do this most efficiently in a particular set of conditions, or **habitat**. Two species cannot share the same niche in the same geographic location, because the more efficient will succeed, the other will become extinct. These three concepts of species are complementary, and generally in agreement except for variable or closely related species.

Species isolation mechanisms

Members of different species are prevented from interbreeding or exchanging genetic material by **isolation mechanisms**. These must evolve in the process of speciation. The list of isolation mechanisms below is based on an original list by Dobzhansky.

Premating isolation mechanisms

Premating isolation mechanisms prevent interbreeding and are efficient because resources and effort in reproduction are not wasted producing inviable or sterile offspring.

(i)   **Behavioral (sexual, ethological) isolation.** Individuals choose members of their own species and reject other species as mates. In species with external fertilization this behavior is expressed by the gametes.

(ii)   **Spatial (geographic, ecological or habitat) isolation.** Individuals of different species do not meet when mating because they mate in different locations. The separation must be considered on the scale of the motility of the organism, and includes different continents, ecosystems, food plants, or different dryness, acidity or temperature preferences in the habitat. The effects are identical whether the cause is genetically determined selection of habitat or passive geographical separation of isolated populations on different islands, lakes, mountains, etc. Sedentary insects on one tree may be effectively isolated from a population on a similar tree 100 m away.

(iii) **Temporal isolation.** Different species mate or flower at different times of the day or year.

(iv) **Mechanical isolation.** The genitalia or flower parts prevent copulation or pollen transfer respectively.

(v) **Isolation by different pollinators.** Related plant species may attract different pollinators.

(vi) **Gametic isolation.** In externally fertilized species, gametes show selective attraction and/or acceptance. Sperm may only swim towards eggs of their own species. In plants, styles may prevent entry of incompatible gametophytes. In internally fertilized species the male gametes may not survive in the sex ducts of the other species (compare to i).

Postmating isolation mechanisms

Postmating isolation mechanisms reduce the viability and/or fertility of hybrids produced by matings between members of two species. This is very inefficient because reproductive effort is wasted producing useless offspring. This prevents the two species coexisting in the same habitat (prevents sympatry) except as migrants, because the rarer species predominantly meets the other (wrong) species when looking for a mate, and so produces mostly low fitness heterozygous hybrid offspring. The rarer species therefore faces greatest selection (Section D3).

(i) **Hybrid inviability.** Hybrid zygotes are inviable or have reduced viability. This could be because the products of genes from the two parental species do not interact together correctly.

(ii) **Hybrid sterility.** One or both sexes of the F1 hybrids are sterile or have reduced fertility. This can be caused by sequence changes from random mutations. When *Saccharomyces* DNA sequences diverge by more than approximately 0.6% (which probably takes about 300 000 years isolation) mismatch repair systems start to prevent meiotic synapsis because they do not identify chromosomes as being homologous. If pairing fails, the chromosomes do not segregate properly into gametes and aneuploidy is caused. More rapid isolation can occur following fixation of chromosomal rearrangements in one of the isolated populations. These can also cause sterility in heterozygous hybrids because of failure of chromosomes to segregate properly in meiosis (Sections C3 and D6).

(iii) **Hybrid breakdown.** F2 or backcross hybrids are inviable, sterile, or have reduced viability or fertility. This occurs because F1 individuals with one complete set of chromosomes from each parent are viable, but after meiotic assortment the gametes do not carry a full genome. This will occur if there are small translocations so some genes are no longer on homologous chromosomes (Section D6).

Speciation

Speciation is the process of forming two or more species from a single species. Study of this splitting arrangement is termed **cladistics**. The 'family tree' of descent of related species is a **phylogeny**, and study of this subject is called **phylogenetics**. Separate species can only form if: (i) gene flow between populations of one species is restricted to allow them to diverge, and then (ii) a component of the divergence produces a complete barrier to interbreeding, called a mating isolation mechanism. The initial barrier can be either spatial or temporal. Changing climate may isolate populations on mountain peaks or in remnant lakes, or rare migrants may colonize new areas such as the Galapagos and

Hawaiian islands (Figure 1). Geographic isolation is termed **allopatry** and gives rise to **allopatric speciation**.

Disruptive selection (Section D2) in a polymorphic species may select for two species to form if the intermediate hybrid is less fit than its parents. If the two types continue to interbreed, however, it is difficult to see how they can diverge sufficiently for two species to arise.

Sometimes selection may produce ecotypes living in adjacent habitats in the same geographic area. This is termed **sympatry**. There are some suggested cases of sympatric populations evolving into separate species (e.g. fish in the same lake and insects living on different species of food plants in the same habitat). This is called **sympatric speciation**, however it is increasingly being discovered that the patches of habitats involved may be so distant compared with the motility or dispersal of the organisms when mating that they are effectively micro-geographically isolated. These isolated populations are incipient species.

Random genetic change by mutation and drift over a long period (probably exceeding about 300 000 years) will produce enough divergence in the DNA for mismatch detection in meiosis in hybrids to prevent recombination and chiasmata. The resulting nondisjunction will produce inviable gametes and cause hybrid infertility. This makes speciation inevitable if populations are separated for long enough. Divergent selection may produce more rapid changes increasing the effect. If two populations evolve adaptations to different environments then interbreeding may produce poorly adapted hybrids. There is then selection for premating isolation mechanisms to prevent interbreeding. This can only happen if the postmating isolation mechanisms are strong enough for selection to increase the premating isolation mechanisms faster than gene flow through hybrids dilutes the necessary differences. Selection for premating isolation also occurs when there is any postmating isolation for whatever reason (usually genetic differences).

Secondary contact

When conditions change, the isolated and diverged populations may meet again; this is termed **secondary contact** at a **contact zone**. The outcome then depends upon the degree to which postmating isolation, premating isolation, and ecological (niche) divergence has arisen while they were isolated. If they interbreed freely and hybrids have normal fitness, the populations will merge, producing a **hybrid swarm** (Figure 1c, populations B and C). If interbreeding produces unfit hybrids, selection will act to **reinforce** premating isolation mechanisms, because any individual that correctly mates with its own type will gain an advantage. Those individuals mating incorrectly will produce unfit hybrid offspring. The locally rarer type has a disadvantage because it is more likely to mate mistakenly with the (wrong) majority type and produce low-fitness offspring.

If postmating isolation is weak, and genetic exchange (gene flow) through hybrids occurs, the genetic differences will be diluted rather than reinforced at the contact zone, and divergence cannot occur. Premating isolation will not develop, but the two populations will live **parapatrically**, side-by-side. They will meet in a **hybrid zone** in a region of low population density (i.e. poor habitat), supported by immigration from both populations (Figure 1c, D vs. B/C). Such hybrid zones will be stable for thousands of years if the environment is stable. Genetic loci which do not produce hybrid unfitness will be exchanged between populations, but genes which are unfit in hybrids cannot cross the barrier. If hybrid inviability or sterility is effectively complete there will be no gene flow, and divergence can continue to reinforce premating isolation. When premating isolation

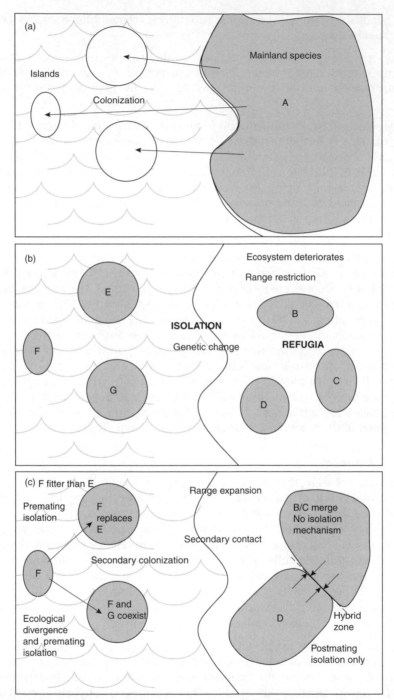

Figure 1.   Production of five species from one. (a) A mainland species colonizes three offshore islands. (b) Climatic change separates the mainland population into three refugia, and the six isolated populations diverge. (c) On future expansion and secondary contact the outcome depends on mating isolation and ecological divergence. B and C have no divergence and merge. B/C has postmating isolation from D, so B/C and D meet at a hybrid zone. F has premating isolation from E and G. It uses the same niche as E and is more efficient, so F replaces E. Ecological differences have evolved between F and G so they can coexist.

is effective, interbreeding stops, and populations which are ecologically different can coexist in one geographic region (Figure 1c, F and G), however if they still use the same niche then the more efficient species will displace the other (Figure 1c, F and E).

Phylogenetic patterns

In nature there is a complete gradation between **geographic races** with no reproductive isolation, through **subspecies** with partial hybrid sterility and **semispecies** with some premating isolation, to **sibling species** which appear to be very similar, but which do not interbreed. Note that quite different species may appear identical to humans, while races which interbreed freely may appear to be very different.

Speciation need not be a dichotomous event, one species splitting to two then four, and so on. All the algorithms used in phylogenetics will artificially enforce a dichotomous series of events, but this may give a very misleading impression. It is likely that unspecialized species with a wide geographic and habitat distribution will survive in many isolated **refugia** during adverse conditions. For example, the ice ages caused ice caps on mountain ranges and separated the populations in adjacent valleys on either side. Each isolated population then started to diverge from the others, becoming an incipient species ready to spread when the climate improved. When a catastrophe destroys many species (e.g. a comet strike killing off the dinosaurs), descendants of surviving species can colonize many new habitats during an **adaptive radiation** when conditions improve. The populations isolated in each newly occupied habitat may simultaneously diverge into many new species. This produces clusters of new sibling species in a very short geological time and confuses attempts to create phylogenies.

As a single species lineage changes through time, it will produce a series of **chronospecies** in the fossil record, but identification of these is arbitrary, and often influenced by the patchy availability of fossils.

D8 Polyploidy

Key Notes

Introduction

Polyploidy is the state of having more than two complete sets of chromosomes. The base haploid number is x, diploids have a diploid number $(2n) = 2x$, hexaploids have $2n = 6x$ and a haploid number $n = 3x$. Polyploidy can cause problems in meiotic segregation if the multiple copies of chromosomes do not divide evenly. Odd-number polyploids are sterile because they cannot divide their chromosomes evenly into two cells at meiosis (it is impossible to divide three or five exactly by two). Polyploidy is rare in animals, but 47% of flowering plants (angiosperms) are polyploid. Vegetative reproduction helps plants through the initial sterile period, and they are less severely affected than animals by chromosomal imbalance. Polyploidy is usually lethal in animals but has happened in evolutionary history. There are many parthenogenic polyploid animal species whose reproductive cycle avoids sterility. The sudden change in chromosome number and consequent hybrid sterility caused by polyploidy can act as postmating isolation mechanisms, causing instant speciation.

Autopolyploids

Autopolyploidy is caused when the chromosome number doubles in an individual (e.g. mitotic separation fails). They are usually infertile because there are more than two copies of each chromosome per cell and they may not pair and segregate evenly in meiosis.

Allopolyploids

Allopolyploids are formed from hybrids between closely related species. The different sets of chromosomes are sufficiently different in DNA sequence not to pair between sets in meiosis. The chromosome number doubles accidentally by failed mitosis so that each cell has two complete sets of chromosomes from each parent, and these pair like with like, restoring fertility.

Introgression

Crosses between species produce a hybrid with a higher level of polyploidy than the parent species because of the doubling required to restore fertility. Genomes introgress from low ploidy species into higher ploidy species, but chromosomes from high polyploids cannot return to diploids.

Polyploid complexes

A cluster of related species (e.g. grasses or magnolias) can hybridize in many combinations producing many polyploid species. Further hybridization leads to more polyploids with higher ploidy, and the eventual loss of diploid species. Eventually a few highly polyploid species remain with no apparent related diploid species.

Characteristics of polyploids	Polyploids tend to be large with large cells. This makes them attractive as food, with a soft texture and high yield. They also produce large flowers. The sterility of triploids is used to produce seedless fruit. The multiple genomes in polyploids tend to stabilize the genotype and phenotype because segregation of extreme genotypes is rare.	
Related topics	(B4) Chromosomes	(D7) Species and speciation
	(D6) Chromosome change in evolution	

Introduction

Polyploidy is the state of having more than two complete sets of chromosomes. (It can be confused with aneuploidy which is an incorrect number of particular chromosomes.) Species with three sets are called triploid, four sets tetraploid, six sets hexaploid, and so on. To avoid confusion, the base (original haploid) number may be called x, and the current haploid chromosome number n, so a hexaploid which behaves as a diploid has a diploid number $2n = 6x$, haploid number $n = 3x$ (e.g. bread wheat).

Polyploid cells have difficulty in meiosis unless they can behave as diploids, so that each gamete gets a balanced set of chromosomes. This requires chromosomes to pair in twos at meiosis. Only even-number polyploids are fertile, because odd numbers of chromosomes cannot be divided in two at meiotic reduction division. Polyploidy is usually lethal to animals. The few polyploid animal species are hermaphrodite (e.g. some earthworms and planarians), or parthenogenic (e.g. some beetles, moths, crustaceans, fish, and salamanders). Parthenogenic animals usually double their chromosome number by replication without division just before meiosis, then pair identical sister chromosomes, and separate them in meiosis to restore the original karyotype. This avoids problems of segregation.

Plant genomes are much more tolerant of changes in chromosome number and 47% of all flowering plants are polyploid. Polyploidy is important as a speciation mechanism in plants because it can prevent interbreeding in a single step. A new polyploid can only produce fertile offspring if its gametes fuse with a gamete of the same ploidy. For example, a new tetraploid produces $2x$ gametes. If these fuse with a haploid gamete from a diploid plant the offspring will be sterile triploids. Fertility is restored by a chance failure of mitotic division that creates a hexaploid cell. Many plants are able to reproduce vegetatively which may allow them to survive the sterile phase until fortuitous doubling occurs. They also have the advantage of not having a differentiated germline. Any dividing cell whose descendent clone could become a flowering shoot and produce gametophytes could undergo chromosome doubling; its seeds would produce fertile polyploids.

Polyploidy has been important for providing spare copies of genes which can evolve new functions. Comparisons of whole genomes provides evidence of complete genome duplication (tetraploidization) in *Saccharomyces* yeasts and also of doublings early in the development of multicellular animals and again in early tetrapods. There are also numerous duplications of small chromosome segments. Usually, one of the two copies of each duplicated gene either mutates and is lost, or rapidly diverges and acquires a new function.

Autopolyploids

Autopolyploids have double the normal chromosome number, but all the chromosomes come from the same species, often the same individual. This can arise by a failure of mitosis or from a diploid gamete produced by a failure of the second meiotic division. Autopolyploids are usually sterile because the three or more homologous chromosomes will not form bivalents (pairs) at meiosis but rather multivalents (three or more synapsed chromosomes). This does not lead to even reduction at division, and gametes will not contain complete sets of chromosomes. Experimentally produced autopolyploids are always less fit than their diploid parents.

Allopolyploids

Allopolyploids have a hybrid origin. They are produced by fusion of gametes from related species. This is common in plants. The two sets of chromosomes in the diploid hybrid may be sufficiently different from each other that they cannot pair correctly at meiosis. This prevents them from consistently orientating towards different poles at the first meiotic division and segregation may be random. Gametes then contain too many copies of some chromosomes, and not enough copies of others, resulting in gamete failure or the formation of inviable zygotes. An accident of mitosis failing to separate the daughter mitotic nuclei can double the chromosome number in that cell, forming an allotetraploid. Each chromosome can now pair with its own duplicate, so the allotetraploid behaves like a normal diploid, but the gametes are now $2x$ 'diploid.' This can be seen in the reconstructed evolution of wheat (Figure 1). Two species, *Triticum monococcum* (*AA* genomes) and *Aegilops speltiodes* (*BB* genomes) each with seven pairs of chromosomes ($x = 7$) formed a hybrid (*AB*). This doubled its chromosome number to 14 pairs ($2n = 28$) becoming tetraploid *Triticum duococcum* (*AABB* genomes). Now the two *A* sets could pair together and the two *B* sets could pair together, producing balanced *AB* gametes. This specific homologous pairing *A-A* and *B-B* (rather than *A-B* pairing, which is termed homoeologous) requires activity of a particular genetic locus (called *Ph* for pairing homology) on the long arm of chromosome 5B. This activity is only found in cultivated wheats, it has never been found in its wild ancestor. Later, an *AB* gamete (14 chromosomes) from *T. duococcum* ($2n = 28$) fused with a haploid (*D* genome) gamete from *Aegilops squarrosa* ($2n = 14$) to give a triploid hybrid (*ABD*), which doubled up its chromosomes to give the fertile hexaploid *AABBDD* ($2n = 42$), *Triticum aestivum*. This produced bread wheat, one of the most important human food plants. It contains two complete genomes from each of its three diploid ancestors.

Introgression

Introgression is the movement of genes from one race or species into another. It is possible for hybrids to arise between species with different levels of ploidy but the outcome is usually an overall increase in ploidy because the chromosome number doubles to restore fertility. This causes introgression because genes from the low ploidy species enter, or introgress into, the higher ploidy species, but the process cannot simply reverse. Genes from high ploidy species cannot return to low ploidy species.

Polyploid complexes

The ploidy in a group of hybridizing species steadily increases because each new hybrid doubles its chromosome number to restore fertility. Such a group is called a polyploid complex. The *Magnoliaceae* provide a good example (Figure 2). Diploid species exist with haploid numbers from seven to 10 chromosomes (7–10 pairs), and extinct species are

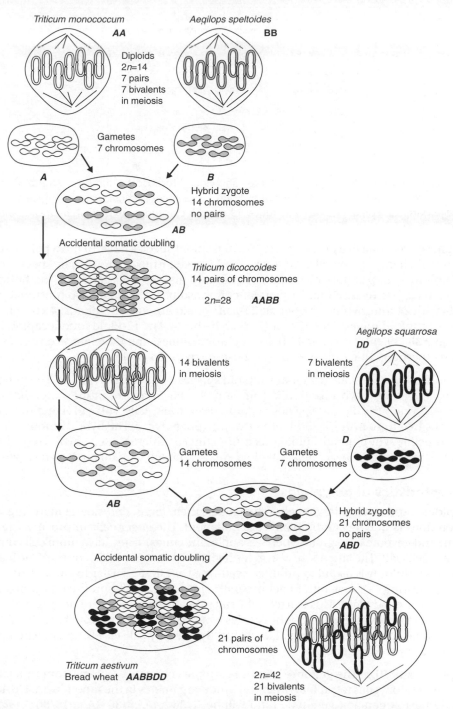

Figure 1.   The evolution of hexaploid bread wheat *T. aestivum* by repeated hybridization and polyploidization. (Note: some authorities use different *Aegilops* species names.)

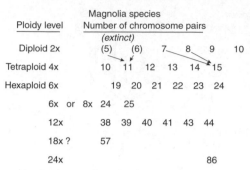

Figure 2.   The magnolia polyploid complex showing chromosome numbers. Numbers in brackets are for species that have not been found, presumed extinct. Arrows suggest two possible parental combinations to produce 11 and 15 chromosome pairs in hybrid offspring.

thought to have had five and six pairs. The tetraploid species have 10 to 15 pairs of chromosomes. There are several ways these could have formed. The 12-pair species could combine two six-pair genomes or a five-pair and a seven-pair genome. The hexaploid species have between 19 and 24 pairs of chromosomes. There are also presumed octaploids with 24 and 25 pairs. Note that 24 pairs could arise as tetraploids: $4 \times 6 = 24$, or by different triploid routes: $3 \times 8 = 24$, or $(7 + 8 = 15) + 9 = 24$. The 12-ploid (duodecaploid) species have 38, 39, 40, 41, 43, or 44 pairs of chromosomes, and there are two species with 57 pairs (perhaps 18-ploid) and 86 chromosome pairs (24-ploid).

Polyploid complexes start with many diploid species and a few polyploids. The polyploid species spread and hybridization is common. As the complex matures after 500000 to 10000000 years the diploid species become rarer, eventually only surviving in isolated locations free from hybridization, or as extreme phenotypes. The hybrid complex is then said to decline. The original diploids become extinct or differentiated beyond recognition, and eventually only a few highly polyploid species remain with no obvious relatives.

Characteristics of polyploids

Polyploids tend to have larger cells. This makes them larger and slower maturing. Most garden flowers and most crop plants are polyploid. The larger cells improve the texture of fruit and vegetables (e.g. strawberries, apples, potatoes). They have more cell contents and less cell wall. The larger size also gives higher yields (e.g. wheat, cotton). The sterility caused by triploidy is useful to produce seed-free fruit that is easier to eat (e.g. bananas) or better tasting (e.g. less bitter cucumbers). Triploid F1 can be produced by crossing a tetraploid and a diploid. This is useful for sugar beet where seeds are not wanted, but is impossible for inbreeding species or species where the seed is the crop (e.g. wheat). Triploid bananas are propagated vegetatively but the sterility causes a problem when trying to breed improved varieties.

Polyploidy **stabilizes the genome** because extreme genotypes segregate out much more rarely. A heterozygous tetraploid *AAaa* will produce gametes in the ratio $1/6$ *AA*:$4/6$ *Aa*:$1/6$ *aa*. The next generation genotype ratio will be: $1/36$ *AAAA*:$8/36$ *AAAa*:$18/36$ *AAaa*:$8/36$ *Aaaa*:$1/36$ *aaaa*.

Only $1/18$ of the progeny are extreme genotype, compared with $1/2$ for a diploid monohybrid cross. If a new hybrid polyploid happens to arise in an environment to which its genotype is well adapted its progeny will tend to maintain the adaptation. Polyploids, however, may be slow to evolve because they cannot respond rapidly to selection.

D9 Phylogeography, molecular clocks, and phylogenies

Key Notes

Evolution by divergence

Evolution occurs by changes in allele frequencies over time. Species form when populations diverge genetically to the extent that they cannot interbreed. They then continue to evolve and diverge independently. Degree of divergence for unselected characters is a measure of time since separation. Shared characteristics which evolved in a common ancestor are said to be homologous (e.g. mammals' back legs). Structures that have evolved similar functions separately are analogous (e.g. bats' wings and birds' wings). Evolving to be more similar from different starting points is called convergent evolution (e.g. whales and fish). Related species share homologous structures from their common ancestors, but do not share characters arising after separation.

Populations

Differences between populations within a species can show patterns of migration and colonization, and indicate degrees of divergence preceding speciation. Divergence may be selected. Dark skin in humans from equatorial regions may reduce damage from sunlight, and pale skin aids photosynthesis of vitamin D by sunlight in northern lands. There are large differences in blood group frequencies in different human populations, which may be due to selection by disease or to a drift and founder effect in small colonizing tribes.

Ring species

A ring species has an extended continuous range around an obstacle, but the populations at the extreme ends of the range are sufficiently diverged to be different species where they meet. An example is a boreal gull whose global range overlaps in Europe as the herring gull *Laurus argentatus* and the lesser black-backed gull *L. fuscus*. This shows that distance alone is an effective isolating mechanism.

Molecular clocks

The rate at which amino acid changing mutations accumulate in genes for specific proteins tends to be constant over time. Similarly, changes in ribosomal RNA sequence, synonymous changes in coding sequences (silent mutations), and changes in noncoding sequences all have particular rates. These can all be used as molecular clocks, faster diverging sequences for more recent events.

	The degree of divergence between two species reflects the duration in time of their independent evolution. There is debate about whether the clock rate is faster at times of rapid evolution (adaptive radiations) and slower during stasis, and also about the effects of generation time and number of germline cell divisions.
Phylogenetics	Arranging species in order of increasing divergence gives a phylogenetic tree which represents evolutionary history. Species within a group are compared with a distantly related outgroup to provide an ancestral root to the tree. Animal mitochondrial DNA evolves rapidly and does not recombine, so is excellent for relatively recent divergences. Conserved proteins (e.g. cytochrome *c*) and ribosomal RNA are useful for studying the whole period of life on earth. Mitochondria and Y chromosomes do not recombine, and, in mammals, follow the maternal and paternal lines respectively. Each could be traced back to a single ancestor, yet many different versions exist in the population. Multiple alleles of nuclear genes exist and recombine, so the phylogeny of a species is the sum of the phylogenies of the DNA within it.
Related topics	(D2) Evolution by natural selection (D5) Neo-Darwinian evolution: selection acting on alleles	(D7) Species and speciation (D10) Human evolution

Evolution by divergence

Evolution is a change in allele frequencies over time, with different changes in different populations causing them to become separate species. We can compare populations to discover how much variation is expected between members of one species (within species) and compare species to see how different they have become. The extent of the differences between populations or species indicates how long they have been diverging, which gives the length of time since they became separated. On the other hand, similarities in some characteristics between populations or species suggest that they share a common ancestor in whom those characteristics first evolved. The more recently two species diverged, the more they should have in common. Such characteristics in descendants are then said to be **homologous** (e.g. back legs of mammals). This is different from **analogous (analogy)**, where species evolve similar structures independently (bats' wings and birds' wings), a process called **convergent evolution**. There is still room for confusion. For example, all mammals, reptiles, and birds have homologous forelimbs descended from the front legs of an early amphibian ancestor, and particular fins on a fish before that. However, while bats' wings and birds' wings are homologous as forelimbs, they are analogous as wings, because they evolved as wings independently. The last common ancestor of bats and birds did not have wings.

Populations

Comparisons between the populations of a species show the degree of divergence within species before speciation. Similarities and differences can reveal migration patterns, changes in selection pressure across the species range, and chance divergence due to drift. The problem is to identify which cause produces which effect. Humans are selected for dark skin in equatorial climates (presumably to avoid sunburn), and for pale skin in less equatorial regions, supposedly so that sunlight can penetrate the skin and synthesize vitamin D, thus avoiding rickets. Some support for this comes from observing that although all native Americans entered the continent from North East Asia and Alaska, they have distinctly darker skin color in the dry sunny regions (e.g. California/Nevada, and the Andes). Another human example is sickle cell anemia, selected for by the tolerance of heterozygotes for malaria. Both malaria and the allele for sickle cell hemoglobin are most frequent in equatorial West Africa.

Blood group distributions are more difficult to explain. The I^B allele (of the ABO blood groups) shows a very uneven global distribution (Figure 1). I^B did not exist in native Americans apparently having failed to enter America or Australia from Asia. This may have been due to founder effect, the relatively small band of colonizers losing I^B by chance drift. Alternatively these populations may have become isolated from Asia when the I^B allele was very rare. The patchy distribution of I^B in the rest of the world may be due to drift or to selection by disease. The A and B antigens are saccharide groups on the surface of red blood cells and many bacteria and some viruses also have these. A large range of diseases appear to have a more severe effect on individuals with particular blood groups. For example, I^A confers susceptibility to smallpox in unvaccinated populations, and may have been selected against. This initial selection or drift may have been followed by a population explosion in some tribes, and migration or invasion, sometimes on a large scale, may have carried the alleles across continents. The Mongol hordes who invaded Eastern Europe and the Middle East in the twelfth and thirteenth centuries have been proposed to explain the frequency of I^B in these areas, but cannot explain the high frequency in West Africa. Such hypotheses are easy to produce but almost impossible to test.

Ring species

When a species covers a sufficient range, the populations at the extremes may diverge sufficiently to be different species. This is apparent when the range circles an uninhabitable

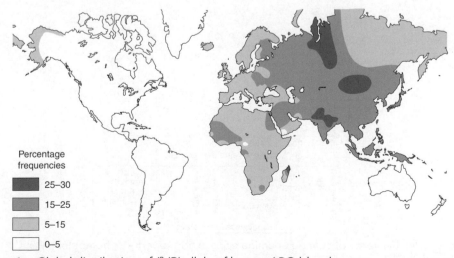

Percentage
frequencies

■ 25–30

■ 15–25

□ 5–15

□ 0–5

Figure 1.   Global distribution of I^B (B) allele of human ABO blood group.

obstacle, and the ends overlap forming a ring species. An example is a boreal (northern) gull which surrounds the Arctic. In Europe the ends of the range overlap as the herring gull *Laurus argentatus* and the lesser black-backed gull *L. fuscus*. Thus distance alone restricts gene flow across large population ranges sufficiently for a widespread species to be many incipient species if the center of the range is removed. The tendency of many bird species to nest and mate near their own birthplace may restrict gene flow despite large seasonal migrations.

Molecular clocks

Mutations occur at a relatively constant rate determined mainly by mistakes during DNA replication. A proportion of these are not deleterious (they may be neutral), and survive, increasing in frequency by drift or selection. This proportion depends upon how tightly the particular protein (or RNA) product is conserved by selection (Section D2). As the differences between populations or species accumulate, they give a guide to the degree of divergence between them. Some proteins (e.g. fibrinopeptide) change relatively rapidly and are useful for studying closely related species, while others change slowly (e.g. cytochrome *c* and ribosomal RNA) and can be used to follow divergence from the earliest living organisms (Figure 2). Because the rate of change in any particular sequence is approximately constant it can function as a **molecular clock**.

Clocks must be calibrated, and this relies on fossil evidence to say when the taxonomic lines being compared became separated. In hemoglobin the rate of amino acid change is about 1.2 changes per amino acid site per 10^9 years (or 1.2×10^{-9} site^{-1} year^{-1}). The value for fibrinopeptide is higher at 8.3×10^{-9} (less conserved) and for histone H4 is only 0.01×10^{-9} (highly conserved). Changes in protein sequence seem to occur at the same rate in mice and whales, per year rather than per generation. Rates of change in noncoding regions and synonymous changes in coding sites of DNA are higher, and appear to be (per generation) faster in mice than in whales. This suggests a considerable degree of selection on proteins to change at a constant time-dependent rate as the environment changes. If correct, it supports the hypothesis that selection, rather than drift of neutral mutations, is the cause of most amino acid changes. There is some doubt about the constancy of

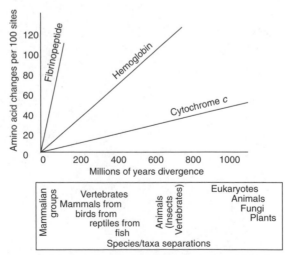

Figure 2.   Relative rates of change of amino acids in fibrinopeptide, hemoglobin, and cytochrome *c*.

molecular clocks. They may run faster at times of rapid evolution such as the adaptive radiation of mammals following the extinction of the dinosaurs and many other animals at the end of the Cretaceous period. They may also slow during periods of stasis. The molecular clock suggests that the main groups of mammals diverged substantially earlier than 65 million years ago, but the first fossils of these groups are unequivocally younger than this. There are more mutations per generation in species with long generation times, and males have higher mutation rates than females. There are more cell divisions in spermatogenesis than in oogenesis, so mutation rate per year and the molecular clock rate may depend on the number of mitotic divisions in the germline cells.

Phylogenetics

A phylogeny arranges species to reflect relatedness by descent from a common ancestor. Closely related species descended from a recent common ancestor are grouped together. The evolutionary distance between them is represented graphically. Phylogenetics is a statistical science in itself. The product is a tree-, bush-, or star-shaped diagram (Figure 3). One problem is to find the original common ancestral state, or root, which is likely to be extinct, but can be inferred. This can be done by using an **outgroup** species which is equally distantly related by time and descent to them all: for example, by using a fungus or plant to root the cytochrome *c* tree for animals (Figure 3). Faster mutating sequences (e.g. mitochondrial DNA) are used to construct phylogenies of more closely related species. Fast-mutating sequences cannot be used to go back far in evolutionary time because the mutations effectively randomize the sequence. Reverse mutations restore the original sequence at some sites, and comparisons become error prone, or impossible.

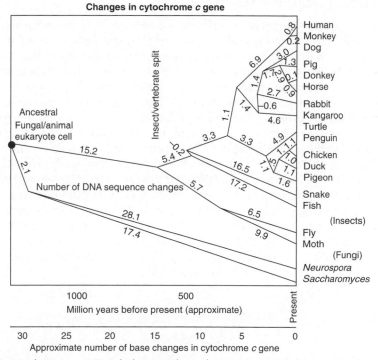

Figure 3.   An evolutionary tree (phylogeny) based on amino acid changes in the cytochrome c gene. Numbers show average number of amino acid changes along each branch. Fungi are used to root the tree.

Animal mitochondrial DNA (mDNA) evolves much faster than nuclear DNA, presumably because it is a small genome under less selection for accuracy, so DNA synthesis is more error prone. This makes it useful for examining relatively recent events. It also has the advantage of only being inherited from females (uniparentally) so different lineages do not recombine, and each lineage carries its own history. Analysis of human mDNA suggests that it all comes from a single ancestral sequence (one woman, hence 'Eve') who lived around 140 000–290 000 years ago (Section D10). Note that it is a mathematical certainty that any particular mutation or unrecombined segment of DNA can be traced back to a single origin if we can follow it far enough, so 'Eve' is not controversial. All the other lineages became extinct, probably by chance rather than by selection. There is some controversy over Eve's place of residence. The most favored suggestion is that she lived in Africa. The two pieces of evidence for this are: (i) most branches of the phylogeny are represented there; and (ii) the most divergent types of mDNA are found there. This suggests that parallel female lineages have been living in Africa longer than anywhere else. Most Europeans are descended from seven founder females, with little divergence. The data do not clearly separate individuals from Europe, Africa, and Asia and is probably not adequate to construct the tree absolutely accurately. There is too much divergence for this one character to be conclusive on this timescale.

Cytochrome *c* is a conserved protein that can be used to compare all eukaryotes (Figure 3). The cytochrome *c*-based phylogeny confirms our view that mammals are one related group and birds are another, as was supposed from comparative morphology and common sense. The whole evolutionary tree of animal evolution is very well supported. Perhaps the most surprising observation is that the two most divergent species by far are two species of fungus, both ascomycetes, that separated twice as long ago as the ancestors of mammals separated from the ancestors of insects.

Some **problems with phylogenies** can be seen in Figure 3. If mutation rates were the same in all lineages then each would have diverged by the same amount and the numbers on the branches for the cytochrome *c* tree would all add up to the same value. It appears that *Neurospora* has a mutation rate 50% faster than that of *Saccharomyces*. All sequence changes, including these amino acid changes, must happen as individual discrete events, but averaging algorithms (numerical techniques) used to construct the tree produced averaged fractional changes. For example there is one amino acid difference in cytochrome *c* between humans and rhesus monkey, and it occurred in the human lineage, not the monkey, which still has the sequence of our common ancestor. This is shown as 0.8 human, 0.2 monkey in the tree. The algorithms typically give a dichotomous branching pattern. This is correct for mDNA, but not for nuclear genes which can recombine, and it is misleading for species because they exist as populations and can split simultaneously into many isolated species. There is an important distinction between the phylogeny of an individual sequence, which can be traced to its origin by mutation in a particular individual, and the phylogeny of a species which includes all the sequences existing in the population, some of very great age. This is compounded by recombination in nuclear genes, which shuffles and obscures lineages.

D10 Human evolution

Key notes

Overview of human evolution

Our branch of the primate evolutionary tree separated from other groups about six million years ago. The fossil record shows an African origin. A number of prehuman hominid species existed, including *Australopithecus* and later several *Homo* species. *Homo sapiens* has probably only existed for 190 000 years. Early studies were completely dependent upon analysis of fossils but recently DNA sequence analysis has become a vitally important tool.

DNA sequence analysis and human evolution

In theory it should be possible to make reliable inferences about the spread of human populations throughout the world by analyzing DNA polymorphisms in current native populations. Since new mutations arise at specific times and are carried only by descendants of the individual in whom they arose, it is possible to develop phylogenetic trees that show how groups of individuals are related. Meiosis causes problems with this approach as it causes polymorphisms to be shuffled by recombination. Mitochondrial and Y chromosome DNA sequences are haploid and do not undergo meiosis and are used to trace human evolution. DNA polymorphisms are detected by PCR in human and recent fossil material.

Human migration throughout the world

Homo erectus and *Homo sapiens* are the only hominid species to have migrated out of Africa. Analysis of DNA polymorphisms shows that African populations are older than any other human populations. The same approach can show the patterns of ancient migrations to colonize different parts of the world such as the Pacific islands. In addition, these methods can be used to show that all humans have a common male and female ancestor, the Y chromosome Adam and the mitochondrial Eve, and give approximate dates for when they existed. Nearly all genetic variation in humans is found in all racial groups. Humans have limited variation in mitochondrial DNA suggesting that a population bottleneck event took place about 90 000 years ago.

Alternative theories of recent human evolution

DNA evidence is largely consistent with the theory that *H. sapiens* arose in Africa and subsequently spread to the other continents, replacing the pre-existing *H. erectus* species. This is known as the **Replacement Theory**. An alternative theory, the **Multiregional Hypothesis** suggests that *H. sapiens* arose from *H. erectus* in separate locations of Africa, Asia, and Europe. Some evidence from early Asian fossils suggests an intermediate state between *H. erectus* and *H. sapiens*, or at least some interbreeding between the two groups. From studies of mitochondrial DNA there is evidence that this did not happen between Neanderthals and *H. sapiens*. The multiregional theory would put the date of common ancestors for modern humans back to about 1.8 million years ago. The two hypotheses have been used to support different views of racial differences.

Related topics	(B3) Eukaryote genomes	(C7) Genes in eukaryotic
	(C3) Meiosis and	organelles
	gametogenesis	(D9) Phylogeography, molecular
		clock, and phylogenies

Overview of human evolution

The only human species currently alive is the species to which we belong, *Homo sapiens*. This is a recent branch of the order Primates that diverged from a precursor species along with our nearest relatives, the chimpanzees, about six million years ago. The fossil record is unambiguous in that all the earliest hominid fossils have been found in Africa, indicating that our earliest ancestors were African or East African in origin.

Fossil evidence points to the existence of a number of early hominid species of a genus known as *Austropithecus* that lived up to 1.5 million years ago. Overlapping in the later part of this period were a number of species of the genus *Homo*. The earliest of these was *Homo hablis*, followed by *Homo erectus*, which persisted in various forms until very recently. The earliest *H. sapiens* fossils date to about 190 000 years ago, and considerable controversy has surrounded their classification, but *H. sapiens* and *H. erectus* clearly coexisted for a long period. The evolution of *H. sapiens* as a chronospecies (Section D7) from *H. erectus* was a slow process, taking around a million years from the first *H. erectus*, during which time African hominids went through intermediate stages. At the start of this period *H. erectus* occupied Africa and Asia, and was more diverse than modern humans. By the end, one African population had become *H. sapiens*. Initially, this subject could only be investigated by comparative anatomical studies of fossil skeletons or, more often, small fragments of skeletons, unearthed in excavations. This leads to many problems, as often a species is known by only a few or even a single artifact. In many cases it is not possible to ascertain with confidence whether two fossils represent separate species or simply demonstrate variation within a single species. The fact that fossils are rare in many parts of the world means that it is difficult to draw definitive conclusions about how *H. sapiens* became distributed throughout the continents. These uncertainties mean that the scientific data can be used to support conflicting theories about the origin of modern humans.

Analysis of DNA in current human populations and in some more recent fossil material has allowed the development of novel approaches to the problem that are not dependent on the serendipitous discovery of human fossils. DNA sequence polymorphisms in existing human populations and in DNA extracted from fossil bones hold out the possibility of producing definitive data on the degree of genetic relatedness between groups of the present human population and our relatedness to earlier human species.

DNA sequence analysis and human evolution

In essence, the use of differences in DNA sequence (polymorphisms; Sections B3 and D4) allows a very simple method to investigate how our species has evolved over time and how it has become distributed across the world. Since mutations are passed on from parent to child and new mutations occur at specific points in time, it should be possible by analyzing DNA from large numbers of people to produce a phylogenetic tree showing degrees of relatedness between groups of humans. A simple example of this type of analysis is shown in Figure 1. Members of an early population migrate from Island 1 to Island 2. On Island 2 a mutation (A) occurred at a specific locus and eventually was fixed for the whole population. In a subgroup of this population that colonized Island 3 a new mutation (B) arose. In time it also becomes fixed in the population of this island.

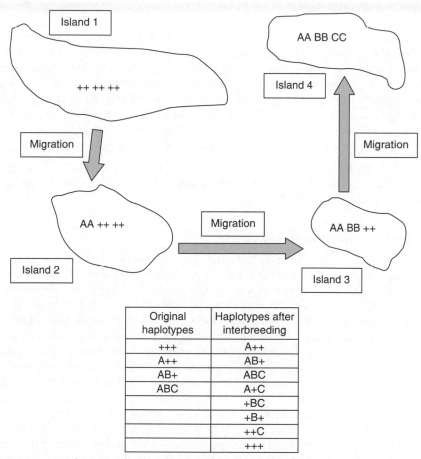

Original haplotypes	Haplotypes after interbreeding
+++	A++
A++	AB+
AB+	ABC
ABC	A+C
	+BC
	+B+
	++C
	+++

Figure 1.   As a population progressively migrates from island to island genetic differences have time to build up and become fixed in each population. Thus an individual can be identified as having a genetic origin to one island even if subsequently population mixing has taken place. Meiosis will have no effect on the genotypes if there is no interbreeding between individuals from different islands. However if interbreeding takes place between individuals from different islands novel haplotypes will arise. This will mask the original pattern.

If the process was repeated and another migration took place then a third new mutation (C) might become fixed in the final population. In theory it would be possible, even if considerable mixing of the populations subsequently took place, to identify the location of an individual's ancestors from analysis of their DNA.

However, because of the effect of meiosis in recombining alleles (Section C3), if inter-breeding took place between individuals from different islands, novel combinations of the three pairs of alleles would be found in the subsequent population and analysis of origins or patterns of migration would become very complex. In addition, due to the dip-loid nature of our genotype, when studying a number of polymorphisms it is not possible to determine which haplotype is inherited from each parent.

To utilize DNA analysis effectively in human evolution studies we require DNA that is present in a haploid state and does not undergo meiosis. Fortunately there are two exam-ples of DNA in human cells that meet these criteria. Mitochondria contain DNA which is haploid and does not undergo recombination (Section C7). It is passed on from mothers to all their children of either gender without any alteration except for rare mutational events. The rate of mutation is relatively high compared to chromosomal DNA but low enough to be able to establish patterns of relationship within human populations. The other example is the Y chromosome. This is passed on from fathers to their sons, and the Y chromosome, except for the small pseudoautosomal region, does not take part in meiosis as it has no homolog with which to pair. Although some research is being carried out with chromosomal DNA sequences, mitochondrial and Y chromosome DNA occupy the major focus in genetic studies of the evolution of *H. sapiens*.

Normally DNA is analyzed using PCR (Section E2), and in the case of fossil remains this is always the case as the DNA of fossil bone is highly degraded into short fragments that would be unsuited to any other technique.

Human migration throughout world

There is no indication that the early hominid species expanded beyond Africa. However over 200 fossil remains of *H. erectus* have been studied and there is clear evidence that they spread from Africa to Asia and Europe. *H. erectus* first appeared about 1.8 million years ago and their descendants existed up to approximately 30 000 years ago. *Homo sapiens* is of much more recent origin, with the oldest fossils being dated to 190 000 years ago. This species also clearly migrated from eastern Africa to the Middle East and onward to all of the continents. The DNA evidence supports and extends the fossil record in a number of ways. Firstly, it is clear that the variation in mitochondrial DNA sequences is greatest in native African populations. This would be expected if our species originated in Africa. If the mutation rate is constant then more mutations will appear in older pop-ulations than in recently established populations. Secondly, by looking at the patterns of mitochondrial mutations in living populations it has been possible to infer many of the routes that early humans used in moving out of Africa and colonizing the rest of the globe. Examples of this include the populating of the Pacific from South East Asia, and the migration from Asia into the Americas. Y chromosome studies have progressed more slowly due to early difficulties in obtaining polymorphic markers, but are now confirming the mitochondrial data.

In addition to charting the migration of modern humans out of Africa, studies of DNA polymorphisms have allowed us to look for common ancestors for all humans alive today. From the comparison of mitochondrial DNA polymorphisms there is good evi-dence that all humans can be traced back to a single female ancestor. Calculations based on the rate of mutation of mitochondrial DNA suggest that this individual, known as the **Mitochondrial Eve**, lived approximately 150 000 years ago. Similar logic suggests that the current European population is descended from only seven different women. Studies using the Y chromosome identify a putative **Y chromosome Adam** who would have lived some 100 000 years ago. It must be stressed that the mitochondrial Eve and Y chromosome Adam were not the originators of our species but simply the most recent common female and male relative for individuals alive today.

Studies of chromosomal DNA sequences have supported the idea that we are a closely related species. Almost 90% of the variation in chromosomal alleles found in humans is common to all racial groups. This indicates that modern racial groups have arisen only very recently in the evolution of humans.

DNA analysis can also provide evidence of events that would be impossible to interpret from paleontological data. Humans show much less variation in their mitochondrial DNA sequences than other members of the Primate order. It is thought that this is evidence of a population **bottleneck** event (Section D4) when the human population was reduced to very low numbers about 90 000 years ago.

Alternative theories of recent human evolution

As described above there is considerable evidence from DNA sequence data that *H. sapiens* evolved in Africa and progressively spread out from that area to colonize the whole of the world. In doing so they replaced the populations of *H. erectus* from Africa, Asia, and Europe, although it is clear that both species coexisted at specific sites for long periods. This theory is known as the **Replacement Hypothesis**, or the 'Out of Africa' hypothesis. There are, however, other theories of the emergence of modern humans, and one with considerable support, particularly from fossil evidence is the **Multiregional Hypothesis**. This also suggests that human life originated in Africa with the emergence of *H. erectus*, but differs in how this species evolved into *H. sapiens*.

The fossil record clearly shows that *H. erectus* migrated into Asia and Europe. In eastern Asia fossils of Peking Man and Java Man represent descendants of the early *H. erectus* migrants. The recently discovered 'Hobbit' fossils from Indonesia may also be in this group. If so, *H. erectus* would have persisted until 18 000 years ago. However, there is much controversy regarding the origin of these fossils.

The Multiregional Hypothesis argues that *H. sapiens* arose separately from *H. erectus* in different locations across the world, and that the similarity of all modern humans is due to interbreeding (gene flow) between these groups. If this were the case then the common ancestors of modern humans would have originated approximately 1.8 million years ago and not 150 000 years ago as in the Replacement Hypothesis. Support for the Multiregional Hypothesis comes mainly from fossil remains. In some cases in Asia fossils that appear to be intermediate between *H. erectus* and *H. sapiens* have been discovered, suggesting a progressive change in morphology. Alternatively, these may represent interbreeding between the two species.

One group of fossils that has been intensively studied is the Neanderthals. They existed in southern Europe and eastern Asia from 200 000 to 30 000 years ago, and are thought to be more closely related to humans than any other hominid species. Comparison of mitochondrial DNA extracted from Neanderthal fossils with data for the current human population shows that they differ by about 27 base substitutions. This suggests that Neanderthal mitochondria are not present in the human population, that Neanderthals were an evolutionary dead end and did not contribute genes to modern humans through a continuous female line. Additionally, extensive Y chromosome analysis of Chinese populations shows no evidence of non-African genes, arguing against the Multiregional Hypothesis.

However, recent sequencing of Neanderthal chromosomal DNA suggests that some Neanderthal DNA sequences may be found in modern humans. Current best estimates of Neanderthal DNA in humans are between 0 or 0.5 and 4%. At present this evidence is based on few samples and there are problems caused by contamination and decomposition of the original DNA making it difficult to interpret the sequences. (Section E3).

The two hypotheses can be extrapolated into social perspectives. The Replacement Hypothesis fits well with social attitudes that relate to racial equality. The Multiregional Theory has been used to support attitudes that promote difference between racial groups, because it puts the splitting of humans into distinct regional (racial) groups much further back in time, hence making them less closely related.

E1 Using sequence specificity to study nucleic acids

Key Notes

Sequence-specific reactions in nucleic acids

Central to nearly all modern techniques to study nucleic acids is the fact that specific base sequences can be identified in reannealing experiments.

DNA reannealing

This is a process in which double-stranded DNA in solution is split into its two separate strands and then allowed to reform its double-stranded nature. Splitting the strands is usually carried out by raising the temperature. Because G:C-rich DNA is more stable than A:T-rich DNA it is more difficult to separate. The temperature at which a DNA becomes 50% single stranded is known as the melting temperature (Tm). If the temperature is lowered the strands will come together and complementary sequences will reanneal. At low levels of stringency much of the double-stranded product will have mismatches, but higher stringency conditions will remove these from the reaction.

Southern blotting

This technique detects specific DNA sequences and is used to analyze gene structure. DNA is digested with a restriction enzyme and the fragments are separated by size on an agarose gel. The gel is blotted by capillary action and the DNA fragments are transferred to a membrane where it is hybridized with a radiolabeled probe. The membrane is washed to remove unbound probe. Exposure of the washed membranc to X-ray film produces bands corresponding to hybridizing DNA fragments whose lengths can be estimated from their position on the membrane.

In situ hybridization

This technique detects expression of mRNAs in intact cells and can be used to identify individual cells expressing a gene within a tissue containing different cell types. Labeled probe is added to thin tissue slices on microscope slides. The probe enters the cell cytoplasm and hybridizes to mRNA. Normally the probe is labeled so that it can be detected by fluorescence, but other methods can be used. The technique can be adapted to identify sequences in chromosomal DNA.

DNA chip technology

DNA chips are arrays of single-stranded DNA organized in spots on small glass or silicon wafers. Very large numbers of different DNA sequences can be included in an array. Fluorescently labeled target DNA is hybridized to the DNAs on the chips. Of the DNA sequences on the chip, only those that are complementary to target sequences can hybridize. These are detected by their fluorescence. The technique can be adapted to use with cDNAs as the labeled target, allowing gene expression patterns to be studied. Chip technology has had a major effect because of the large number of analyses that can be carried out in a short period of time.

Comparative genome hybridization   Comparative genome hybridization uses DNA chips that are organized to represent chromosomes. Competitive hybridization is used to identify deletion or amplification of chromosomal areas in cancer tissue.

Related topics
(A1) DNA structure	(F2) Genetic screening
(A7) Regulation of gene expression	(F3) Genes and cancer
(F1) Genetic diseases	(G1) Genetics in forensic science
	(G2) Biotechnology

Sequence-specific reactions in nucleic acids

The fact that DNA and RNA are both polymers of specific sequences of nucleotides has proved most useful in developing techniques to analyze these molecules, understand their cellular functions and also to manipulate them in experimental situations. The key to these processes is the recognition of specific base sequences by complimentary nucleic acids in annealing reactions or by proteins.

Annealing of complimentary nucleic acid strands is the basis of the blotting and chip technologies described below and the annealing of short single-stranded DNA sequences, primers, to specific regions of DNA or RNA molecules that subsequently act to prime various polymerases (Section E2) is the basis of PCR and reverse transcription.

Experimental analysis or manipulation of RNA molecules is difficult as the single-stranded molecules are easily degraded in experimental conditions. For this reason RNA is usually now converted to DNA through reverse transcription (Section E2).

Many classes of proteins will bind to specific DNA sequences *in vivo* but the most important group for analysis and manipulation of DNA are restriction endonucleases, usually referred to as restriction enzymes. These bacterially derived enzymes have the property to bind specific palindromic DNA sequences and then cut the double-stranded molecule within the sequence that is bound. They can be used to cut DNA into appropriate-sized fragments to analyse by gel electrophoresis or use in cloning DNA. Their role in the latter is described in detail in Section E4.

DNA reannealing

This was one of the first techniques used to analyze DNA sequences and currently underpins almost all molecular DNA technologies. Essentially the process consists of splitting a DNA double helix in solution, into its two separate strands (sometimes called denaturing) and subsequently allowing the strands to come together to reform the helix (reannealing), a process often referred to as DNA hybridization.

DNA helices are held together by hydrogen bonds (Section A1) between complementary base pairs. The simplest way to destroy these bonds is to raise the temperature of the solution, allowing the two strands to separate. The temperature at which 50% of the DNA is single-stranded is known as the melting temperature (Tm). G:C base pairs share three hydrogen bonds and are more stable than A:T pairs that share only two. For this reason the base composition of any DNA will determine its Tm, with G:C-rich DNA having higher Tm values. Other factors, such as the salt concentration of the buffer solution and the presence of helix-destabilizing compounds such as formamide, also affect Tm values. High pH can also be used to cause DNA strands to separate.

When DNA strands are separated, lowering the temperature allows reannealing. Rapid cooling does not permit the strands time to reanneal resulting in a solution of single-stranded DNA, but if the temperature is too close to the Tm reannealing occurs very slowly. A compromise often used is to allow reannealing to take place at 25°C below the Tm.

Reannealing occurs through random collisions between single-stranded DNA molecules in solution. Strands with complementary base sequences will form energetically stable double-stranded structures after collision. However, the process is more complex than it may appear. Because similar base sequences will occur several times throughout a genome it is also possible to get partially double-stranded molecules being produced. Partially double-stranded structures (mismatches) can be avoided by increasing the stringency of the reannealing conditions. This usually means increasing the temperature of the reaction.

When appropriate stringency is applied in reannealing, all regions of the genome should reanneal at the same rate; however, as noted in Section B3, satellite DNA is made up of sequences that are repeated up to a million times in a genome, so the strands are much more likely to meet by diffusion than single-copy sequences. This class of DNA sequence was first detected in reannealing experiments because it reannealed well before the remainder of the genome.

Southern blotting

This technique, named after its inventor Ed Southern, is used to detect the presence or alteration of DNA sequences in an organism. The key event is the reannealing/hybridization of labeled copies of the DNA of interest, the probe DNA, to DNA isolated from an organism, the target DNA. In Southern blots the probe is in solution and the target DNA immobilized on a filter (Figure 1).

DNA for analysis is purified from cells and is obtained as large fragments of chromosomal DNA, typically 20 000 bp or more in length. The first step involves digestion of the DNA with restriction enzymes (Section E4). This produces thousands of DNA fragments ranging in size from just a few to several thousand bases. The DNA fragments are then subjected to electrophoresis in an agarose gel. This separates them according to size with the large fragments near the top of the gel and the small fragments near the bottom. After electrophoresis, the gel is soaked in an alkaline solution where the high pH splits the DNA into single strands. The next stage involves transferring these fragments from the gel to a membrane made of nylon (or sometimes nitrocellulose) where they become accessible for analysis using a probe. Transfer is achieved by capillary action, hence the term blotting. This process is now often carried out by electroblotting, which is quicker. A replica of the pattern of fragments in the gel forms on the membrane. The membrane is treated either by baking it at 80°C or by exposing it to ultraviolet radiation to attach the single-stranded DNA fragments firmly.

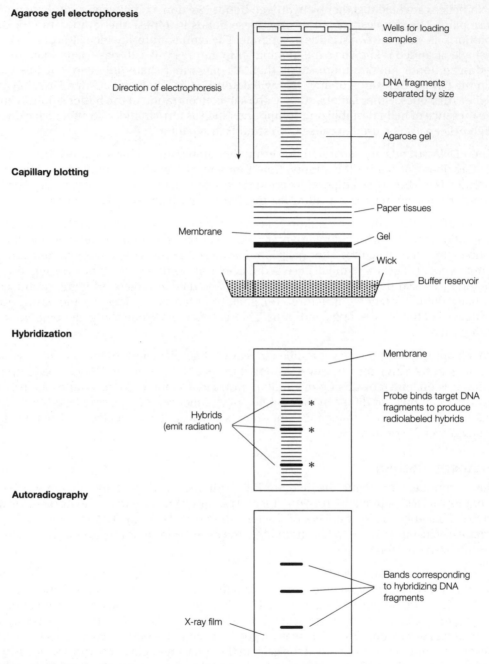

Agarose gel electrophoresis

Wells for loading samples

Direction of electrophoresis

DNA fragments separated by size

Agarose gel

Capillary blotting

Paper tissues

Membrane

Gel

Wick

Buffer reservoir

Hybridization

Membrane

Hybrids (emit radiation)

Probe binds target DNA fragments to produce radiolabeled hybrids

Autoradiography

Bands corresponding to hybridizing DNA fragments

X-ray film

Figure 1. Southern blotting.

The next stage involves incubating the membrane with a labeled single-stranded probe. This is known as hybridization and is carried out at a temperature and a salt concentration that favors the formation of hybrids between the probe, which is in solution, and fragments of DNA bound to the membrane whose sequences are complementary to the probe. The membrane is then washed with a buffer to remove probe that is bound nonspecifically so that only labeled probe bound to target sequences remains. Washing conditions can also be used to increase stringency. If the probe is labeled with radioactivity, the membrane is placed in the dark against a sheet of X-ray film. Development of the film reveals one or more dark bands which correspond to the position of fragments hybridized to the probe. The length of the fragments can be calculated from their position relative to marker DNA molecules of known length. In this way the structure of individual genes can be characterized in terms of the number of hybridizing fragments and their sizes. One typical example of the use of Southern blots is to show alterations in oncogenes in cancer cells where the pattern of bands will be altered or the intensity will be increased if the gene has been rearranged or amplified (Section F3).

A similar technique known as northern blotting can be used to analyse RNA. This was the original technique used to analyze RNA. Its use has now been largely superseded by the RTPCR-based technologies (Section E2). In northern blotting the RNA molecules are electrophoresed in agarose gels to separate them by size. As was the case for Southern blotting, the gel is blotted and the membrane is hybridized with a radioactive probe specific for the RNA being investigated. The membrane is then washed and exposed to X-ray film. Development of the film reveals usually a single band corresponding to the RNA recognized by the probe. The position of the band can be used to estimate the length of the mRNA and the intensity or darkness of the band is a rough measure of how much of the mRNA was present in the original cells. However, accurate quantification requires real-time PCR (Section E2). The current major use of northern blots is to confirm the size of an RNA molecule.

In situ hybridization

This technique is different from other hybridization methods in that it is used to detect nucleic acid sequences present in intact cells. The main use of *in situ* hybridization is to identify expression of specific mRNAs in individual cells present in tissues containing a number of different cell types. For example, *in situ* hybridization can be used to show that insulin mRNA is produced only by the β cells of the pancreas. Thin slices of tissue, known as sections, are placed on a microscope slide. Probe is added to the cells on the microscope slide and is taken up into the cells where it hybridizes to target mRNA forming hybrids. Probes may be labeled in a number of ways. Usually they are labeled so that they can be detected by fluorescence: the technique is then known as fluorescence *in situ* hybridization (FISH). Alternatively they can be labeled with enzymes that catalyze the formation of colored precipitates. The former allows better definition of the cytology but precipitate methods do not require specialized microscopes for analysis.

The method is also applicable to detecting DNA sequences in nuclei and chromosomes.

DNA chip technology

DNA chip (microarray) technology employs essentially the same basic procedures as Southern blotting. However, it allows large numbers of genes to be analyzed in very short periods of time. In this case the single-stranded probes are attached in an organized matrix on a small glass or silicon wafer. Several hundred thousand probes can be arrayed on one chip. The target DNA is labeled with a fluorescent marker after being isolated from tissue, made single-stranded and hybridized to the probes on the chip. Unhybridized DNA is washed away and the chip is read under a laser light. Only probe DNAs that have hybridized to the target DNA will fluoresce. The identity of each probe on the matrix is known and thus the target DNA can be checked for the presence of any desired DNA sequence.

	Sequences on the array	Mutation	A/A	A/B	B/B
1	ACCTGTCGATTG**A**TCTTAACGTTGA	None			
2	ACCTGTCGATTG**C**TCTTAACGTTGA	C			
3	ACCTGTCGATTG**G**TCTTAACGTTGA	G			
4	ACCTGTCGATTG**T**TCTTAACGTTGA	T			
5	ACCTGTCGATTG TCTTAACGTTGA	Deletion			

6	TGGACAGCTAAC**T**AGAATTGCAACT	None			
7	TGGACAGCTAAC**C**AGAATTGCAACT	C			
8	TGGACAGCTAAC**G**AGAATTGCAACT	G			
9	TGGACAGCTAAC**A**AGAATTGCAACT	A			
10	TGGACAGCTAAC AGAATTGCAACT	Deletion			

Figure 2. Detection of single nucleotide polymorphism by DNA microarray.

Allele A	5'ACCTGTCGATTG<u>A</u>TCTTAACGTTGA 3'
	3'TGGACAGCTAAC<u>T</u>AGAATTGCAACT 5'
Allele B	5'ACCTGTCGATTG<u>C</u>TCTTAACGTTGA 3'
	3'TGGACAGCTAAC<u>G</u>AGAATTGCAACT 5'

The sequences on the left-hand side of the figure show the 10 DNA sequences that would be used to identify SNPs arising in the central base of the sequence indicated in bold. The right-hand of the figure represents the array after hybridization. If the DNA array was screened with DNA containing allele A, two spots, 1 and 6, corresponding to the exact complementary sequence of the coding and non-coding strands, would form hybrids. For allele B the spots would be 2 and 8. For an A/B heterozygote all four spots would form hybrids.

Stringency can be controlled as in Southern blotting. Chips can be designed and made using different sets of DNA probes depending on what they are required to analyze. They can be used for many purposes including identifying DNA from infecting bacteria, parasites or viruses in samples taken from patients in clinical situations or to differentiate between different alleles of genes in population studies (Figure 2).

Gene expression patterns can also be studied using DNA chips. In this case the probes would be made up of a large number of different genes. mRNA is extracted from the tissue of interest, converted to cDNA and labeled with a fluorescent marker by RTPCR (Section E2). After hybridization only genes that were expressed in the tissue will fluoresce. Comparisons can be made between different tissues of normal cells and tumors. In this case the mRNA from a tumor is labeled with a red fluorochrome and mRNA from normal tissue with a green fluorochrome. When both are hybridized on to the chip they compete for the target sequences. Targets that are only expressed in the tumor fluoresce red, those only expressed in normal tissue fluoresce green and those expressed in both are orange. Sophisticated detectors can determine the ratio of red to green fluorescence for any target to determine the relative expression of the gene in both samples.

DNA chips have had a major impact on genetic research because of the high numbers of samples that they can analyze and the large number of probes that can be used simultaneously.

Comparative genome hybridization

The concept of using DNA from different sources, labeled with different fluorochromes to compete in hybridization reactions has proved very adaptable. Comparative genome hybridization uses a DNA chip that carries a large number of targets representing a complete genome arranged to show the linear organization of chromosomes. Differentially labeled target DNA from two sources, for example a tumor (red) and normal (green) tissue, are hybridized to the chip. Spots on the chip that fluoresce green indicate sequences that are deleted in the tumor and those that fluoresce red represent areas that have been amplified in the tumor. Most spots will fluoresce orange indicating no change. This approach allows candidate tumor genes to be identified (Section F3).

E2 PCR and related technology

Key Notes

Polymerase chain reaction

Polymerase chain reaction (PCR) allows specific DNA sequences to be copied or amplified over a million-fold in a simple enzyme reaction. Some of the sequence at each end of the region of DNA to be copied must be known to enable primers for the DNA polymerase to be synthesized. Target DNA is amplified by 20–40 cycles of DNA synthesis. Each cycle has three stages carried out at different temperatures: (i) denaturation – the reaction is heated to above 90°C to separate the strands of the double helix; (ii) annealing – the reaction is cooled to 40–60°C to allow the primers to bind to the single-stranded template DNA; (iii) extension – the reaction is heated to 72°C where the polymerase is most active and the target DNA sequence is copied. Each molecule of target DNA acts as a template for the synthesis of new DNA in the next cycle. This leads to a rapid increase in the amount of target DNA with successive cycles.

Reverse transcription

To make RNA more stable for further analysis it can be converted to complementary or copy DNA. This is carried out using the enzyme reverse transcriptase. The double-stranded DNA molecule produced can be used in PCR, sequencing or cloning procedures. RTPCR is used to detect specific RNA molecules in cell extracts.

Real-time PCR

Real-time PCR measures numbers of specific RNA molecules present in cells or tissues. RNA is converted to cDNA and this is amplified by PCR. The rate of increase in PCR product is monitored using fluorescently labeled probes that only fluoresce when bound to the amplified DNA. The rate at which the cDNA is amplified is directly proportional to the initial concentration of the RNA molecule in the reaction.

Related topics

(A1) DNA structure	(E3) DNA sequencing
(A4) DNA to protein	(F1) Genetic diseases
(A7) Regulation of gene expression	(F2) Genetic screening
	(F3) Genes and cancer
(B1) Concepts of genomics	(G1) Genetics in forensic science

Polymerase chain reaction

The polymerase chain reaction (PCR) is a powerful and widely used technique. It allows specific DNA sequences to be copied from genomic DNA in a simple enzyme reaction. The only requirement is that some of the DNA sequence at either end of the region to be copied is known. DNA corresponding to the sequence of interest is copied or amplified by PCR more than a million-fold. Sufficient DNA is obtained for detailed analysis or manipulation of the amplified gene.

Each PCR has four key components (Figure 1).

- **Template DNA.** This contains the DNA sequence to be amplified. The template DNA is usually a complex mixture of many different sequences, as is found in genomic DNA, but any DNA molecule that contains the target sequence can be used. RNA can also be used for PCR by first making a DNA copy using the enzyme reverse transcriptase (Section B7).
- **Oligonucleotide primers.** Each PCR requires a pair of oligonucleotide primers. These are short single-stranded DNA molecules (typically 20 bases) obtained by chemical synthesis. Primer sequences are designed to anneal by complementary base-pairing to opposite DNA strands on either side of the sequence to be amplified.
- **DNA polymerase.** A number of DNA polymerases are used for PCR. All are thermostable and can withstand the high temperatures (up to nearly 100°C) required. The most commonly used enzyme is Taq DNA polymerase from *Thermus aquaticus*, a bacterium present in hot springs. The role of the DNA polymerase in PCR is to copy DNA molecules. The enzyme binds to single-stranded DNA and synthesizes a new strand complementary to the original strand. DNA polymerases require a short region of double-stranded DNA to initiate synthesis of a new strand. In PCR, this is provided by the oligonucleotide primers which create short double-stranded regions by binding on either side of the DNA sequence to be amplified. In this way the primers direct the DNA polymerase to copy only the target DNA sequence.
- **Deoxynucleotide triphosphates (dNTPs).** Each reaction requires supplies of the four dNTPs (dATP, dGTP, dTTP, dCTP) which are used by the DNA polymerase to synthesize new DNA.

PCR proceeds as a series of cycles. Each cycle involves three stages (**denaturation**, **primer annealing**, **elongation**) which take place at different temperatures.

Denaturation

The reaction is heated to greater than 90°C. At this temperature the double helix is destabilized and the DNA molecules separate into single strands capable of being copied by the DNA polymerase.

Primer annealing

The reaction is cooled to a temperature that allows binding of the primers to the single-stranded DNA without permitting the double helix to reform between the template strands. This process is called **annealing**. The temperature used varies (typically 40–60°C) and is determined by the sequence and the number of bases in the primers.

Extension

This stage is carried out at the temperature at which the DNA polymerase is most active. For Taq, this is 72°C. The DNA polymerase, directed by the position of the primers, copies

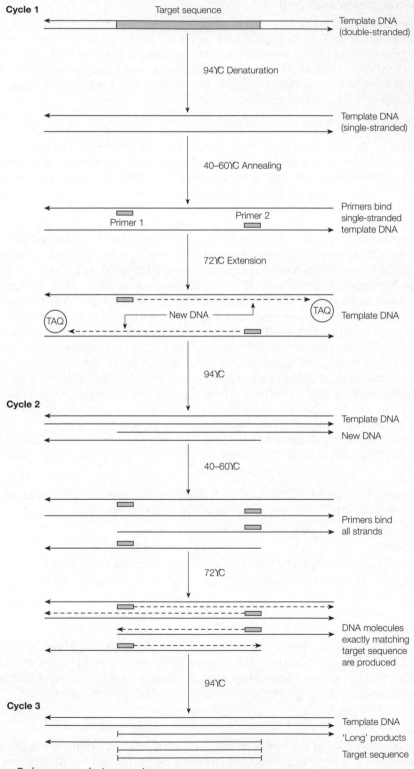

Figure 1. Polymerase chain reaction.

the intervening target sequence using the single-stranded DNA as a template: 25–35 cycles are carried out depending on the abundance of the target sequence in the template DNA. To deal with the large number of separate incubations needed, the PCR is carried out using a microprocessor-controlled heating block known as a **thermal cycler**. In the first cycle, DNA molecules are synthesized which extend beyond the target sequence. This is because there is nothing to prevent the DNA polymerase continuing to copy the template beyond the end of the target sequence. However, in subsequent cycles, newly synthesized DNA molecules which end with the primer sequence act as templates and limit synthesis to the target sequence so that the amplified DNA contains only the target sequence.

PCR is used both to analyze genes and RNA molecules after RTPCR (see below), and also in constructing sequences to be cloned or sequenced.

Reverse transcription

RNA molecules are capable of being analyzed directly but they are much more easily degraded during handling than DNA. For this reason most analyses are carried out after RNA has been copied to **complementary or copy DNA** (cDNA). Conversion to DNA also makes it available for amplification by PCR, cloning, and sequencing. Copying an RNA molecule into DNA is carried out with the reverse transcriptase enzyme used by retroviruses to convert their genome from RNA to DNA (Section B7). In common with other DNA polymerases, the enzyme requires a primer. In the case of mRNA a primer of poly-T is used which is complementary to the poly-A tail at the 3' end of the mRNA molecule (Section A4). Reverse transcriptase then copies the RNA to make a double-stranded DNA/RNA hybrid. The RNA strand is degraded and the new single-stranded DNA copied by DNA polymerase to form a double-stranded cDNA copy of the original RNA. To copy other RNA molecules or discrete regions of mRNAs specific primers have to be designed and synthesized. When RNA is converted to cDNA standard PCR protocols (see above) are used to produce multiple copies of the sequence. This procedure is known as **reverse transcription polymerase chain reaction** (RTPCR). RTPCR is now the standard method to detect if a specific RNA molecule is present in extracts of tissues.

Real-time PCR

There are many instances where it is important to assay accurately the number of copies of a particular RNA molecule in cells or tissues. In medicine this may mean assessing the level of a viral infection. In cell biology it may mean monitoring the change in level of a specific mRNA during embryogenesis. For accurate quantitative estimation a technique known as **real-time PCR** has been developed. Real-time PCR involves, as its name implies, monitoring the amount of DNA product at every cycle of the PCR reaction. This means that the whole procedure is very rapid and can be completed in a few hours.

To quantify RNA sequences it is necessary to copy the RNA to cDNA using reverse transcriptase. The number of molecules of the double-stranded DNA product of this reaction can then be quantified.

A number of methods are available to monitor accurately the production of new copies of the target DNA molecule during the PCR reaction. All of these require a modified thermal cycler which is adapted by incorporating a spectrophotometer to measure fluorescence during the PCR reaction. The amount of DNA synthesized in the PCR reaction is measured at each cycle, using probes labeled with fluorescent tags that are only capable of fluorescing when annealed to their target sequence (Figure 2). Specialized chemical

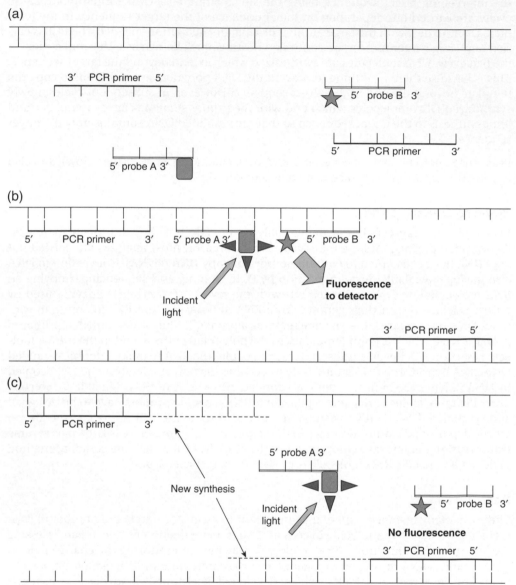

Figure 2.   Real-time PCR. (a) Prior to denaturation: the double-stranded target DNA molecule, two single-stranded PCR primers and the two probes are shown. Probe A is tagged with a fluorochrome (dark rectangle) and probe B with a different fluorchrome (star). After the denaturation step of the cycle target DNA becomes single-stranded. (b) On lowering the temperature for the annealing phase, the primers and probes hybridize with their complementary sequences in the target DNA. In the presence of exciting wavelengths of light, fluorochrome A emits fluorescence (black arrowheads) over a short range. Because the two probes are held close together in the hybrid this fluorescence can reach the fluorochrome (star) on probe B and induce it to emit its own fluorescence. It is this wavelength that the detector is programmed to detect. (c) During the extension phase, both probes detach from the hybrid. The fluorochromes are no longer close enough together to interact and fluorochrome B ceases to emit its fluorescence; these probes are available to bind in the next cycle of PCR.

processes are required to design these probes. The rate of increase of fluorescence is directly proportional to the number of molecules of the specific RNA present at the start of the reaction. Thus accurate measurement of RNA levels can be calculated. By using different fluorescent tags it is possible to quantify more than one RNA species in each reaction and to include standards of known concentration to act as controls.

The technique can also be used to determine the number of molecules of a DNA sequence present in tissues in the same way. In this case the initial reverse transcriptase step is omitted.

E3 DNA sequencing

Key Notes

The first sequencing methods

DNA sequencing has been the most important technological advance in the history of genetics, providing data at the highest resolution. The problems of sequencing biological polymers are first to isolate pure molecule and second to develop methods of sequencing. Initial methods using successive fragmentation and deduction allowed 90 nucleotide tRNAs to be sequenced. Chain elongation techniques dramatically improved the technology.

Plus–Minus method

Sanger and Coulson discovered that when a chain was extended with one nucleotide at low concentration the chain often terminated immediately before that base. The length of the chains produced could be measured by electrophoresis and gave the positions of each nucleotide. The method was used to sequence the entire 5386 nucleotide ϕX174 bacteriophage genome which was published in 1977.

Maxam and Gilbert chemical sequencing

Maxam and Gilbert used chemical methods to cut purified single-strand DNA at specific bases or base combinations. Measuring the range of lengths from the original end to each cut gave the distance to that base. The chemistry was easier than extension methods then available.

Sanger dideoxy chain termination methods

Sanger developed a more reliable method of terminating chains during synthesis by incorporating a low concentration of dideoxynucleotide triphosphates in the polymerization mix. These lack a 3′ OH group so cannot be extended. Using four reactions, one with dideoxyribose, for each base and comparing length from the 5′ end gives the positions of each base.

Cloning and polymerase chain reaction provided pure samples

The development of restriction enzymes and cloning techniques allowed DNA fragments to be inserted into vectors where they replicated asexually and could be excised and purified in large quantities. The polymerase chain reaction could be used to amplify enough copies to sequence directly from genomic DNA if the flanking sequences for priming were known. Unknown DNA fragments can also be amplified directly without cloning after primer adaptors are ligated to the ends.

Fluorescent chain termination

The chain termination method was modified by attaching different fluorescent dyes to the 3′ blocked nucleotides.

One reaction was used for all four bases. As the strands migrated by electrophoresis past a laser and detector, the color indicated the base at that position. The process was completely automated and used to reference sequence the human genome.

Massively parallel sequencing

Massively parallel sequencing is the latest system for automated sequencing. A single DNA molecule with adapters ligated to the ends is isolated, attached to a solid bead or surface and amplified by PCR so that all the millions of copies are also localized in a region a few microns across, enabling many reactions in a very small area. A primed extension reaction is then performed in such a way that light is emitted according to the base being added. The exact form of reaction depends on the detection technique being used. The lengths read tend to be low, making it difficult to align fragments from large genomes unless there is already a sequence to align them to, but a single machine can read millions of bases in a day.

Photo-detection techniques

All four bases may be added in one reaction tube, 3' blocked so that only one base can be added, and labeled with different fluorochromes. The color is read at each step, then the fluorochrome is chemically removed and the 3' hydroxyl group unblocked so that the next base can be monitored. In Pyrosequencing® each single normal nucleotide triphosphate is added in turn and light is emitted if it is inserted at the next position, the light being proportional to the number of bases where there is a short mononucleotide run. A camera or other photo-detector records the light and a computer program reads off the sequence.

Applications of massively parallel sequencing

Applications of massively parallel sequencing include sequencing thousands of humans to get a good indication of population genetic variation; sequencing DNA from fossil bones, including Neanderthals, and identifying hominid sequences from the mass of bacterial DNA; and sampling DNA from environmental samples of soil and seawater to catalog unique sequences which indicate unique species, particularly of bacteria and viruses that are otherwise unknown.

Related topics

(A1) DNA structure
(B1) Concepts of genomics
(D1) Population genetics and evolution

(D2) Evolution by natural selection
(D9) Phylogeography, molecular clocks, and phylogenies

The first sequencing methods

DNA sequencing is providing massive amounts of genetic information at the maximum possible resolution, the individual bases in the combination that makes the genome. It is by far the biggest source of factual genetic data and has opened a world of understanding and enquiry into the mechanisms of genetic processes unimaginable when Watson and Crick proposed their theory of DNA structure in 1953. Although purely technical, its contribution to the science of genetics equals the discovery of chromosomes or the role of DNA and the genetic code.

The two main problems of determining the sequence of biological polymers, protein or nucleic acids are obtaining sufficient quantities of pure molecules of a size that can be analysed by current sequencing techniques, and developing a method of identifying specific units (amino acids or nucleotide bases). The history of DNA sequencing and genomic studies is the progressive improvement and development of the techniques for achieving these objectives.

The first attempts at sequencing involved purifying a particular, relatively small molecule, such as a tRNA, then rounds of breaking it into smaller overlapping fragments, purifying these and repeating the procedures until the chemical composition of each fragment could be determined. The sequence of the larger combinations was then deduced from the smaller derivatives. For example, a trinucleotide fragment that could be cleaved to give TC plus A or AT plus C must be ATC in that order. These techniques could sequence about 50 nucleotides, and three would be sufficient to complete a 90-nucleotide tRNA with overlaps.

Chain extension by polymerase (Section A1) was used by Weissmann and co-workers in 1969 to deduce the sequence of a RNA bacteriophage β. They used a replicase (polymerase) that started at the 3′ end, then they pulse-labeled the growing chain with radioactive nucleotides. Another technique for DNA used incorporation of one base as a ribose nucleotide which allowed chains to be cleaved by alkali at that base.

Plus–Minus method

The Plus–Minus method of sequencing was published by Sanger and Coulson in 1975. The method used chain extension by DNA polymerase to produce chains with the same 5′ end and a defined 3′ end. Single-stranded bacteriophage φX174 DNA was replicated from a manufactured primer. One of the four nucleotide triphosphate bases used for incorporation was radioactively labeled and was in a very low concentration so synthesis often terminated at the preceding base because of a shortage of substrate. This produced chains whose length corresponded to the position of that base. Four separate reactions were used, one with a low concentration of each base. The products could be separated by electrophoresis and gave a complete sequence for about 80 nucleotides. The technique was used to sequences almost all of the 5386-nucleotide φX174 genome by 1977, the first complete DNA sequence ever produced.

Maxam and Gilbert chemical sequencing

Maxam and Gilbert published a chemical sequencing method in February 1977 which was briefly popular. This used purified DNA directly. The 5′ end was labeled with $[\gamma^{32}P]$ATP. The newly synthesised DNA was then chemically modified at specific bases to weaken the phosphodiester chain. Purines (A+G) were removed with formic acid, guanines methylated with dimethyl sulfate, pyrimidines (C+T) methylated with hydrazine and cytidines alone by hydrazine with salt to protect the thymines. The levels of reaction were kept low to average one modification per molecule, and the weak points cleaved with hot piperidine. The four reactions were separated by length side-by-side on an acrylamide gel. The length of the molecules from 5' label to cleavage point allowed the entire sequence to be deduced (Figure 1). The technique was later developed to demonstrate the protection footprints where DNA-binding proteins bound to specific nucleotide sequences and protected the bases from modification.

Sanger dideoxy chain termination methods

Sanger's dideoxynucleotide chain termination method was published in December 1977 as an improvement on the Plus–Minus method. Dideoxyribose (dd) has no 3' hydroxyl group so the chain cannot be extended because this is where the next nucleotide triphosphate should be added (Figure 2). A low concentration of one dideoxynucleotide will cause a small proportion of chains to terminate at that base. In the reaction containing ddC, chains will terminate in a C. Running the four reactions out side-by-side by electrophoresis allows the sequence to be read directly (Figure 3).

Cloning and polymerase chain reaction provided pure samples

At the same time that chain termination technology became available, restriction enzymes also became readily available (Section E1). These allowed DNA of unknown sequence to be cut producing sticky ends that fit into specially made cloning sites in vectors (Section E4). Subcloning into the single-stranded DNA phage M13 produced single-stranded template, and primers were available complementary to the vector upstream of the insert, allowing the insert to be sequenced from its first base. By 1983 these techniques producing ^{32}P-labeled products became readily available and affordable in many genetics labs. Thermostable DNA polymerase *Taq* from *Thermus aquaticus* revolutionised working with DNA. The polymerase chain reaction (Section E2) can be used with unique primers complementary to the ends of a known sequence to amplify that sequence directly from total genomic DNA. Sufficient copies can be produced to sequence without the need to clone it into a vector. In the most sensitive techniques *Taq* PCR allows sufficient exponential amplification of a single molecule to determine its nucleotide sequence. A second advantage of using *Taq* for the sequencing reaction at high temperature, around 75°C, is to keep the DNA denatured and single-stranded in regions which tended to form double-stranded DNA by foldback pairing. This allows sequencing of 'difficult' sequences

Sequence	A+G	G	C+T	C
C			—	—
T			—	
C			—	—
A	—			
C			—	—
C			—	—
A	—			
G	—	—		
T			—	
G	—	—		
T			—	
C			—	—

Figure 1. Maxam and Gilbert chemical cleavage sequencing.

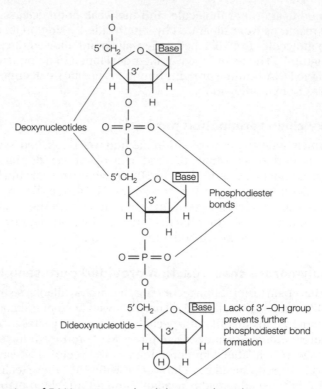

Figure 2. Termination of DNA sequence by dideoxynucleotides.

impossible at lower temperatures. Recombinant *Taq* polymerase has been produced to improve the accuracy of replication.

Fluorescent chain termination

This method became the basis of automated sequencing using dideoxynucleotides with a fluorochrome attached, a different color for each nucleotide. All the reactions are run in one tube and the products separated by capillary electrophoresis. As the chains run out of the capillary in order of increasing length, an exciting laser is shone on them and the color of the resulting fluorescence is measured (Figure 3). The order of colors automatically indicates the sequence of bases. This is fast and can handle 96-well or larger reactions trays automatically without the need for potentially dangerous radioactive label or the need to develop radiographic film. It provided the human genome sequence between 1990 and 2003, although it took several more years to clean up, interpret, and annotate the data. It is still the main method for sequencing specific individual DNA fragments accurately. The method is capable of giving 600 to 800 bases accurately, which makes it relatively easy to align overlapping sequences.

Massively parallel sequencing

Massively parallel sequencing is the descriptive name of the latest (third) generation of sequencing techniques that can sequence millions of single DNA molecules without any prior cloning or amplification. The fastest systems can produce 400 million bases of sequence data in 10 hours. Their limitation is that the new methods mostly produce

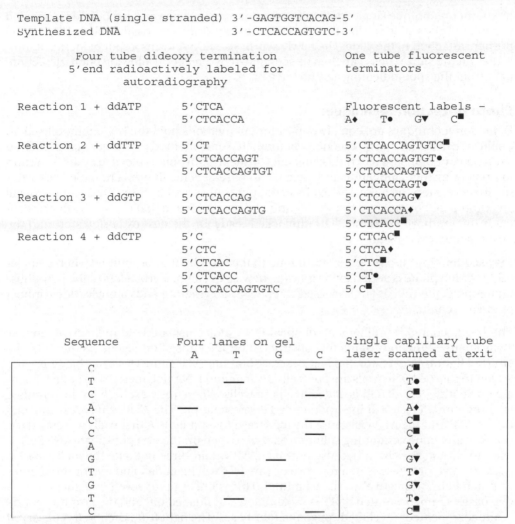

Template DNA (single stranded) 3'-GAGTGGTCACAG-5'
Synthesized DNA   3'-CTCACCAGTGTC-3'

	Four tube dideoxy termination 5'end radioactively labeled for autoradiography	One tube fluorescent terminators
Reaction 1 + ddATP	5'CTCA 5'CTCACCA	Fluorescent labels – A♦ T● G▼ C■
Reaction 2 + ddTTP	5'CT 5'CTCACCAGT 5'CTCACCAGTGT	5'CTCACCAGTGTC■ 5'CTCACCAGTGT● 5'CTCACCAGTG▼ 5'CTCACCAGT●
Reaction 3 + ddGTP	5'CTCACCAG 5'CTCACCAGTG	5'CTCACCAG▼ 5'CTCACCA♦ 5'CTCACC■
Reaction 4 + ddCTP	5'C 5'CTC 5'CTCAC 5'CTCACC 5'CTCACCAGTGTC	5'CTCAC■ 5'CTCA♦ 5'CTC■ 5'CT● 5'C■

Sequence	Four lanes on gel				Single capillary tube laser scanned at exit
	A	T	G	C	
C				—	C■
T		—			T●
C				—	C■
A	—				A♦
C				—	C■
C				—	C■
A	—				A♦
G			—		G▼
T		—			T●
G			—		G▼
T		—			T●
C				—	C■

Figure 3. Comparison of dideoxy termination and fluorescent dye termination technology.

much smaller read lengths than capillary sequencing which makes it much more difficult to combine overlapping fragments unless there is a completed reference sequence available to align them against. There are several commercial methods. In a typical technique, DNA is fragmented to a suitable length. Read lengths from different techniques currently vary between about 30 bases up to around 800 bases. Paired adapter sequences are ligated (Section A1) to the ends of the fragments to act as binding sites for a primer-pair. Each end needs a different adapter because each strand has to have a different primer and only one strand must attach to the bead for sequencing. One strand of DNA is fixed to a minute bead by the adapter at one end. Each bead needs to capture a single DNA molecule to produce a usable pure sequence. The beads are individually trapped in PCR reaction mixture in water droplets in an emulsion in oil and the DNA is replicated repeatedly to coat the bead with copies of one strand. Each bead with attached pure DNA is then transferred to a flow cell used for a microscopic sequencing reaction (Figure 4).

In another technique single molecules with adapters on each end are bound by one end to a treated glass slide at a low density. They are then amplified in a thin film by PCR using primers attached to the slide. The newly synthesised DNA forms a spot of identical molecules on the slide. These can then be sequenced by a reaction allowing photo detection, light from the spot indicating addition of a base.

Photo-detection techniques

Detection techniques are equally important. All methods hold the DNA immobilised on a solid substrate and pass reagents over them. In some methods the sequencing reaction is carried out one base at a time using nucleotide triphosphates blocked at the 3' carbon to prevent further elongation and attached to base-specific fluorescent molecules. After the reaction step the color of each bead or spot is observed by a sensitive camera using an automated computer system to identify the base added to that chain. The fluorescent tag is then removed and the 3'OH unblocked ready for the next extension step, and the next reaction cycle begins.

Pyrosequencing® is a propriety technique that causes light to be emitted when a nucleotide triphosphate is added to the growing new DNA chain. 'Pyro' refers to the pyrophosphate molecule (PPi, $P_2O_7^{4-}$) released when DNA polymerase adds a nucleotide triphosphate to a chain.

The beads coated in millions of identical DNA molecules are held individually in tiny wells about 29 microns diameter constructed on the ends of optical fibers that take the emitted light to a detector. Smaller beads containing ATP sulfurylase and luciferase are added to the well. The wells are flow cells. Primer and DNA polymerase are added. A single nucleotide (e.g. dCTP) is added to all the wells. Where the next base on the template is G the polymerase will incorporate dCTP releasing pp. The ATP sulfurylase converts this to ATP and the ATP will cause the luciferase to emit light. Other wells will stay dark. (An ATP analogue containing sulfur instead of oxygen on the first phosphate, dATPαS, is used to extend the chain because normal dATP would directly excite the luciferase.) If there are two or three Gs in a row, two or three Cs will be added and two or three times as much light, respectively, will be emitted. This is linear up to seven or eight consecutive bases. Unincorporated dCTP is broken down or flushed out and the next nucleotide is added, repeatedly cycling through the four bases. The light from each well is observed

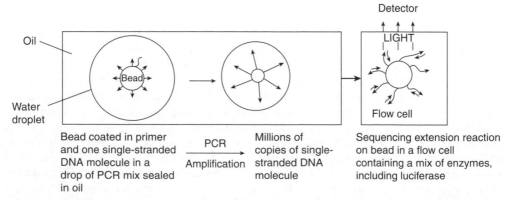

Figure 4. Bead-linked amplification of a single molecule and automated sequencing.

simultaneously by a camera and the sequence is determined automatically. By sequencing single molecules this method can detect the number of copies of each gene and the frequency of sequence variants in a mixed, heterozygous, or polyploidy sample.

Applications of massively parallel sequencing

Massively parallel sequencing produces millions of bases of sequence per day, often in short lengths that require very intensive computerised analysis. One use is to sequence the genomes of very many humans in the **1000 Genome Project**. The new sequences are aligned on the known 'human genome'. Because individual strands of DNA are sequenced, the technique identifies heterozygosity and also copy number. If a small region is duplicated on a chromosome so that three copies are present in a diploid, then there will be three sequences from that region for every two of the rest of the genome. This gives a much better indication of the variability present in the population. Screening a large number of samples of cancer tissue can reveal the frequency of specific mutations and suggest which mutations are most significant. This may reveal less common but important mutations, and the larger samples are more likely to exclude spurious false-positive associations.

DNA from Neanderthal bones some 30 000 years old has been sequenced. Most of the sequences were contaminating bacterial DNA, but the sequences with a close similarity to human represent the original Neanderthal. The technique sequences short single DNA sequences so is extraordinarily sensitive to contamination. There were two initial sequencing attempts. One found no sequences identical to modern humans, the other found large amounts of similarity, most or all of which was contamination, and there is as yet no conclusive evidence either way (Section D10).

Sequencing from environmental samples such as seawater or soil can indicate the number of species of bacteria and viruses present. The data are all recorded on databases as, for example, a catalog of ribosomal RNA variants. The presence of a particular known bacterium or virus at any of the sample sites can be checked by screening the database for species-specific sequences. A cause for caution is that every sequencing error appears as a new sequence, implying a new population variant or new species. It should be recalled that around half the protein-coding 'genes' in the initial release of the human genome turned out to be errors.

E4 DNA cloning and transfection

Key Notes

DNA cloning

This technique allows individual DNA sequences to be isolated from complex mixtures and copied, permitting detailed analysis and manipulation. The DNA to be cloned is recombined with vector DNA and introduced into host cells where it is copied. Recombinant vector is purified from cultures of host cells and the cloned DNA can be recovered for analysis.

Restriction enzymes

Recombining DNA molecules for cloning depends on the use of bacterial restriction enzymes which cut DNA molecules at palindromic sequences and produce sticky ends. Two DNA molecules cut with the same restriction enzyme can be joined by complementary base-pairing between the sticky ends and covalently linked using DNA ligase.

Plasmids

These are small circular DNA molecules found in bacteria that are frequently used as cloning vectors. Plasmids are easily purified and confer antibiotic resistance to host bacteria allowing easy identification of recombinants. Linearized plasmid is recombined with foreign DNA and the recombinant plasmid is transformed into bacteria. Colonies containing recombinant vector are isolated from agar plates containing antibiotic, and vector containing the cloned DNA is purified from bacterial cultures. Early plasmids such as pBR322 contain twin antibiotic resistance genes that allow identification of recombinants. Later plasmids such as pUC identify recombinants by blue/white selection based on disruption of the *lac Z* gene by insertion of the foreign DNA. Additional plasmid modifications include multiple cloning sites, phage promoter sequences for *in vitro* transcription and the ability to express cloned sequences as protein.

Lambda (λ) phage

Bacteriophage λ which infects *E. coli* has been adapted as a cloning vector. The central portion of the phage DNA is deleted and can be replaced with foreign DNA. Recombinant phage DNA is packaged *in vitro* into capsids which infect *E. coli* producing plaques on agar plates. λ vectors are used to construct genomic libraries. These are collections of recombinant phage containing cloned sequences representative of an entire genome. Libraries can be screened for sequences of interest by hybridizing plaque lifts with a probe. cDNA libraries are constructed from mRNA and contain clones representing expressed sequences only. Expression libraries are cDNA libraries that allow screening with antibodies.

Cosmids

These cloning vectors resemble plasmids but contain λ phage *cos* sequences which allow them to be packaged into λ capsids. Packaged cosmids infect *E. coli* and are replicated. Cosmids do not contain λ genes and so produce bacterial colonies instead of plaques. Cosmids can accommodate large inserts up to about 44 kb.

Yeast artificial chromosomes (YACs)

These vectors use eukaryotic host cells and replicate in the same way as host cell chromosomes. They contain features required for chromosome replication including an origin of replication, a centromere and telomeres. YACs can accommodate inserts of up to two megabases and can be used to construct chromosome maps. They have been largely superseded by bacterial artificial chromosomes and P1 artificial chromosomes, which are much more stable.

Plant cloning vectors

Most plant cloning vectors are based on the Ti plasmid isolated from *Agrobacterium tumefaciens* which causes crown gall disease in plants. Part of the Ti plasmid can integrate into the host cell chromosome and is used to carry useful genes into the plant genome.

Transfection

Several methods can be used to transfer cloned DNA into animal or plant cells. Chemical methods include DEAE dextran or coprecipitation with calcium phosphate. Liposomes can also be used to carry DNA across membranes. Physical methods of transfection include microinjection, electroporation, and gene guns. Transfection can be transient or stable where the DNA is integrated into the host's chromosomes.

Related topics

(B1) Concepts of genomics   (B6) Bacteriophages
(B2) Prokaryote genomes   (G2) Biotechnology
(B3) Eukaryote genomes   (G3) Transgenics
(B4) Chromosomes

DNA cloning

Large genomes, such as those of mammals, are estimated to contain about 20 000 genes. In DNA isolated from mammalian cells, individual gene sequences are therefore present in only very small amounts. DNA cloning is a powerful technique that allows specific DNA sequences to be separated from other sequences and copied so that they can be obtained in large amounts, permitting detailed analysis or manipulation. An important use of DNA cloning is to isolate new genes allowing them to be investigated and characterized.

All DNA cloning experiments are based on the construction of **recombinant DNA** molecules. This involves joining different DNA molecules together. The DNA molecule to be cloned (often a fragment containing a gene of interest) is inserted into another, usually circular, DNA molecule called a **vector**. The recombinant vector is introduced into a **host cell**, usually the bacterium *E. coli*, where it produces multiple copies of itself. When the host cell divides copies of the recombinant vector are passed on to daughter cells. Large amounts of the vector are produced which can be purified from cultures of the host cells and used for analysis of the foreign DNA insert.

Restriction enzymes

The ability to join different DNA molecules together for cloning is dependent on the use of enzymes from bacteria called **restriction endonucleases**. These enzymes cut DNA molecules at specific sequences, usually of 4–8 bases. The sequences recognized are **palindromes**. This means that the sequence is the same reading 5'→3' on both strands. Each enzyme has a specific target sequence. For example the enzyme **EcoRI**, which is obtained from the bacterium *E. coli*, will cut any DNA molecule that contains the sequence GAATTC. Other restriction enzymes recognize different sequences. The cut made by restriction enzymes is usually staggered such that the two strands of the double helix are cut a few bases apart (Figure 1a). This creates single-stranded overhangs called **sticky ends** at either end of the cut DNA molecule. Some restriction enzymes cut the DNA leaving the

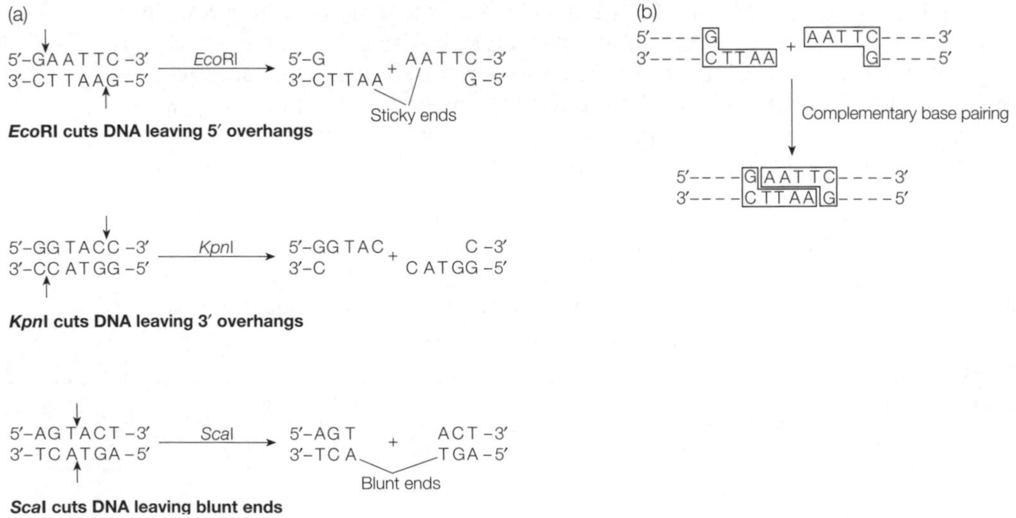

Figure 1. (a) Action of restriction enzymes on DNA. (b) Complementary base-pairing between sticky ends.

5' end overhanging and others leave a 3' overhang. A few restriction enzymes cut both strands of the double helix at the same position creating what is known as a **blunt end**. Two DNA molecules cut with the same restriction enzyme can be joined together by complementary base-pairing between the sticky ends (Figure 1b). Although the molecules are joined by the sticky ends, they are not covalently linked. However, the enzyme DNA ligase can be used to catalyze the formation of a phosphodiester bond between the two DNA molecules, thus permanently recombining them. A number of systems exist for cloning DNA molecules based on the use of different types of vector. Each has its own individual characteristics and uses. Some of the more important one are reviewed below.

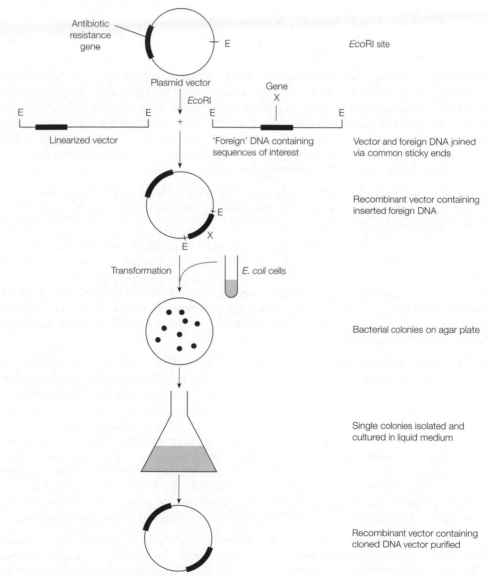

Figure 2. Cloning with plasmids.

Plasmids

These are small, circular DNA molecules present in bacteria. They occur in addition to the main bacterial chromosome and can replicate autonomously (Section B2). Plasmids were the first cloning vectors to be developed and have been used since the 1970s. Plasmid cloning involves a number of steps (Figure 2): first, the plasmid is digested with a restriction enzyme that cuts it at a single site converting it from a circular molecule into a linear molecule with sticky ends. The foreign DNA to be cloned is also digested with the restriction enzyme to produce the same sticky ends. When the plasmid and the foreign DNA are mixed, molecules of plasmid become joined to molecules of foreign DNA via their common sticky ends and circular recombinant plasmids are obtained. DNA ligase is then used to covalently join the two. The recombinant plasmid is introduced into host bacteria (usually *E. coli*) by a process called **transformation**. The transformed bacteria are spread on agar plates and bacterial colonies derived from individual cells that have taken up the recombinant plasmid are grown. Individual colonies are picked and cultured in liquid medium. Large amounts of plasmid can then be purified from the cultures and the cloned DNA recovered for analysis.

Several features of plasmids make them especially suitable as cloning vectors. Their small size (usually about 3 kbp) makes them easy to purify from bacterial cultures allowing the cloned DNA to be recovered easily. In addition, plasmids often contain genes encoding proteins that make the bacteria resistant to antibiotics such as ampicillin and tetracycline. By growing colonies on agar plates containing antibiotic, it is possible to isolate bacteria that have taken up the plasmid during transformation because only these will be resistant to antibiotic and will be able to grow.

Plasmid cloning vectors were initially based on naturally occurring plasmids. These have gradually been replaced by improved vectors whose DNA sequences have been altered to include features useful for cloning.

One of the earliest plasmid vectors to be developed was **pBR322**. This plasmid contains two genes that confer resistance to the antibiotics ampicillin and tetracycline. During cloning, foreign DNA is inserted into the tetracycline gene, thereby inactivating it. Transformed bacteria containing recombinant plasmid could therefore be identified by being resistant to ampicillin but not to tetracycline. Bacteria which had taken up a plasmid that did not contain foreign DNA but had simply been re-ligated to itself could be identified by being resistant to both antibiotics.

The pBR322 plasmid was followed by the **pUC** series of vectors, which allowed identification of colonies containing recombinant plasmid by a method called **blue/white selection** (Figure 3). This method relies on the presence of a gene called *lac Z* which encodes the enzyme β-galactosidase and is located on the plasmid at the point where the foreign DNA is inserted. Bacteria that contain the intact plasmid synthesize β-galactosidase which acts on a synthetic substrate called **X-gal** (5-bromo-4-chloro-3-indolyl-β-D-galactopyranoside) to produce a colored product. When colonies are grown on agar plates containing X-gal they take on a blue color. However, when foreign DNA is inserted into the pUC plasmid, the *lac Z* gene is disrupted and β-galactosidase is no longer produced. As a result, colonies containing recombinant plasmid remain white when grown on X-gal and are easily distinguished from colonies containing re-ligated vector which are blue.

Another useful feature of pUC vectors is that the sequence of part of the *lac Z* gene is modified to create a series of clustered restriction enzyme sites. This is called the **multiple cloning site (MCS)**. Its purpose is to create extra flexibility during the cloning procedure by allowing the foreign DNA to be inserted at any one of several restriction sites (Figure 4).

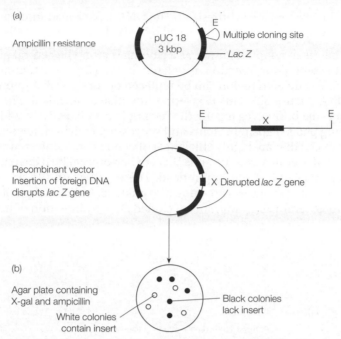

Figure 3. (a) Insertion of foreign DNA inactivates *lac Z* gene. (b) Recombinant colonies appear white on agar plates containing X-gal.

Other useful plasmid modifications include the presence of promoter sequences from bacteriophages inserted on either side of the MCS which allow *in vitro* transcription of the inserted foreign DNA by RNA polymerase. This feature is useful for producing RNA probes from cloned sequences. Some plasmids are modified to allow cloned sequences to be translated into protein. These are known as **expression vectors**.

Lambda (λ) phage

Bacteriophages (phages) are viruses that infect bacteria. They consist of a nucleic acid genome inside a protective protein coat called a capsid. The bacteriophage λ which infects *E. coli* has a linear double-stranded DNA genome (Section B6). The phage attaches itself to the surface of a bacterium and injects its DNA into the cell. Inside the cell, the λ DNA is

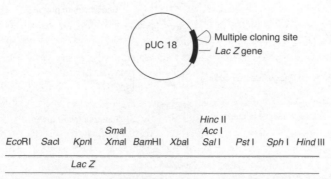

Figure 4. Multiple cloning site of the pUC18 vector.

copied and capsid proteins are synthesized. The DNA is packaged into capsids and new phage particles are produced that are released by lysis of the infected cells.

Lambda phage has been adapted for use as a cloning vector. The central portion of the λ DNA, that is not essential for infection, is deleted, leaving 5' and 3' fragments known as **arms** (Figure 5a). The deleted region can be replaced by foreign DNA to produce recombinant phage DNA (Figure 5b). This is inserted into phage capsids *in vitro* by a process called **packaging** which involves mixing the recombinant phage DNA with a **packaging extract** containing phage capsid proteins and processing enzymes. Recombinant phage particles are produced that are highly efficient at infecting *E. coli*. Infected cells are spread on an agar plate and produce a continuous sheet of bacteria called a **lawn** which contains small clear areas about the size of a pin head. These correspond to areas of lysis produced by infection with phage and are known as **plaques**. Individual plaques can be isolated and used to generate large amounts of cloned DNA by infection of fresh cultures of *E. coli*.

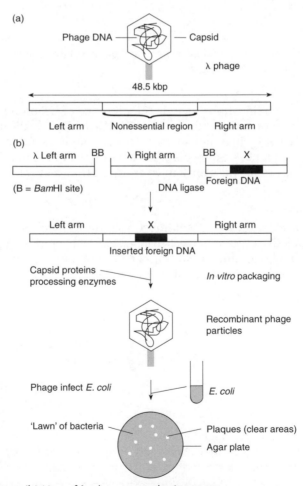

Figure 5. (a) λ phage. (b) Use of λ phage as a cloning vector.

The main advantage of λ as a cloning vector is that the size of the fragments that can be cloned is much larger than for plasmids. Lambda vectors can accommodate fragments up to 25 kb as compared to less than 10 kb for plasmids. The ability to clone larger fragments has led to the development of the main use of λ vectors which is the construction of **DNA libraries**. A library is a collection of recombinant phage which together contains clones representative of all the DNA sequences present in the genome of an organism. Because λ can accommodate relatively large fragments of DNA, many fewer clones are required to represent an entire genome than would be required if plasmids were used. **Genomic libraries** are constructed from DNA purified from cells which has been broken randomly into fragments of around 20 kb, either by digestion with a restriction enzyme or through physical shearing by pipeting or sonication. The fragments are ligated to the λ arms and are cloned at random. Many thousands of plaques are produced, each of which contains a different cloned sequence. Plaques corresponding to a cloned sequence of interest can be identified by screening the library with a probe using a procedure called a **plaque lift**. This involves taking an agar plate containing plaques and laying a sheet of special nylon membrane on top of it. Some of each plaque adheres to the membrane and a replica of the pattern of plaques on the plate forms. The membrane is then treated with alkali to denature the DNA in the plaques and is hybridized to a DNA probe labeled with radioactivity. After washing away unbound probe, exposure of the membrane to X-ray film produces a series of black dots corresponding to the position of plaques containing the desired sequence. These can then be isolated from the agar plate and used to obtain the cloned DNA.

Libraries can also be produced using RNA. The enzyme reverse transcriptase is used to convert the RNA into complementary DNA (cDNA; Section E2) which can then be cloned in the same way as for genomic libraries. Libraries made this way are called **cDNA libraries** and contain clones that are representative of the genes that are active in the cells used to isolate the RNA. Thus, a cDNA library from liver cells will have many different clones from a library derived from lung cells or kidney cells because the active genes in each cell type will be different. cDNA libraries have the advantage that the cloned sequences do not contain introns. This greatly simplifies the characterization of cloned genes because the complete coding sequence of the gene may be present in a single clone. **Expression libraries** are a type of cDNA library in which the cloned sequences are translated into protein by the host bacteria. This allows the library to be screened using antibodies specific for the protein encoded by the cloned sequence.

Cosmids

This type of vector combines features found in plasmids and λ phage. Cosmids contain all the normal features found in plasmids, including a MCS and genes conferring antibiotic resistance, but also include sequences found in λ called cos **sequences**. These occur at either end of the λ DNA molecule and are responsible for its insertion into the phage capsid. The presence of *cos* sites on cosmids allows them to be packaged into phage capsids. Cloning with cosmids combines features associated with the use of both λ and plasmids as cloning vectors (Figure 6). Cosmid DNA is cleaved with a restriction enzyme and ligated to foreign DNA. The recombinant cosmid is then packaged into λ capsids and used to infect *E. coli*. Cosmids do not contain any λ genes and so do not form plaques after infection. Instead, infected cells are grown on agar containing antibiotic and resistant colonies containing recombinant cosmids are obtained which can be propagated in the same way as plasmids. Cosmids have the advantage of being able to accommodate larger inserts. Because cosmids are small, typically 8 kb or less, and the λ capsid can accommodate up to 52 kb, inserts of up to 44 kb can be cloned.

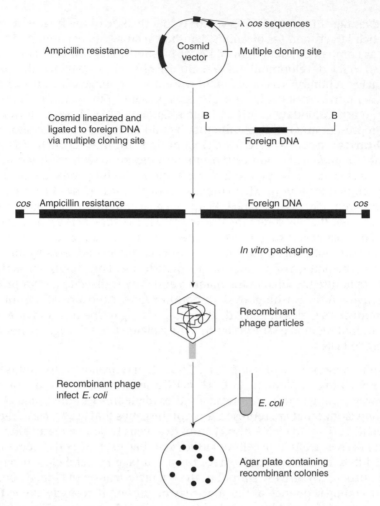

Figure 6. Cloning with cosmid vectors.

Yeast artificial chromosomes (YACs)

These vectors represent an approach to gene cloning which uses eukaryotic host cells with a vector that is replicated in the same way as a host cell chromosome. YACs contain all the essential features of a chromosome required for its propagation in a yeast cell including an **origin of replication**, a **centromere** to ensure segregation into daughter cells and **telomeres** to stabilize the ends of the chromosome (Figure 7). Very large DNA molecules up to two megabases can be cloned using YACs. This is significant because individual clones are large enough to encompass an entire mammalian gene. YACs have been widely used to construct maps of parts of the human genome by identifying clones containing adjacent regions of the genome. More recently, YACs have been largely superseded by two related types of vector, bacterial artificial chromosomes (BACs) and P1 artificial chromosomes (PACs) which have uses similar to those of YACs. These can accept somewhat smaller DNA inserts, up to 350 kb, but are much more stable than YACs. They have become the mainstay of modern genome projects (Section B1).

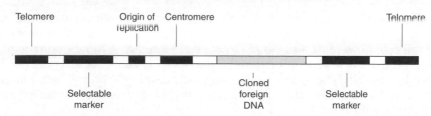

Figure 7. Structures of a yeast artificial chromosome (YAC).

Plant cloning vectors

Most vectors for cloning with plants as the host organism are based on the **Ti plasmid** which occurs in a soil bacterium called *Agrobacterium tumefaciens*. The bacterium invades plant tissue and causes a cancerous growth called a **crown gall**. During infection part of the Ti plasmid, called the T-DNA, integrates into the plant chromosomal DNA. Cloning vectors based on the Ti plasmid use this ability of T-DNA to carry useful genes into the plant genome; such genes may confer useful features to the plant such as resistance to disease (Section G3).

Transfection

Cloned genes are inserted into animal or plant cells by a process known as **transfection**. This is the basis of much biological research and of particular importance in gene therapy (Section F4) and the production of transgenic plants and animals (Section G3). A number of methods are utilized to transfect DNA depending on properties of the cells to be transfected or whether the aim is to study the expression of the gene over a short period (transient transfection) or to produce cells where the DNA is integrated into the chromosomes and stably retained by the cells.

Several methods involve close association of DNA with chemicals that assist its passage across cell membranes. The earliest transfection experiments utilized either **calcium phosphate** or **DEAE-dextran**. DNA reacts with calcium chloride in the presence of phosphate. The resulting precipitate is added to animal cells in tissue culture and the cells take it across their membranes. The precipitated form may give some protection to the DNA molecules. This procedure is suitable for producing transient and stable transfections. DEAE-dextran is a cationic polymer that binds to the negatively charged DNA molecules. Again the complex is taken up by cells; however this technique is only useful for transient transfections. Other cationic polymers have also been found to be successful.

More recently, a number of artificial **liposomes** have been developed to carry DNA through cell membranes. The lipid content passes easily through the lipid of the cell membrane. These systems are very efficient and can be used for the production of stable and transient transfections.

Alternatives to chemically mediated transfection include physical processes that damage the cell membrane. The simplest of these is microinjection where the cloned gene is directly injected into the nucleus of the recipient cell. This requires sophisticated apparatus and only relatively few cells can be injected.

Electroporation is a process in which DNA is added to cultured cells and the mixture exposed to a short electrical pulse. This disrupts the membranes sufficient to allow DNA to enter the cell; however it also causes much cell mortality. A more dramatic physical approach is the use of a '**gene gun**.' Here DNA associated with colloidal gold particles is fired directly into cells. The approach was originally developed for use with plant material as it is capable of transfecting DNA directly into living plant tissues, but may now be applied to work with animals.

E5 Bioinformatics

Key Notes

Overview

Bioinformatics is the collection, storage, analysis, collation, and use of biological information. The data contain mapping and phenotype information, nucleotide and amino acid sequence, but also protein structure, function, and expression data. The data come from cDNA (RNA), local and genomic DNA sequencing, amino acid sequences from polypeptides, and structural information from X-ray diffraction and nuclear magnetic resonance. There are also results of earlier analyses. The objective is to forecast structure and function from sequence data to improve medical treatments and drug design quickly. Sequence data are also very useful for constructing phylogenies and investigating evolutionary relationships.

Databanks

The information is stored logically in databanks which include databases and the tools to access the data. These are accessed via the Internet. Typical questions arise when a new sequence or structure is tested against the database to look for similarities to the deposited data. Information about the matching entries in the database may be applicable to the new sequence or structure. Some databank URLs are in the text.

Techniques of alignment

Algorithms are designed to test for similarities between two sequences (pairwise alignment). In a perfect match the two would align side by side. Amino acid sequences are better than nucleotide sequences. They are more conserved, and with 20 amino acids against only four bases, there are fewer chance matches between amino acids.

The tools used

BLAST (Basic Local Alignment Search Tool) is currently the principal pairwise alignment search program. Multiple alignments are possible, and iterative searches can combine data from different partial matches to identify distant relationships. PSI-BLAST (Position Sensitive Iterated-BLAST) uses matches from pairwise searches to refine subsequent searches and can detect alignments between more divergent sequences. Comparison of protein structure can detect great conservation even after extensive evolutionary divergence. DALI (Distance-matrix ALIgnment) can detect structural similarities when very little amino acid identity remains.

Predicting structure

The DNA sequence determines the amino acid sequence (primary structure) of the polypeptide or protein. This in turn influences the secondary structure (alpha helix, beta

	pleated sheet, flexible loops, and turns). These structures fold into the tertiary structure, mainly energetically driven by hydrogen bonding both between amino acid residues and between the amino acids and water molecules. Hydrophilic regions are excluded from solution by the hydrogen bonding between the water molecules and are forced into the center of the protein, or into lipid membranes.	
Predicting binding and function	The structure of the protein, and particularly the distribution of its electron cloud, is responsible for the protein's function as a catalyst and in binding to other molecules. Drugs can be designed to interact with specific parts of specific proteins to produce specific effects if pharmacologists have sufficient knowledge of the protein's structure. Bioinformatics hopes to speed the path from sequence or structural data to the production of efficient and profitable drugs.	
Comparative bioinformatics	The wealth of data available, including mRNA and protein expression, allows comparison between cells and tissues in an organism including cancer and developmental stages, between individuals for population genetics and epidemiological purposes, and between species for evolutionary purposes	
Related topics	(B1) Concepts of genomics (B2) Prokaryotic genomes (B3) Eukaryote genomes (D9) Phylogeography, molecular clocks, and phylogenies	(E1) Using sequence specificity to study nucleic acids (F1) Genetic diseases (F3) Genes and cancer (G2) Biotechnology

Overview

Bioinformatics is the collecting, collating, and analysis of large amounts of biological data in order to make useful deductions. This started as bibliographic databases about pheno-types and genetic mapping of the loci responsible. Accumulating local and genomic DNA sequences, cDNA (i.e. mRNA ESTs) and amino acid sequence data required computer storage and analysis. The current process combines the nucleic acid sequence and amino acid sequence data with structural and functional data from known polypeptides and proteins in order to deduce the structure and function of the products of other genes that are only known by their sequence. At the same time, evolutionary relationships between different proteins are revealed. The dogma is that DNA sequence determines protein sequence which determines protein structure and thus protein function. In essence, computer search algorithms are used to detect similarities between sequences of unknown function and sequences whose function is understood in order to make useful predictions. The ultimate goal is to be able to go from a nucleic acid or amino acid sequence to a detailed three-dimensional structure of the product, where the position of each atom is predicted accurately, down to the orbits of its electrons and charge distributions, and to then identify all molecules which will interact with it, including drugs. The product considered most frequently is a protein, but may be DNA or RNA, both of which have secondary structure and may be direct targets for drugs in the broad sense to control synthesis of a particular protein, for example by antisense RNA.

The greatest current problem is predicting protein folding. Many large polypeptides have numerous possible shapes, and chaperone molecules direct their folding as they leave the ribosome. X-ray crystallography and nuclear magnetic resonance are used to discover structure empirically. Currently, structure can only be predicted reliably for sequences similar to those already worked out.

The accumulated sequence data is very useful for comparing species and investigating the differences (and similarities) that have developed. Sequences data, in particular, is the raw material for constructing evolutionary phylogenies (Section D9).

Databanks

Information is held in databanks where it is stored in a logically organized way with the tools to find what you need. It is usually accessed via the Internet using world-wide-web technology. The data consist of DNA and protein sequences (genomics and proteomics, respectively), RNA expression and splicing patterns (transcriptomics), protein expression patterns (proteomics again), nucleic acid and protein structures and functions. These data are annotated with notes about their source, form and function, including where they are expressed and any other useful information, or references to that information. There are also bibliographic databanks to find publications on a subject, and databanks linking databanks.

A typical query is 'I have a new sequence or structure, are there sequences or structures like it in the database which will predict a function for mine?' This extends to asking 'What other proteins or structures interact with mine?' This will become an important part of the development of medicines, as drug molecules are designed to interact with biomolecules to counteract disease conditions.

This is a small selection of useful URLs (uniform resource locators). Most have links to other sites. It may save time and expense if you can use a mirror site nearer to your own location. OMIM, On Line Mendelian Inheritance in Man, collates human genetics (http://www.ncbi.nlm.nih.gov/omim/). It is the prime site for information on genetic diseases, even where the gene is not positively identified. Entrez is an American site providing entry to, and searches in, a wide range of databases from pubmed, a literature database, to molecular databases (http://www.ncbi.nlm.nih.gov/entrez/). The Wellcome Trust Sanger Centre, Cambridge, UK, provides a link to genome databases (http://www.Sanger.ac.uk/) including Ensembl, a European equivalent to Entrez (http://www.ensembl.org/). One of the oldest repositories for DNA sequences is GenBank with over 15 million sequences; (http://www.ncbi.nlm.nih.gov/GenBank/GenbankSearch.html). SWISSPROT is the main protein sequence database with more than 533 049 sequence entries (Nov. 16, 2011), half of which were added in 2008 and 2009 (http://www.expasy.org/sprot). Two smaller protein databases are PFAM which specializes in linking proteins into families, and SMART which specializes in identifying protein domains and protein interactions. Protein Data Bank contains protein structures (over 75 000 on Nov. 15, 2011) and a useful education resource (http://www.rcsb.org/pdb/).

Techniques of alignment

The primary identification requires algorithms to detect matches between two sequences and to give a probability of finding such a match by chance in random sequences. If it has

```
              A    C    A    C    T    G    A    T    G
        1                   C    T    G
        2              C         T    G
        3              C                        T    G
        4              C    T                   G
        5         C              T              G
```

Figure 1. Possible sequence alignments, longest and best first.

a very low probability of occurring by chance, it is probably real. Both nucleotide and amino acid sequences can be compared, but the amino acid sequence is preferred because: (i) there are 20 amino acids compared to four nucleotides so chance matches are rarer; (ii) the sequence is shorter, saving computing time; (iii) amino acid sequences are more conserved and hide silent (synonymous) base changes which do not change the amino acid (Section A3). Consider a sequence CTG being compared to a sequence ACACT-GATG. Five possible alignments are shown in Figure 1. The first example with the bases CTG adjacent and perfectly matched is the least likely to occur by chance. The last example, matching one base at a time, is meaningless.

If alignments are plotted on a matrix, matches fall along or parallel to the diagonal. Figure 2 shows the sequence ACACTGATG aligned both to itself, and also to a related sequence ACGACAGATG containing mutations, an insertion that offsets the alignment and a base change which causes a gap. The numbers show the length of uninterrupted alignment: higher numbers indicate a better match, more likely to be the result of both sequences being descended from the same ancestral sequence. A dot plot demonstrates the alignments visually. Notice that duplications produce offset alignments parallel to the diagonal, and the insertion causes a false alignment of three bases. The algorithm has to piece these matched regions together to find the most likely alignment. It is also possible to create databases of protein structure and interactions, and to align these in proteins so diverged that most of their amino acids are now different.

The tools used

FastA (for Fast Alignment) is a pairwise alignment program devised by Lipman and Pearson, and established the standard format for sequence data. The Basic Local Alignment Search Tool (BLAST) algorithm was developed by Altschul *et al.* and has the benefit of

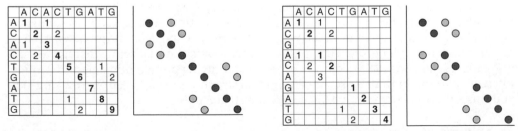

Figure 2.   Matrix alignment of the sequence acactgatg with itself, and with a sequence ACGACAGATG differing by two mutations, an insertion at 3 and a T→A at 5/6. The 'dot plot' shown beside the matrix is much more impressive for long sequences.

some very fast computer implementations. These programs compare the test sequence to each individual database sequence independently in turn. Position Sensitive Iterated-BLAST (PSI-BLAST) matches individual sequences first to identify patterns in the different comparisons, and then uses these matches to refine and repeat the search. PSI-BLAST is much better than BLAST at detecting similarities between diverged proteins. Searches can be based on a combined sequence generated from individual matches. In these multiple alignments, allowances can be made for a weighted match between different amino acids with similar properties appearing at the same position. A problem with this is that effects of chance misalignments in an early search, and gaps introduced to allow for different insertions, can accumulate and reduce the quality of the output.

Many structural and functional units are conserved. Secondary databases of consensus profiles or motif sequences have been devised weighted for the expected frequency of an amino acid, or type of amino acid, at each position (e.g. PFAM and SMART). Searching a test sequence against such motifs can identify matches and likely function quickly and allow rapid refinement of searching. Matching such weighted profiles or motifs can also connect widely divergent but related proteins in a manner impossible by exact pairwise alignment. DALI (Distance-matrix ALIgnment) is a program developed by Holm and Sander that looks at the interactions between regions of proteins. As mutations change the amino acids, the structure is often preserved, so the replacement amino acid residues still make contact. DALI identified structural similarities between mouse adenosine deaminase and the bacterial proteins phospho-diesterase from *Pseudomonas diminuta* (only about 13% amino acid identity) and also urease from *Klebsiella aerogenes*.

Predicting structure

Predicting the structure and function of a protein from its sequence requires the use of information derived experimentally from X-ray crystallography and nuclear magnetic resonance of known proteins. This information must be related to the sequence of the new protein. The amino acid sequence is the **primary structure**. The polypeptide chain comprises a repeating nitrogen-carbon-carbon (NCC) chain (NCCNCCNCC—). The peptide bond [C(O)N] is usually flat and in *trans* (meaning the next carbons in the chain are on opposite sides, Figure 3a) except proline which is *cis*. The other single bonds can rotate, but their conformation is restricted by stearic (three-dimensional) interactions between the side chains. There are two main **secondary structures**. The most stable is an **alpha helix** (Figure 3b) where successive amino acids coil around with their side chains projecting outwards. There are 3.6 amino acids per turn and each amino acid is 0.15 nm (nanometers) above the next. The structure is stabilized by hydrogen bonding between the oxygen of one peptide residue and the hydrogen on the nitrogen of the fourth peptide residue below it. (A 'residue' is one unit in the polymer, i.e. one amino acid.) The other stable structure is a **beta pleated sheet** (Figure 3c). The polypeptide forms a linear zigzag with alternate side chains projecting up and down. These linear zigzags line up side-by-side in parallel or antiparallel to form a beta pleated sheet, again by hydrogen bonding between peptide oxygen and hydrogen atoms. Flexible regions form loops and bends between these structured domains. The folding and alignment of helix, sheet and loop domains creates the **tertiary structure** of a protein (Figure 3d). The specific properties of a polypeptide come from the side chains on the amino acids (Table 1) which affect the stability of the folded polypeptide. Glycine has a single hydrogen atom side chain, which poses little stearic hindrance and allows exceptional flexibility in the polypeptide chain. Some side chains are polar, charged either acid (-COO$^-$) or basic (-NH3$^+$) and tend to form hydrogen or ionic bonds to each other or hydrogen bonds to water molecules. If they bond to the main chain they end a region of helix. Other aromatic or aliphatic side

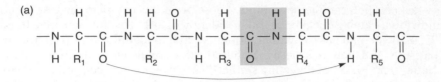

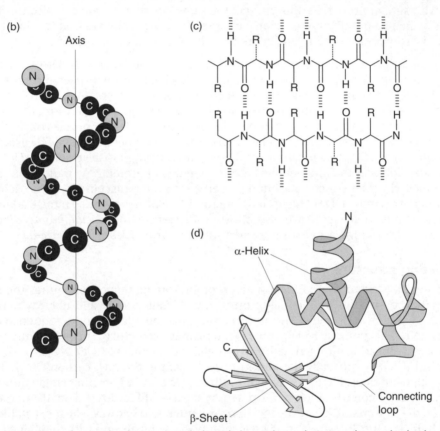

Figure 3. (a) Part of peptide chain, peptide bond shown boxed. Arrow shows the hydrogen and oxygen hydrogen-bonded in an alpha helix. (b) Alpha helix showing only the backbone. (c) Beta pleated sheet. (d) Tertiary structure showing schematic helix and beta sheet domains. From McLennan *et al.* (2012), *Instant Notes in Molecular Biology,* 4th Edn. Garland Science, Abingdon.

chains are hydrophobic. Hydrogen bonding between water molecules squeezes out these side chains which are forced together like oil droplets into the interior of the protein. An alpha helix surrounded by hydrophobic residues may become embedded in a membrane. Hydrophilic side chains associate with water molecules and so tend to be on the surface of proteins. An important function of alpha helix and beta sheet is that it allows hydrogen bonding between the hydrophilic parts of the peptide backbone when it is buried in a hydrophobic protein core. Most proteins are only marginally stable, and two or three unbound hydrophilic sites would stay at the surface and bond to water molecules. The final protein must be stereochemically possible (all bond angles and physical size must be accommodated), all polar atoms must be hydrogen bonded, enough hydrophobic

Table 1. Amino acid codes and properties

Amino acid	Code 3-letter	Code 1-letter	Properties
Alanine	Ala	A	Nonpolar, hydrophobic
Arginine	Arg	R	Polar, hydrophilic
Asparagine	Asn	N	Polar, positive charge (basic)
Aspartic acid	Asp	D	Polar, negatively charged (acidic)
Cysteine	Cys	C	Polar, hydrophilic
Glutamic acid	Glu	E	Polar, negatively charged (acidic)
Glutamine	Gln	Q	Polar, hydrophilic
Glycine	Gly	G	Nonpolar, hydrophobic
Histidine	His	H	Polar, positive charge (basic)
Isoleucine	Ile	I	Nonpolar, hydrophobic
Leucine	Leu	I	Nonpolar, hydrophobic
Lysine	Lys	K	Polar, positive charge (basic)
Methionine	Met	M	Nonpolar, hydrophobic
Phenylalanine	Phe	F	Nonpolar, hydrophobic
Proline	Pro	P	Nonpolar, hydrophobic
Serine	Ser	S	Polar, hydrophilic
Threonine	Thr	T	Polar, hydrophilic
Tryptophan	Trp	W	Nonpolar, hydrophobic
Tyrosine	Tyr	Y	Polar, hydrophilic
Valine	Val	V	Nonpolar, hydrophobic

surfaces must be buried for stability, and the core must be densely packed for Van der Waal's forces to make it stable. Van der Waal's forces are the interactions between electrons of adjacent atoms. They interact to hold atoms together as they touch, but also resist compression and maintain volume.

Predicting binding and function

When the three-dimensional physical structure and the charge distribution of the electron cloud of a complete protein is known, its reactive and binding properties can be estimated. This can be used to design drugs (small molecules that interact with the protein to modify or block its function) to be specific for a single protein type, and to have the desired effect. An example is the interaction of aspirin with prostaglandin cyclooxygenases (COX). One enzyme, COX-1, is expressed in the stomach, and its inhibition causes unwanted side effects. COX-2 is involved with inflammation. COX-2 has a small amino acid, valine, where COX-1 has the larger isoleucine. By designing a molecule shaped like aspirin, but with atoms projecting to bump into the isoleucine, a drug was produced which inhibited COX-2, reducing inflammation, but which was unable to bind COX-1, so avoiding the side effects.

Comparative bioinformatics

An early strength of genetics as a biological tool was the comparison of mutants with normal phenotypes to deduce how individual components of a biological organism worked. Now it is possible to go to databases and compare DNA sequences and protein expression patterns in detail at almost any levels from cells and tissues in an individual through individuals in populations and species to cross-species differences. Differences and similarities between developing embryos (Section C11) can elucidate both cell interactions and the evolution of body form (Evo-devo). Comparison between individuals in a human population and correlation with disease can highlight mutational variation associated with disease susceptibility including cancer (Section F3). Comparison of human populations around the world has shown how humans have evolved into a range of ecotypes as they spread from the hot tropics to colder polar regions and encountered new diseases. The paths of human migration within Africa, and from Africa throughout the rest of the world can be traced in the genes evolving along the track (Section D10). Sampling of all the ribosomal RNA in an environment such as soil, water, deep sea, or ice cores, even cores drilled out of rocks, can allow an accurate sampling of the bacterial species present giving the first real entry to bacterial ecology comparable to the ecology of multicellular organisms. The quantity of data is such that the ingenuity and computing power needed to analyze it is becoming a limiting factor.

F1 Genetic diseases

Key Notes	
Genetic diseases	These are a diverse group of disorders caused by mutations and chromosome abnormalities. Three categories occur. (i) Single gene defects are caused by mutations which result in the synthesis of a defective protein. Mutations may be inherited or arise *de novo* in the parents' germ cells. (ii) Chromosome disorders involve loss or gain of chromosomes or alterations in chromosome structure. Most chromosome disorders arise in the parents' germ cells. (iii) Multifactorial disorders include many common diseases and most congenital abnormalities. They are caused by complex interactions of genes and environmental factors.
Inheritance	Three patterns of inheritance occur. In autosomal dominant disorders inheritance of a single mutant allele causes the disease to occur. Affected individuals are heterozygous and offspring have a 50% chance of being affected. In autosomal recessive disorders both alleles must be mutated for the disease to occur. Heterozygous individuals who have a single mutated allele are carriers. For carrier parents, one in four offspring is affected, one in four is normal, and one in two is a carrier. In X-linked disorders, males are affected and females are usually carriers. For female carriers, 50% of male offspring will be affected and 50% of female offspring will be carriers.
Mutations in single gene disorders	Mutations that cause single gene disorders are diverse. Several types of point mutation occur. Missense mutations alter the encoded amino acid with varying effects on protein function. Nonsense mutations produce stop codons which terminate translation early, usually with a serious effect on protein function. Frameshift mutations result from the insertion or deletion of bases which alter the reading frame, again with a serious effect on protein function. Splice site mutations alter signal sequences at exon-intron boundaries causing abnormal splicing. Promoter mutations affect gene transcription. In some instances a short sequence of bases, usually three, expand *in situ* to cause mutations. Gross mutations include gene deletions, insertions, and rearrangements.
Single gene disorders	These are a diverse group of disorders with variable outlook for sufferers. Cystic fibrosis is an autosomal recessive disorder found mostly in Western European populations. The *CFTR* gene encodes the cystic fibrosis transmembrane conductance regulator protein. This is mutated in cystic

fibrosis. Thick mucous secretions accumulate in the lungs of cystic fibrosis sufferers causing chronic lung disease. Around 2000 *CFTR* gene mutations are known. The Δ508 mutation is found in about 75% of patients. **Huntington's disease** is an autosomal dominant disorder which causes progressive degeneration in the central nervous system. Symptoms appear in late adult life and gradually worsen resulting in death from dementia about 15 years later. The defective *huntingtin* gene contains a triplet repeat sequence which is expanded in Huntington's patients. **Duchenne muscular dystrophy** is an X-linked recessive disorder in which affected males are immobilized by progressive muscle weakness eventually causing death by the age of 20. The defective DMD gene encodes a protein called dystrophin which has a structural role in muscle cells. **Hemophilia** is an X-linked recessive disorder which affects the blood coagulation system. Affected males suffer prolonged bleeding following an injury. Two forms occur, known as hemophilia A and B which result from defects in the *Factor VIII* and *Factor IX* genes. Many different mutations have been identified. A recurrent mutation caused by recombination between a sequence in intron 22 of the *Factor VIII* gene and homologous sequences outside the gene is found in many patients with severe hemophilia A. The **hemoglobinopathies** are a group of disorders caused by defects in the structure and synthesis of hemoglobin. Sickle cell disease is caused by a missense mutation in the β-*globin* gene causing a glutamine to valine substitution at codon 6. The abnormal hemoglobin, HbS, causes sickling of red blood cells leading to anemia and tissue ischemia. The **thalassemias** are a group of disorders in which globin chain synthesis is defective. α thalassemia is caused by deletion of α-*globin* genes. β thalassemia is caused by a range of β-*globin* gene mutations which disrupt β-*globin* mRNA synthesis. **Charcot-Marie-Tooth Disease** results in muscle wasting. This can be caused by point mutation or by duplication of a gene on chromosome 17.

Identifying genes causing predisposition to single gene disorders

Linkage analysis has been successfully used to assign genes which cause disease to specific regions of the genome. Once these regions have been delimited all genes that lie in them are regarded as possible candidates. Comparisons of DNA sequences between individuals who suffer from the disease and controls can identify which genes are causative for the disease. Several factors including incomplete penetrance can hamper this approach. If the defective protein that causes the disease is known the relevant gene can be identified from the mRNA.

Multifactorial disorders

Diabetes is a good example of a multifactorial disease. The disease may be associated with particular families, but because of the importance of environmental factors it does not show a Mendelian pattern of inheritance. Two methods are used to identify genes involved in this type of disease. Genome-wide association detects alleles that are found more commonly in sufferers than in the whole population. This has shown that diabetes type 1 is associated with specific alleles of the human *HLA* locus. Analysis of affected sib pairs in families which display a history of diabetes can identify regions of the genome containing a gene involved in the disease. A number of genes involved in diabetes have been identified by these methods.

Related topics	(A5) DNA mutations	(C10) Sex and inheritance
	(A8) Epigenetics and chromatin modification	(F2) Genetic screening
	(B4) Chromosomes	(F3) Genes and cancer
	(C8) Quantitative	(F4) Gene therapy

Genetic diseases

These are a diverse group of conditions which result from gene mutations and chromosome abnormalities. Disorders with a genetic basis fall into three categories:

- *Single gene defects* (also known as **Mendelian disorders, monogenic disorders**, or **single locus disorders**). These are a group of diseases caused by the presence of a single mutated gene. The mutation alters the coding information of the gene such that it either produces protein which is defective or fails to produce any protein at all. The resulting protein deficiency is responsible for the disease symptoms. The gene mutation may be passed between generations from parents to children or may arise spontaneously (*de novo*) in a germ cell (sperm or ovum) of a parent which gives rise after fertilization to a child who carries the mutation in every cell.
- *Chromosome disorders.* These are conditions caused by the loss or gain of one or more chromosomes or by alterations in chromosome structure. Most chromosome disorders arise *de novo* in the parents' germ cells but examples of inherited chromosome disorders also exist. Abnormalities relating to the number of chromosomes may involve the presence of multiple copies of each chromosome (**polyploidy**) or the gain or loss of individual chromosomes (**aneuploidy**; Section B4). Structural chromosome abnormalities result from chromosome breakage and may involve the deletion, duplication or rearrangement of chromosome segments.
- *Multifactorial disorders.* These include many common diseases such as diabetes, asthma, hypertension, and coronary artery disease as well as most congenital malformations. They are influenced by several genes in complex ways. This means that the diseases are associated with specific families but do not show a defined Mendelian inheritance pattern (Sections C1 and C2). The genes involved respond to specific environmental triggers so the phenotype represents a complex interaction of multiple genes and environmental factors.

During the last 20 years, there have been enormous advances in our understanding of single gene disorders. The information in the following sections refers to some of these disorders.

Inheritance

Single gene disorders are passed on between generations from parents to their children. Three patterns of inheritance occur: **autosomal dominant**, **autosomal recessive**, and **X-linked** (Table 1). Every human cell contains 22 pairs of homologous chromosomes known as **autosomes** and a pair of X or Y **sex chromosomes**. Females have two X chromosomes and males have a single X chromosome and a Y chromosome. Duplicate copies of genes occur on chromosome pairs and are known as **alleles**.

In **autosomal dominant** disorders, the inheritance of a single mutated allele is sufficient for an individual to be affected by the disease. Affected individuals have one normal allele and one mutated allele and are **heterozygous**. The child of an affected individual will have a 50% chance of inheriting the mutated allele and being affected by the disorder themselves (Figure 1a).

In **autosomal recessive** disorders, two mutant alleles (one from each parent) must be inherited for an individual to be affected by the disease. Affected individuals are said to be homozygous for the mutant allele. Individuals who inherit a single mutant allele are heterozygous and are not affected by the disease but are carriers and may pass their mutant allele on to their children. When both parents are heterozygous carriers of an autosomal recessive disorder, one in four of their children will be affected by the disorder, one in four will be normal, and one in two will be carriers (Figure 1b; Section C1).

Table 1. Single gene disorders

Disorder	Frequency per 1000 births	Pattern of inheritance	Mutated gene	Characteristics
Hemophilia A	0.1	X-linked	*Factor VIII*	Abnormal bleeding
Hemophilia B	0.03	X-linked	*Factor IX*	Abnormal bleeding
Duchenne muscular dystrophy	0.3	X-Linked	*Dystrophin*	Muscle wasting
Becker muscular dystrophy	0.05	X-linked	*Dystrophin*	Muscle wasting
Fragile X syndrome	0.5	X-linked	*FMR1*	Mental retardation
Huntington's disease	0.5	Autosomal dominant	*Huntingtin*	Dementia
Neurofibromatosis	0.4	Autosomal dominant	*NF-1,2*	Cancer
Thalassemia	0.05	Autosomal recessive	*Globin* genes	Anemia
Sickle cell anemia	0.1	Autosomal recessive	β-*globin*	Anemia, ischemia
Phenylketonuria	0.1	Autosomal recessive	*Phenylalanine-hydroxylase*	Inability to metabolize phenylalanine
Cystic fibrosis	0.4	Autosomal recessive	*CFTR*	Progressive lung damage and other symptoms

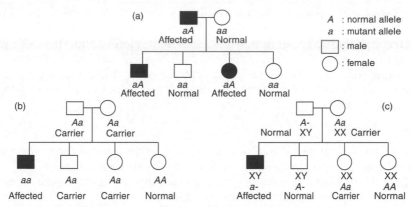

Figure 1. (a) Autosomal dominant. Inheritance of a single mutated allele (*a*) results in disease. (b) Autosomal recessive. Affected individuals have two mutant alleles (*aa*). Heterozygotes (*Aa*) are carriers. (c) X-linked. For female carriers, 50% of male offspring are affected and 50% of female offspring are carriers.

In **X-linked disorders**, the mutated gene is present on the X chromosome. As males have a single X chromosome, inheritance of a mutated allele is sufficient for the disease to occur. Affected males are said to be **hemizygous**. Females have two X chromosomes and usually remain unaffected since most X-linked disorders are recessive, although the mutations will be expressed in half of their cells because of X inactivation (Section A8). Females may be carriers of X-linked disorders with 50% of their male offspring affected and 50% of female offspring being carriers (Figure 1c; Section C9).

Mutations in single gene disorders

Two types of mutation occur in single gene disorders: **point mutations** which involve single base changes and **gross mutations** which involve alteration of longer DNA sequences (Section A5). For each disorder, the type of mutation present varies. In addition, individuals affected by the same disorder may carry different mutations of the same gene.

A number of different types of point mutation exist:

- *Missense mutations.* These are base changes that alter the codon for an amino acid resulting in its replacement with a different amino acid. The effect the mutation has on the encoded protein depends on the importance of the amino acids being replaced and the nature of the substitute amino acid (charged/uncharged, acidic/basic, hydrophilic/hydrophobic, large/small).
- *Nonsense mutations.* These are base changes that convert an amino acid codon to a stop codon resulting in premature termination of translation and the production of a shortened protein. Nonsense mutations usually have a serious effect on the encoded protein, especially if they occur near the 5′ end of the gene.
- *Frameshift mutations.* These mutations result from the insertion or deletion of one or more bases causing the reading frame of the gene to be altered and a different set of codons to be read downstream (3′) of the mutation. Frameshift mutations usually have serious consequences for the encoded protein, especially when they occur near the 5' end of the gene.
- *Splice site mutations.* These mutations alter the signal sequences for splicing that occur at the 5' and 3' ends of exons leading to a failure of RNA transcribed from the

mutated gene to be spliced properly. Mutations also occur in introns that result in the creation of new splice sites, again causing splicing to occur abnormally.

- *Promoter mutations.* These mutations occur infrequently and affect the way gene transcription is regulated, often reducing or eliminating expression of the gene.
- *Triplet expansion.* In these mutations a triplet of bases undergoes expansion *in situ*. This can cause problems by altering the coding sequence of a gene or by interfering with gene expression.

Gross mutations include:

- *Deletions.* These mutations involve a loss of part of the normal gene sequence. Deletions vary greatly in size from just a few bases to the entire gene sequence.
- *Insertions.* These mutations involve the insertion into the normal gene sequence of a segment of DNA from elsewhere in the genome.
- *Rearrangements.* These mutations involve reciprocal movements between segments of gene sequence and sequences elsewhere in the genome.
- *Duplications.* In these cases tandem duplications of a stretch of DNA cause a gene to be present twice and this results in increased expression of the gene thus altering the phenotype.

Single gene disorders

To date, over 10 000 single gene disorders have been identified in humans. Most are rare but together they affect between 1 and 2% of the population. The following is a list describing some of the more important and common single gene disorders. A more complete description is provided in some of the articles listed in the Further Reading section.

Cystic fibrosis

Cystic fibrosis (CF) is an autosomal recessive disorder. It is common in individuals of Western European origin where it affects about 1 person in 3000. A much lower incidence is seen in non-Europeans. In patients with CF, thick mucous secretions accumulate in the lungs which block the airways and cause persistent infections. Fibrous tissue forms in the lungs as a result of the infections and leads to permanent lung damage. Antibiotics and physiotherapy are used to lessen the effects of the mucous secretions but most patients die as a result of their chronic lung disease by the age of 40. The only effective treatment at present is a heart-lung transplant. Mucous secretions also affect the pancreas blocking the pancreatic ducts and preventing release of pancreatic digestive enzymes. However, this can be treated with dietary supplements of pancreatic enzymes. The liver and reproductive tract may also be affected.

The defective gene in cystic fibrosis is the *CFTR* (CF Transmembrane conductance Regulator) gene on chromosome 7. This encodes a large cell membrane protein which acts as a chloride channel and also regulates other chloride, sodium, and potassium channels. The normal CFTR protein secretes chloride ions from cells into mucus and alters the sodium chloride balance reducing the viscosity of mucous secretions.

The CF mutation database has 1903 different *CFTR* gene mutations listed but a single mutation of European origin involving deletion of the three bases of codon 508 (Δ508) occurs in about 75% of CF patients. This mutation has a very dramatic effect on phenotype as the polypeptide is not processed properly in the Golgi apparatus and no protein is inserted into the membrane. The Δ508 mutation can be identified in a simple PCR-based test which is used to identify heterozygous carriers of the mutation. The other CF mutations identified include missense, frameshift, nonsense point, and deletion mutations. Most are rare but a few occur at significant levels in certain populations. For example, a point mutation at codon 542 is seen in 12% of CF patients belonging to the Ashkenazi Jewish population. Certain CF mutations alter the splicing of CF mRNA. These tend to have a less severe phenotype as a proportion of normally spliced mRNA is also produced.

The relatively high frequency of CF mutations in Caucasian populations has led some researchers to suggest that, at some point in the past, CF heterozygotes must have had a selective advantage. Suggestions for the most likely effect include partial tolerance of cholera or salmonella infection or reduced diarrhea caused by lactose intolerance when prehistoric animal herders started drinking milk. Heterozygote advantage is known to explain the high frequency of sickle cell anemia mutations (Section D5).

The inevitably fatal outcome in CF and the current lack of an effective treatment have led to attempts to develop a cure based on gene therapy. This involves delivering a functional *CFTR* gene into the lungs using an inhaler. Some progress has been made and researchers are cautiously optimistic about future success (Section F4).

Huntington's disease

Huntington's disease (HD) is an autosomal dominant disorder which affects the central nervous system (CNS). It has a late onset with symptoms only appearing after about the age of 40. The disease is progressive. Initially the symptoms are mild but gradually get worse over time. Patients with HD suffer involuntary movements such as twitching of the limbs and face and have slurred speech. As the disease progresses, they lose their intellectual abilities and have psychiatric disturbances which eventually lead to dementia and death, usually about 15 years after the initial appearance of the symptoms.

The defective gene responsible for HD is called ***huntingtin*** and is located on chromosome 4. The gene is expressed in many cells in the CNS but the function of its encoded protein is unclear. Recent evidence suggests that the protein interacts with the cytoskeleton and may transport vesicles of brain-derived neurotrophic factor. The defect in the *huntingtin* gene which causes HD is typical of triplet expansion mutations. The gene contains a repeated CAG sequence near the 5' end. In normal individuals, the sequence is repeated 26 times or less. However in individuals with HD the repeat sequence is expanded and occurs more than 40 times. The number of repeats tends to increase in successive generations and the age of onset of the symptoms decreases as the number of repeats increases. At present, there is no known treatment or cure for HD.

Duchenne muscular dystrophy

Duchenne muscular dystrophy (DMD) is an X-linked recessive disorder in which affected males suffer progressive weakening of the muscles. Mutations that produce a protein with some remaining function cause a milder form known as **Becker muscular dystrophy** (BMD). DMD affects about 1 male in 3500. Symptoms first appear between the ages of 3 and 5 and worsen with time. Increasing muscle weakness affects patients' mobility and most are confined to a wheelchair by the age of 11. There is no effective cure for DMD and most patients die from heart or lung failure before the age of 20.

The gene responsible for DMD is present on the X chromosome. At 2.3 million base pairs, it is one of the largest genes in the human genome. About two-thirds of the mutations found in DMD patients are deletions. The size and the location of the deletions vary but two regions of the DMD gene, one involving the first 20 exons and the second near the center of the gene, are affected more often. Mutations identified in the remaining one-third of DMD patients are point mutations which result in reduced or absent gene expression. These include nonsense, frameshift, splicing, and promoter mutations.

The DMD gene encodes a large protein called **dystrophin** which has a structural role in muscle cells. Dystrophin is found close to the cell membrane and acts as link between actin protein molecules on the inside of the cell and laminin protein molecules on the outside. Absence of dystrophin causes gradual degeneration of muscle cells.

Gene therapy offers the only realistic hope of a future cure for DMD. Transgenic mice which lack dystrophin have been used to test methods of delivering a synthetic *dystrophin* gene into muscle cells but effective gene therapy for patients remains a prospect for the future only.

Hemophilia

Hemophilia is an X-linked recessive disorder which affects the blood's ability to clot. Individuals with hemophilia bleed for prolonged periods following an injury. There are two types of hemophilia (A and B) which have similar symptoms but are caused by separate genetic defects. **Hemophilia A** affects 1 in 5000 males and is more common than **hemophilia B** which affects 1 in 40 000.

There is a wide variation in the severity of the disease among hemophilia sufferers. Some patients are mildly affected and suffer bleeds only rarely. Others are severely affected and may bleed spontaneously into muscles and joints causing severe pain and swelling. Multiple bleeds may cause a joint to be destroyed and results in serious disability.

Hemophilia A is caused by mutations in the *Factor VIII* gene which is located on the X chromosome. The gene encodes the Factor VIII protein which is a plasma-clotting factor with an important role in the blood coagulation system. Hemophilia B is caused by mutations in the gene for another important coagulation protein known as Factor IX. Both types of hemophilia can be treated very effectively with injections of Factor VIII or Factor IX purified from donated plasma or produced synthetically by genetic engineering.

More than 600 different *Factor VIII* gene mutations have been identified in patients with hemophilia A. About 75% are point mutations with deletion mutations accounting for a further 5%. The severity of the hemophilia in individual patients is directly related to the type of mutation present. Gross mutations and point mutations such as nonsense and frameshift mutations, which severely affect the ability of the gene to produce intact Factor VIII protein, are associated with severe hemophilia, whereas missense mutations, which may only have a minor effect on the encoded protein, tend to be associated with mild disease.

A recurrent mutation is found in about 20% of patients. The mutation is caused by recombination between a sequence present in intron 22 of the *Factor VIII* gene and homologous sequences further along the X chromosome. This results in the *Factor VIII* gene being split into two sections separated from each other by several million base pairs of intervening sequence. The mutation completely disables *Factor VIII* gene function and causes severe hemophilia.

Over 400 *Factor IX* gene mutations have been identified in hemophilia B patients. As for hemophilia A, most are point mutations with some deletion and insertion mutations also found.

Hemophilia A and B are good candidates for gene therapy programs since only a small increase in plasma levels of Factor VIII or IX results in greatly improved blood clotting. Results of tests carried out in animals show that gene therapy for hemophilia is feasible but early human trials have failed to reach sufficient levels when the gene is introduced into muscle and work well but only temporarily in liver (Section F4).

Hemoglobinopathies

The hemoglobinopathies are a group of disorders caused by inherited abnormalities of hemoglobin – the protein in red blood cells responsible for the transport of oxygen. In adults each molecule of hemoglobin (Hb) contains two α-globin and two β-globin polypeptide chains bound to a small molecule called heme. Mutations in the α- and β-*globin* genes are the underlying cause of the hemoglobinopathies.

Two types of hemoglobinopathy occur: (i) abnormal hemoglobins which result from structural defects in the globin polypeptide chains and (ii) thalassemias which are caused by defects in the synthesis of the globin chains.

Abnormal hemoglobins

Abnormal hemoglobins result from *globin* gene mutations that affect the structure of the globin polypeptides. Some of the mutations identified affect the stability of the hemoglobin molecule, causing the globin chains to precipitate inside the red cells, damaging the cell membrane and causing red cell lysis. Other mutations affect the ability of the hemoglobin molecule to carry oxygen. Over 300 abnormal hemoglobins have been identified. Most are harmless but a few are associated with disease and occur at significant levels in certain populations and ethnic groups.

Sickle cell disease is the most common **hemoglobinopathy**. It is caused by a point mutation which results in the substitution of glutamine with valine at codon 6 of the β-globin polypeptide. The abnormal hemoglobin produced is called HbS. The substitution of glutamine with valine makes HbS less soluble than normal Hb and causes the HbS molecules to polymerize. This has the effect of deforming the red blood cells, causing them to adopt a sickle shape. The abnormal red cells are quickly removed from the circulation by the spleen, causing serious anemia. The sickle cells also tend to become lodged in small blood vessels, blocking them and depriving tissues of their oxygen supply (ischemia). Heterozygous carriers of the sickle cell mutation remain healthy but homozygotes suffer a number of serious complications including kidney failure, heart failure, pneumonia, and strokes. The mutation is maintained in the population in areas where malaria is endemic because heterozygotes have a tolerance of malaria.

Thalassemias

The thalassemias are a group of disorders caused by defects in the synthesis of the globin polypeptides. Absent or reduced synthesis of one of the globin chains results in an excess of the other. In this situation, free globin chains, which are insoluble, accumulate inside the red cells and form precipitates which damage the cell, causing cell lysis and resulting in anemia. There are two main types of thalassemia in which synthesis of α- or β-globin is defective. Thalassemia is common in Mediterranean countries, the Middle East, the Indian subcontinent, and in South-East Asia where malaria is (or was) present but the sickle cell allele is absent.

In humans, two copies of the α-*globin* gene (four alleles) are present on chromosome 16. In α **thalassemia**, deletion of one or more of the four α-*globin* alleles occurs causing reduced or absent synthesis of the α-globin polypeptide. The severity of the disorder depends on how many of the four alleles are deleted. In the most severe form, all four alleles are deleted resulting in a complete absence of α-globin synthesis. This form is incompatible with life and results in fetal death due to severe anemia, a condition known as **hydrops fetalis**. Milder forms of α thalassemia in which some α-globin is produced are associated with deletion of 1–3 alleles. In this case, the fetus survives and the affected individuals suffer varying degrees of anemia.

In β **thalassemia**, underproduction of β-globin chains occurs as a result of mutations which affect the ability of the gene to produce β-globin mRNA. Over 100 different mutations are known to cause β thalassemia including promoter, splicing, and nonsense mutations. Unlike α thalassemia, the β form is rarely caused by gene deletion. Individuals who are homozygous for β-*globin* gene mutations are said to have thalassemia major and require multiple blood transfusions due to severe anemia. Patients who are heterozygous for β-*globin* gene mutations are said to have thalassemia minor and usually remain healthy.

Charcot-Marie-Tooth Disease

This is the most common inherited peripheral neuropathy, with a frequency of 1 in 2500 of the human population. It results in progressive wasting of the muscles of the arms and legs. Symptoms are fully expressed by the mid-thirties. The disease is caused by mutation of several different genes, but the most common form (CMT1) is caused by mutation of the *peripheral myelin protein-22* gene (*PMP-22*) situated on the short arm of chromosome 17. Point mutations in this gene have been demonstrated in some cases of CMT1 and in a similar syndrome, Trembler-J, found in mice. However, in many cases of the disease, a **tandem duplication** of approximately 1.5 Mb is observed in chromosome 17. The *PMP-22* gene is completely contained within the duplicated region. The duplication has been shown to be the result of unequal crossing over in spermiogenesis. It is now considered that the syndrome can be the result of either mutation or duplication of *PMP-22* leading to over-expression of the gene.

Identifying genes causing predisposition to single gene disorders

Several methods have been used to identify genes responsible for inherited disease. For those diseases that show clear Mendelian inheritance patterns **linkage analysis** (Sections B1 and C5) is important. This can map the gene to a region bounded by specific SNPs or VNTRs. The next step is to identify all genes in the region. Any genes that have expression patterns that correlate to the tissues that are involved in the disease or that code for proteins that could be involved in the disease phenotype are considered **candidate genes**. The sequence of these genes in individuals who have the disease is compared with controls. Candidate genes with no mutational differences from normal can be excluded. Finally functional assays are needed to unambiguously indentify the causative gene. Apart from mutations in exons, alteration of splice sites (Section A4), or copy number variation (Section A5) can also be indicative.

A number of difficulties can make linkage analysis more difficult. The number and size of families in which the disease is segregating can make it difficult to map the gene accurately, but data from different families can be combined. Some diseases, for example Xeraderma pigmetosum, can be caused by several unrelated and unlinked genes. Similarly misdiagnosis can be a problem where two diseases have overlapping symptoms. In addition, not all individuals who have the appropriate genotype develop the disease. This is known as **incomplete penetrance** and arises where a mutation requires some other environmental or genetic factor for the disease to arise. For example inherited breast cancer has a penetrance of approximately 80%, meaning that one woman in five who carries the dominant mutation in *BRCA1* (Section F3) does not develop the disease. Despite these difficulties linkage analysis has been successful in identifying most single genes that cause predisposition to disease in humans.

In some cases the specific protein that is responsible for a gene is known. Obvious examples are the hemoglobinopathies where disease is due to abnormalities in one of the globin chains. Here the gene responsible can easily be identified by using the mRNA that codes for the protein to identify the gene.

Multifactorial disorders

Diabetes is a classical example of a multifactorial disease. Although it is clear that the disease has a high frequency in many families there is no evidence of any pattern of Mendelian inheritance. The disease is characterized by high sugar levels in the blood. There are two main forms of the disease: type-1 diabetes is characterized by early onset and loss of the Islets of Langerhans in the pancreas, the site of insulin production; type-2 diabetes shows a later onset and is often associated with obesity. The latter form accounts for approximately 90% of the incidence of the disease. Both forms of diabetes are phenotypes that result from complex interactions between several different genes and the environment. Two major methods are used to identify genes that are involved in the development of diabetes. Both of these depend on the availability of large numbers of polymorphic DNA markers such as VNTRs and SNPs (Section B1).

The simplest approach, termed **genome-wide association** (Section B1), compares polymorphic DNA markers in normal and affected individuals. Any alleles that occur more frequently in affected individuals than normal individuals may lie close to genes involved in the disease. In type-1 diabetes this approach has shown that specific alleles of the human *HLA* locus give a significantly increased risk of diabetes. This locus accounts for approximately 50% of the genetic contribution to the disease. Over 30 loci, mainly genes related to the immune system, insulin signaling, and sugar transport and metabolism, are reported as showing association with type-2 diabetes.

The second approach involves families in which diabetes is common. Polymorphic DNA markers in **affected sib pairs** are compared. If a marker is inherited by affected siblings more frequently than expected from normal Mendelian predictions, this provides evidence of probable linkage (Section C5) between the marker and a gene responsible for susceptibility to diabetes. Data from many families are pooled for analysis. This approach has indicated up to 20 possible chromosomal regions involved in type-1 diabetes, although only three genes have been identified so far. In type-2 diabetes a number of chromosomal regions are involved but only one gene, *hepatocyte nuclear factor 4A* on chromosome 20, has been positively identified.

Other approaches to identifying genes involved in diabetes include the use of animal models of the disease and identifying proteins that are important in the production of insulin and meditating its effects.

F2 Genetic screening

Key Notes

Genetic screening

Genetic screening is used to identify otherwise healthy carriers of autosomal recessive and X-linked genetic diseases. It can also be used to predict the development of disease in autosomal dominant disorders which have late onset.

Biochemical markers

Biochemical markers are proteins or metabolites whose plasma levels can be used to identify carriers of recessive genetic disorders. The presence of one normal and one mutated allele in carriers predicts that the marker level should be 50% of normal. The usefulness of biochemical markers is limited by unpredictable variations in levels caused by non-genetic influences.

Screening for disease-causing alleles

Where a specific mutation is known to be inherited in a family the relevant exon can be assessed by sequencing. Alternatively there are a number of allele-specific methods based on primers designed to identify mutant alleles followed by PCR or using DNA microarrays which carry all known mutations for the gene in question. Gene tracking, where the inheritance of a DNA marker closely linked to the mutant allele is monitored over several generations of a family, was widely used before sophisticated techniques for recognizing mutations became available.

Pre-natal diagnosis

Pre-natal diagnosis is used to detect genetic defects in an unborn child. Fetal cells for genetic analysis can be obtained by removing amniotic fluid (amniocentesis) at 16 weeks or sampling the chorionic villi at 12 weeks. Both procedures carry a significant risk of miscarriage. Preimplantation genetic diagnosis involves analysis of DNA from a cell removed from an embryo at the eight-cell stage. Down's syndrome can be identified from fetal cells in a mother's blood.

Ethical issues

Genetic screening raises ethical issues concerning benefits gained by patients. Identification of carriers of recessive disorders provides obvious benefit. Predictive screening of autosomal dominant disorders may be beneficial depending on the availability of treatments. Other issues include informed consent and testing of children.

Related topics

(B3) Eukaryote genomes	(E2) PCR and related
(C1) Basic Mendelian genetics	technology
(C3) Meiosis and gametogenesis	(F1) Genetic diseases
(E1) Using sequence specificity	(G6) Ethics
to study nucleic acids	

Gonotic screening

The purpose of genetic screening is to identify individuals who are carriers of genetic diseases and are at risk of passing the condition on to their children. By providing accurate information about the risk of having an affected child, individuals can make informed choices about parenting.

Inheritance of single gene disorders can be autosomal dominant, autosomal recessive or X-linked (Section F1). In autosomal dominant disorders, the identification of carriers is straightforward since these individuals will already have the disease. However, in autosomal recessive and X-linked disorders, carriers remain healthy and specialized methods are needed to distinguish them from noncarriers. In a few autosomal dominant disorders, such as Huntington's disease, where symptoms of the disease only appear later in life, genetic screening can be used to identify individuals who will go on to develop the disease. A number of different techniques are used to identify carriers of single gene disorders.

Biochemical markers

Carriers of autosomal recessive and X-linked disorders have one normal allele and one mutated allele and, in theory, should have only 50% of the amount of the gene product present in a noncarrier with two normal alleles. In single gene disorders, where the gene product appears as a protein in the blood, carriers can sometimes be identified by measuring the level of that protein. For example, in Tay–Sachs disease (an inborn error of metabolism), levels of the enzyme hexosaminidase A in carriers fall between those seen in normal and affected individuals. In other cases, this approach is less reliable. For example, in female carriers of the X-linked disorder hemophilia A, levels of the coagulation protein, Factor VIII, would be expected to be 50% of that seen in normal individuals. In practice, Factor VIII levels in plasma are influenced by many different factors including female sex hormones (whose levels vary during the menstrual cycle) and proteins that mediate the inflammatory response. As a result, even in normal individuals, Factor VIII levels are subject to unpredictable variations and can only be considered as an indicator of hemophilia carrier status.

In some cases, the biochemical marker is not the gene product but a protein or metabolite whose level is altered by the gene product or as part of the disease process. For example, in Duchenne Muscular Dystrophy, the muscle enzyme creatine kinase leaks into the blood as a result of defects in the muscle membrane. Carriers of DMD often have higher plasma levels of creatine kinase than noncarriers but sometimes the levels overlap and additional investigations must be carried out to confirm the diagnosis of carrier status.

Screening for disease-causing alleles

It is now possible to analyze the entire genome of an individual to identify alterations in DNA sequence that would identify them as a carrier of a genetic disease or as being likely to develop a late onset disease. Although the cost of sequencing DNA is dropping rapidly this approach is not yet economically practical. In most cases an individual will only be concerned about the chance that they will have disease susceptibility alleles at only one or a small number of genes. If there is a family history of the disease and the mutation responsible is known, then it is straightforward to test for the presence or absence of that specific mutation by sequencing the chromosomal region, for example the individual exon in question. There are, however, some difficulties in this approach if the disorder is recessive, because two copies of the relevant sequence will be present and it may be difficult to distinguish the mutant from the normal (Section E3). It may be better to use allele-specific methods to identify the mutation. A number of these are available that depend on hybridization of specific primers constructed to differentiate between mutant and normal sequences, followed by PCR or sequencing reactions. Often it is possible to screen for a number of mutations simultaneously in multiplex reactions by using carefully selected primers. The process can be automated by fluorescently labeling the primers and separating PCR products using the capillary electrophoresis technique to identify each product on the apparatus used for DNA sequencing (Section E3). DNA microarrays are also being used to identify mutations and microarrays are now available which contain all common mutations of a specific gene such as *CFTR*, the gene involved in cystic fibrosis. These methods can also be applied to individuals with no family history of an inherited illness but new or rare mutations might be missed unless the whole gene is sequenced.

Gene tracking was an early approach of screening for specific disease-causing alleles where there was a family history of the disease. Testing was targeted to identify a specific VNTR or SNP marker that was known to be tightly linked to an allele causing disease within the family. This was a very useful development but there was always a chance that recombination had taken place between the marker and the disease-causing allele. This approach is rarely used today and has been replaced by direct screening for mutant alleles.

Pre-natal diagnosis

Pre-natal diagnosis is a procedure used to detect defects in an unborn child and is offered during pregnancy to parents at risk of having a child with a genetic abnormality. The decision to undergo pre-natal testing is a difficult one for parents since, depending on the outcome, they may have to consider termination of the pregnancy. A number of techniques are used for pre-natal diagnosis.

Amniocentesis

This procedure is usually performed around the 16th week of pregnancy and involves using a needle inserted through the wall of the abdomen to remove 10–20 ml of the amniotic fluid which surrounds the fetus. The amniotic fluid contains cells derived from the fetus which can be collected by centrifugation. Fetal DNA can be isolated from the cells for genetic analysis or they can be grown in a flask placed in an incubator to provide material for chromosome analysis or biochemical studies. The disadvantages of amniocentesis include a risk of miscarriage (0.5–1% of procedures) and the possibility of having to consider a termination relatively late in pregnancy.

Chorionic villus sampling

This procedure is carried out around the 12th week of pregnancy and involves removing a sample of the chorionic villi which are derived from the outer cell layer of the embryo. As for amniocentesis, the tissue removed can be used to extract fetal DNA or cultured to provide fetal cells for chromosome analysis or biochemical studies. Chorionic villus sampling has the advantage that it can be carried out earlier in pregnancy but the procedure also carries a significant risk of miscarriage (2–3%).

Preimplantation genetic diagnosis

This new technique allows genetic testing of embryos at a very early stage. The procedure involves removing an egg from the mother and fertilizing it *in vitro*. The fertilized embryo is grown in a petri dish until it reaches the eight-cell stage. At this point, one of the cells is removed and can be analyzed by PCR for genetic defects. If the embryo is normal, it can be implanted into the mother's uterus and will go on to develop normally.

Analysis of maternal blood

A fetus with Down's syndrome carries an extra copy of chromosome 21. Recent research has shown that it is possible to detect elevated levels of chromosome 21 DNA in the mother's blood. This is due to small numbers of fetal cells entering the mother's circulation and sensitive assays can detect the increases in chromosome 21 DNA. This procedure has the major advantage of being noninvasive and may be applied to other trisomies.

Ethical issues

Genetic screening raises a number of specific ethical issues. A more general discussion of ethical issues raised by genetics as a subject is given in Section G4.

The benefits of genetic screening are clearest for individuals affected by autosomal recessive and X-linked disorders. Providing an accurate diagnosis of carrier status allows individuals to make informed choices about having children. The benefits are less clear where genetic screening is used in autosomal dominant disorders to identify individuals who will go on to develop a disease in later life. For example, familial adenomatous polyposis is a disorder in which affected individuals are at risk of developing cancer of the colon. In this case, genetic screening is probably beneficial because affected individuals can be closely monitored for the development of tumors allowing early treatment before the cancer is too advanced. In contrast, the benefits of genetic screening for Huntington's disease, for which there is no effective treatment to delay the onset or progression of symptoms, are less clear.

Other ethical issues raised by genetic screening include informed consent and testing children. It is important that persons undergoing testing for genetic diseases should do so as a free choice without pressure from outside influences such as employers or insurance companies. The issue of testing children is contentious since the child may not fully understand the consequences of the result. Many argue that testing of children should be left until they are old enough to make an informed choice.

F3 Genes and cancer

Key Notes

Cancer as a genetic disease

Several lines of evidence suggest that genes are involved in cancer. Tumor cells are known to have altered numbers of chromosomes and the chromosomes have often undergone rearrangements. Some translocations are specific to certain forms of cancer. Mutagens are potent causes of cancer. In rare cases predisposition to cancer is inherited in families.

Oncogenes

These are derived from protooncogenes, genes that normally play a role in regulating cell proliferation by signal transduction or apoptosis. The protooncogenes are activated to form oncogenes by mutation or over-expression. They were first identified in oncogenic retroviruses, but can be detected in nonviral tumors by transfection into mouse 3T3 cells. Oncogenes in retroviruses are referred to as v-*onc* and those in tumors as c-*onc*. Oncogenes are sometimes amplified in tumor cells.

Cancer-specific chromosome translocations

Some tumors are characterized by specific translocations which can result in increased transcription of an oncogene; this is found in Burkitt's lymphoma. Alternatively the fusion of two protooncogenes may form a novel gene with oncogenic properties; this is observed in chronic myeloid leukemia.

Tumor suppressor genes

Tumor suppressor genes require both copies to be inactivated before tumors can develop. This is usually achieved by mutation of one allele and deletion of the other. The best studied tumor suppressor gene is *p53*, which is mutated in about 40% of human tumors. It has several roles in both regulating the cell cycle and apoptosis. Inactivation of tumor suppressor genes may involve DNA methylation.

Hereditary cancers

Hereditary cancers are caused by the inheritance of mutant alleles of tumor suppressor genes. Their pattern of inheritance is that of autosomal dominants. Mutant alleles of DNA repair genes can also confer inherited susceptibility to cancer.

RNA splicing and cancer

Altered patterns of RNA splicing are commonly found in tumor cells. These can result from mutation within the intron/exon splice site. In addition, where no new mutation is present, changes in cellular splicing processes can produce alternative patterns of splicing of the pre-messenger RNA molecules.

Gene analysis and cancer treatment	DNA chips are now used to identify specific genetic changes in tumors and can indicate specific therapies for individual patients.	
Related topics	(A6) Mutagens and DNA repair	(B5) Cell division
	(A8) Epigenetics and chromatin modification	(F1) Genetic diseases

Cancer as a genetic disease

There are several distinct lines of evidence which indicate that cancer is a disease caused by alterations to genes:

- it has long been known that mitosis is less precise in tumor cells leading to variation in chromosome number between cells of the same tumor. This is referred to as heteroploidy;
- chromosomes in tumors frequently show structural rearrangements. Although most of the rearrangements appear to be random in origin, several cancers have specific rearrangements;
- it is clear that the majority of mutagens are also carcinogens (agents that cause cancer);
- predisposition to either a single or multiple forms of cancer is found to be inherited in some families.

Although this constitutes strong evidence for a major genetic involvement in cancer, it is necessary to identify specifically those genes that are involved before the hypothesis is proven. Three classes of genes have been identified: oncogenes, tumor suppressor genes, and DNA repair genes.

Oncogenes

These were first identified in a group of oncogenic (cancer-causing) animal viruses, the retroviruses (Section B7). These viruses have an RNA genome in the virus particle, but after infecting a cell the genome is converted to DNA by the enzyme **reverse transcriptase**, and integrated into the host cell's DNA. Oncogenic strains of retroviruses differed from nononcogenic strains by the presence of an extra gene. This gene is referred to as an **oncogene**. DNA sequences closely related to viral oncogenes are also found in DNA from normal healthy animals. Oncogenic retroviruses are the result of genetic recombination between nononcogenic viral strains and cellular genes (Figure 1). To differentiate between the viral and the normal cellular copies of the genes the cellular copy is called a **protooncogene**. The viral oncogene differs from the protooncogene in that it is mutated. When present in a retrovirus it is under the control of a powerful enhancer, the viral long terminal repeat (LTR), which increases its rate of transcription. Mutation and/or overexpression is referred to as **oncogene activation**.

Oncogenes are also identified in DNA from tumors where there is no viral involvement. To do this DNA is extracted from tumor cells and transfected into a mouse cell line, 3T3. These cells grow in tissue culture attached to the surface of the culture dish. After they have divided sufficiently to cover the surface of the dish, cells stop dividing; this is known

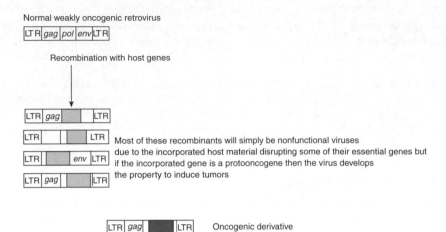

Normal weakly oncogenic retrovirus

Recombination with host genes

Most of these recombinants will simply be nonfunctional viruses due to the incorporated host material disrupting some of their essential genes but if the incorporated gene is a protooncogene then the virus develops the property to induce tumors

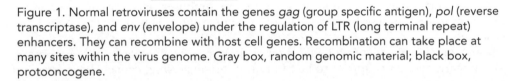

Oncogenic derivative

Figure 1. Normal retroviruses contain the genes *gag* (group specific antigen), *pol* (reverse transcriptase), and *env* (envelope) under the regulation of LTR (long terminal repeat) enhancers. They can recombine with host cell genes. Recombination can take place at many sites within the virus genome. Gray box, random genomic material; black box, protooncogene.

as **contact inhibition**. The presence of an activated oncogene overrides contact inhibition and cells that have been transfected with DNA containing an oncogene continue dividing to form multilayered foci (Figure 2). These foci can be picked, grown on, and the oncogene isolated (human DNA sequences can be identified in a mouse background by the presence of the primate-specific *alu* repeat). Oncogenes which have been activated without viral involvement are referred to as c-*onc* to distinguish them from the viral oncogenes v-*onc*.

Oncogenes are usually named after the retrovirus in which they were first identified or after the tumor from which they were first isolated by the transfection assay. Thus, v-*myc* was first isolated in avian myelocytomatosis virus, and c-*neu* was identified in DNA extracted from a rat neural tumor. The proteins encoded by oncogenes are involved mainly in **signal transduction pathways**. These are multistep systems that receive growth signals at the cell surface and transmit the signal to the nucleus, where transcription of genes involved for cell division is initiated. Some oncogenes are involved in the regulation of **apoptosis**. This is a process in which cells become programed, by response to various triggers, including viral infection or DNA damage, to commit suicide.

Activation of an oncogene results in either constitutive growth stimulation or prevention of cell death, both of which lead to an increase in cell numbers. In some cases oncogene activation arises through gene amplification. Up to several thousand copies of an oncogene can be found in each tumor cell. An example of this is the amplification of c-*myc* in neuroblastomas.

Cancer-specific chromosome translocations

In many cancers specific translocations are detected in almost every tumor. In some instances these are used by pathologists to differentiate between related types of tumor. Hemopoietic tumors such as leukemias and lymphomas are well characterized for translocations. Burkitt's lymphoma has specific translocations of chromosome 8 with

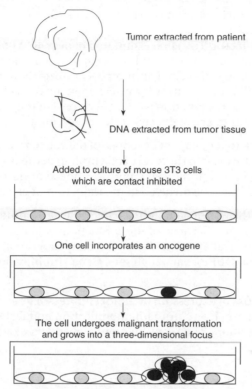

Figure 2. Detection of activated oncogenes in tumor DNA by transfection assay.

chromosomes 14, 2, or 22. The translocation between chromosomes 8 and 14, t(8;14) (q24:q32) is the most commonly observed. This translocation brings the *myc* protoon-cogene, found on chromosome 8, into very close proximity with the immunoglobulin heavy chain gene cluster on chromosome 14. *myc* is activated by being transferred close to the immunoglobulin heavy gene enhancer (Section A7). It is transcribed at a greatly increased rate and the resulting increased level of its protein product stimulates cell division. The translocations to chromosomes 2 and 22 bring *myc* close to the immunoglobulin light chain gene clusters.

Translocations can also cause the 5′ and 3′ ends of two different genes to fuse to form a novel gene. This novel gene may be oncogenic. Examples of this include the so-called **Philadelphia chromosome** in chronic myeloid leukemia, where the translocation t(9;22) (q34:q11) fuses the protooncogene *abl* with the *bcr* gene to form a novel oncogene.

Specific translocations are less well identified for solid tumors. This probably reflects the increased difficulties of obtaining high quality chromosome preparations from these tumors.

Tumor suppressor genes

In genes of this class, both alleles need to be inactivated in order for tumor formation to progress. The mutations are recessive. Inactivation is most frequently achieved by point mutation of one allele, and loss of the other through deletion, although mutation of both alleles is occasionally found. Inactivation of tumor suppressor genes occurs in somatic cells during the development of a tumor. The best studied tumor suppressor gene is *p53*. This is so named because it codes for a 53 kDa nuclear phosphoprotein, which functions as a transcription factor. The gene is mutated in over 40% of human tumors. *p53* is involved in different cellular processes. It has a major role in the transition from the G1 phase to the S phase of the cell cycle. This is a **checkpoint** where cells can arrest during the cell cycle (Section B5). In its native form *p53* prevents the cell moving into S phase, but when phosphorylated this function is lost. This is part of the normal regulation of the cell cycle. Mutant *p53* fails to arrest the cell at this checkpoint.

p53 has also been shown to regulate the response of the cell to DNA damage. This is mediated by two separate pathways: (i) the level of p53 protein in the nucleus is increased after DNA damage – this can prevent the cell entering S phase until the damage is repaired; and (ii) alternatively, *p53* may be involved in the induction of cell death via **apoptosis**.

Both processes reduce the risk of damage to DNA resulting in mutations that could cause cells to develop into tumors. Because of these functions *p53* has been described as 'the guardian of the genome.' Transgenic mice that lack any functional *p53* develop normally but produce spontaneous tumors at a high rate. Gene amplification is found at elevated levels in their cells.

In some instances the loss of expression of tumor suppressor genes has been shown to be due to epigenetic events such as chromatin modification and DNA methylation (Section A8), rather than to mutation. The most frequently observed example of this is the *p16* gene.

Hereditary cancers

A number of rare cancers and a small proportion of the common cancers show clustering within families. These families are said to have a hereditary predisposition to cancer. This may be to a specific form of cancer, e.g. breast cancer, or to a variety of different cancers.

Hereditary predisposition to cancers may arise through germline mutations of tumor suppressor genes. Here all the somatic cells of an individual carry one mutant allele, and there is a greatly increased risk of tumor formation due to inactivation of the single normal allele. Examples in humans include the retinoblastoma gene, *Rb*, on chromosome 13 and a gene for inherited predisposition to breast and ovarian cancer, *BRCA1*, on chromosome 17. The normal allele is inactivated in the cells of the tumor. Familial cancer genes are often found to undergo somatic mutation in sporadic (noninherited) cases of the same cancers.

Hereditary cancers show the inheritance pattern of autosomal dominants, because sufferers pass the predisposition on to half of their offspring (Sections C1 and F1), however mutations in tumor suppressor genes are recessive. This appears to be a paradox. It is explained by the fact that if an individual has only one functional copy of a tumor suppressor gene in every cell then it is highly likely that, in some cells, a second mutation will cause inactivation of the normal functional copy. These doubly mutant cells can then give rise to a tumor.

DNA repair genes are also associated with hereditary cancers. Examples of these include xeroderma pigmentosum, a recessive autosomal condition, in which the affected individuals are highly sensitive to sunlight because they lack an endonuclease involved in DNA repair. Mutant alleles of several different genes can result in predisposition to colon cancer. Two of these are also involved in repair of DNA mismatches (Section A6). The cells of the tumor repair damaged DNA less efficiently than normal cells and produce muta-tions at a high frequency, and some of the mutations will be in tumor suppressor genes or protooncogenes. This is known as a **mutator phenotype**.

RNA splicing and cancer

Pre-mRNA is processed as it matures in the cell. One of the most striking effects is the splicing out of introns (Section A4). In tumor cells this process can take place in an aber-rant manner, resulting in the production of atypical proteins. This can be due to muta-tions in the splice sites leading to the deletion of exons from the mature mRNA. Splice site mutations are a common class of mutation in tumor suppressor genes in cancer cells. Several splice-site mutations have been described for *p53*. In some cases a mutation in an intron results in the creation of a new splice site.

Alternative splicing, where identical RNA molecules are spliced differently at the spli-cosome, is found in normal tissue. However, it is much more common in tumor cells and many examples of tumor-specific splicing patterns have been found. These are common in genes coding for proteins associated with stimulating growth in cells and in oncogenes.

Gene analysis and cancer treatment

DNA chips can be used to identify regions of loss or amplification in tumor tissue and also to detect changes in levels of expression of specific genes. These techniques can be used to suggest specific therapies for individual patients. Chips are available that carry all genes that are known to be involved in development of cancer.

F4 Gene therapy

Key Notes

Overview

It is now possible to treat genetic disorders by the addition of functional copies of the gene responsible for the disorder. This is, at present, limited to treating somatic cells. Germ cell gene therapy is considered unethical. The major problem in developing gene therapies is finding the optimal vector to transfer the therapeutic gene into the recipient's cells. Gene therapy can be carried out on the whole patient (*in vivo* therapy) or on cells which have been removed from the patient and which are then returned after the gene has been added (*ex vivo* therapy). In some disorders the therapeutic gene will have to be targeted to a particular cell type, while for other syndromes several tissues may make suitable targets. In the last decade gene therapy methods have begun to yield clinical successes.

Vectors

Viruses offer the best potential as vectors in gene therapy. The bulk of viral genes are removed to prevent virus replication in the patient. Retroviruses infect most cell types, but only if the cells are dividing. This limits their use to *ex vivo* application. Lentiviruses may prove a more useful alternative. Adenoviruses can also be used but their DNA does not integrate into the target cell's chromosomes and their effect is transient. They also may induce immune response in the patient. An alternative to the use of viral vectors involves encapsulating DNA in liposomes. This causes no side effects, but is much less efficient in transferring DNA to target cells.

Severe combined immunodeficiency syndrome

The genes for adenine deaminase or *IL2RG* can be introduced into hemopoietic stem cells removed from patients with this disease using a retrovirus vector. When the stem cells are reintroduced into patients most show enhanced levels of immune response and this level is stable for several years. A small number of patients have developed leukemia after treatment.

Hemophilia

Hemophilia is a good candidate for gene therapy because the clotting factors simply need to circulate in blood, it doesn't matter which cells make them. Experiments with Factor IX have worked in animal models but only inadequately in human trials.

Cystic fibrosis	Gene therapy for this disease has to be carried out *in vivo* because the major target tissue is the lining of the lungs. Adenovirus vectors have been used. These cause side effects and may be of limited use. An alternative approach is to use liposomes.
Cancer	This is currently the major area of medicine in which gene therapy is being used. There are a number of different approaches. These include introducing genes to kill tumor cells, or to stimulate the response of the host. One potentially useful method involves the use of the herpes simplex thymidine kinase gene to sensitize the tumor cells to drug treatment. Genetically modified T lymphocytes have been used to successfully treat some leukemias.
Vision defects	Gene therapy is currently being used to treat patients with Leber's congenital amaurosis. Some increase in sight has been reported.
Related topics	(F1) Genetic diseases (F3) Genes and cancer

Overview

Major technical advances in molecular genetics made in the past 30 years have enabled us to isolate, clone, and sequence genes from all species including humans. In addition to this the ability to map human genes, in particular those for inherited disorders, has allowed us to identify unambiguously the precise genetic reason for ill health in many individuals. This has raised the question of whether it would be possible to use genetic methods to cure such individuals, or at least to increase their quality of life. This would require the necessary gene to be isolated and targeted into the cells of the affected individual – **gene therapy**. In the case of inherited syndromes there is a choice of approaches: a normal copy of a gene can be introduced into either the germline, or into the somatic cells of the affected individual. The former would prevent transfer of a deleterious allele to the next generation whereas the latter would ameliorate the effects of the allele in the affected individual. **Germline gene therapy** is however regarded as unethical due to the danger of side effects in subsequent generations. Current studies are concerned only with **gene therapy of somatic cells**. In the majority of cases the intention is to introduce a functional copy of a gene into cells which have two nonfunctional alleles, so as to correct the defect. However, in proposed gene therapies for cancer it is often the intention to introduce genes into tumor cells that will induce cell death. Assuming that problems involving cloning of the relevant gene and regulating its expression have been largely solved (Sections A7 and E4), the major obstacle to the development of a gene therapy lies in finding suitable vectors to introduce genes into target cells efficiently while not causing serious side effects. Paradoxically the most promising candidate vectors for gene therapy are derived from viruses that cause disease in humans, precisely because they have evolved efficient systems for transmitting their genomes into human cells.

Gene therapy can be carried out by introducing genes into the patient and hoping they will find an appropriate target cell – *in vivo* **therapy**. This is obviously less efficient, and it is often preferable to remove cells from the patient, manipulate them in cell culture, and then return them to the patient. This is known as *ex vivo* **therapy**.

It is important to realize that targets for therapeutic genes will differ. In certain cases the gene must be delivered to a specific cell type, for example replacement of *globin* genes in inherited hemoglobinopathies in which the gene must be delivered to erythroid cells. In other cases it is only necessary to increase the level of a specific protein, for example a hormone, in the body. Here any cell type that will allow the product to be secreted into the blood system will be a potential target.

Over 1300 clinical trials using gene therapy have been undertaken to treat human diseases, and animal model systems are also used to determine the efficacy of novel strategies. For many years gene therapy seemed only a theoretical possibility. However, in the last decade a number of clinical success stories have emerged.

Vectors

Before a virus can be used as a vector to carry therapeutic genes those genes necessary for viral functions must be removed. This prevents spread of the virus within the body, and reduces the response of the patient's immune system. Viruses deleted in this way are referred to as **gutted viruses**. The desired gene (**transgene**) is inserted into the remaining virus genome along with DNA sequences to promote and regulate its transcription. The construct is then packaged into the appropriate viral coat so that it can infect human cells. Several different virus vectors can be used.

Retroviruses (Section B7) are easily adapted to carry a human gene insert. These have the advantage that they will infect most, if not all, cell types and are efficient at integrating their genome into the host cell. Their main disadvantage is that they will only infect actively dividing cells. For this reason their use is largely limited to *ex vivo* use with cells grown in culture. Another problem is that the inserted gene is often only expressed for a short period, usually less than 2 weeks. A subgroup of the retroviruses, the lentiviruses, which includes the human immunodeficiency virus (HIV), are being developed for use in gene therapy. These can infect nondividing cells and hence are useful for *in vivo* procedures.

One DNA virus group that has been adapted for gene therapy is the **adenoviruses**. Several adenoviruses are associated with common upper respiratory tract infections in humans. Their genome can accept large insertions of human DNA, but they have several major disadvantages, including the fact that their genomes do not integrate into the host cell's genome. This means that they cannot engender a permanent change in any cell they infect. They also induce an immune response which may result in the killing of infected cells. This, however, may be useful in the treatment of tumors.

Other DNA viruses which are under examination as potential vectors include the much smaller adeno-associated viruses (AAVs), Sendai virus, pox virus, and herpes simplex virus, which has a large genome and can thus incorporate large inserts.

All of the viral vectors have drawbacks, either in the efficiency of their delivery system, their triggering of an immune response, or because they are closely related to viruses that may have a capacity to induce tumors. For this reason considerable effort has been put into the development of nonbiological delivery systems in gene therapy. Of these the most promising appears to be liposomes, lipid micels that can be used to carry naked DNA across the cell membrane. The phospholipid binds DNA and because it is of a similar composition to the cell membrane the two can fuse, thus the DNA is carried into the cytoplasm of the target cells. Liposomes have several advantages: they elicit no immune response; they can operate *in vivo* or *ex vivo*; and they can carry any size of DNA fragment. However, the DNA they transfer has a low frequency of integration into the host chromosome.

Severe combined immunodeficiency syndrome

This is a rare, recessive, autosomal inherited disease caused by recessive mutations in either the adenine deaminase (ADA) gene or a cytokine receptor gene on the X chromosome *IL2RG*. The gene that encodes ADA was cloned and initial attempts to treat the disease by gene therapy were first carried out in 1990. The gene was inserted into a retroviral vector, T lymphocytes were removed from the patient, stimulated to grow in culture and transfected with the gene. Cells carrying the functional gene were then grown up, before being transferred back to the patient. This procedure showed success in some cases. However, it was not a cure for the disease: because the T lymphocytes are mature, differentiated cells they were progressively replaced in the body by new cells from the bone marrow.

More recently, the ADA gene has been inserted into patients' hemopoietic stem cells (Section G4) rather than lymphocytes and this appears to lead to long-term improvement in immune response. Similar approaches have also been used with the X-linked disease. Currently over 40 patients have been reported as being successfully treated. Significant success has also been reported with trials involving the *IL2RG* gene.

Care has to be exercised with this approach as a small number of patients have developed cancers after treatment. This may be due to mutations caused by insertion of the gene into a gene involved in growth control (Section F3).

Hemophilia

Hemophilia is a good candidate for gene therapy because the clotting factors circulate in the blood so could be secreted from any cell. Expression of only 3% of normal levels would substantially reduce clinical disease, and 30% levels would be phenotypically normal except in extreme bleeding conditions. The hemophilia B gene, *Factor IX*, has a very small cDNA size, around 1.4 kb, and is easily contained in transfection vectors. The hemophilia A gene, *Factor VIII*, is much larger, over 8 kb, but can either be cut down by removing nonessential amino acids or divided into two parts, one for each chain in the final processed molecule. Successful long-term effectiveness in mice has been achieved using retroviral vectors carrying *Factor IX* perfused through a liver induced to regenerate which stimulates cell division. Results in monkeys, dogs and humans have been less successful. In humans, a trial was run with AAV as vector. Successful transfection of muscle cells produced too little circulating Factor IX. Transfection of liver cells worked well but later the immune system destroyed the transfected cells, so there was only temporary benefit.

Cystic fibrosis

Cystic fibrosis, like ADA, is the result of recessive mutation of a single gene, however the frequency of mutant alleles of this gene is much higher, particularly in Caucasian populations. The defect is in a gene responsible for the transport of chloride ions across the cell membrane. Lack of this function has major effects on the cells that line the lungs and gut. The gene therapy approach was initially similar to that for ADA in that the gene was cloned and engineered into a vector. Because of the cell types that are involved it is not possible to transfect cells *ex vivo*. For this reason adenovirus has been used as a vector. The engineered virus containing the functional gene can be delivered to patients as an aerosol. Some virus particles are able to infect cells of the respiratory tract. Problems in this approach include the difficulty of infecting cells through the mucus lining of the tissue, the fact that adenovirus does not integrate into host cell chromosomes, and possible immunologic reaction to virus-infected cells. An alternative approach is to use liposomes to transfer the gene. This introduces the gene into fewer cells, but may have longer-term effects. At present several early stage clinical trials are being carried out.

Cancer

About two-thirds of all clinical trials involving gene therapy that have carried out have been directed at tumor cells. Given that tumors arise as a result of gene mutation (Section F3) it is theoretically possible to prevent further growth of tumor cells by replacing defective tumor suppressor genes such as *p53* or genes that will allow apoptosis to function normally. Another approach is to use exogenous genes either to induce death in tumor cells or to render them more sensitive to the normal immunological defense mechanisms of the body. Alternatively, genes to make cells more resistant to chemotherapy can be incorporated into the stem cells responsible for hemopoietic cells. When reintroduced into the body, these will make the patient's lymphocytes capable of withstanding higher doses of chemotherapy.

Tumor cells can also be treated with 'suicide' genes that will make them sensitive to drugs. One example of this approach is transfection with the thymidine kinase gene of the herpes simplex virus. This enzyme differs considerably from its human equivalent, and can phosphorylate a wider range of substrates. One substrate that cannot be utilized by human thymidine kinase is gancyclovir. This compound is nontoxic to human cells but its phosphorylated derivatives are highly toxic. Transfer of the herpes thymidine kinase gene to tumor cells thus renders them specifically sensitive to gancyclovir treatment. There are two other advantages in this methodology: (i) thymidine kinase is active only in cells that are replicating DNA and hence nontumor cells are less likely to be affected; and (ii) the phosphorylated gancyclovir molecules, which are produced only in those tumor cells that have taken up the gene, are exported to surrounding tumor cells through gap junctions. The result is that the proportion of tumor cells killed by gancyclovir is greatly increased. This approach has been used with brain and ovarian tumors, and is set out schematically in Figure 1.

Recently a small study has been reported in which two of three patients with chronic lymphocytic leukemia were successfully treated by reintroducing T cells that had been removed from the patient and genetically modified to carry a cell-surface protein that would recognize and kill the malignant B lymphocytes. It was calculated that each engineered cell had killed approximately 1000 leukemia cells, and the cells were retained in the patient for at least 10 months.

Vision defects

Defects in vision in humans are known to result from mutations in a large number of genes. Several of these have been identified and cloned. Mutations of one such gene, *RPE65*, are responsible for Leber's congenital amaurosis. The gene has been inserted into adeno-associated virus vectors which show a high ability to infect retinal epithelium cells and photoreceptors *in vivo*, produce a minimal immune response and give rise to stable

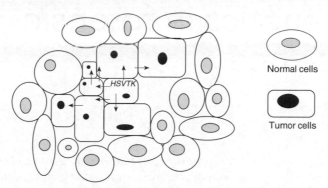

Figure 1. Killing of tumor cells containing the herpes simplex thymidine kinase gene. A single tumor cell which has been transfected with the herpes simplex virus thymidine kinase gene (*HSVTK*) can export the toxic phosphorylated products of gancyclovir to other tumor cells. Normal cells are relatively unaffected because they are not undergoing DNA synthesis

expression of the **transgene**. A number of animal models, both natural mutants and specially designed knock-out mice (Section G3), exist for that and are used to test the efficacy of the gene therapy. Recent clinical trials have shown injection of the vector carrying the *RPE65* gene into humans has resulted in amelioration of the symptoms in a majority of patients.

G1 Genetics in forensic science

Key Notes

Unique correlations

Sexually reproducing organisms are genetically unique. All the individual's cells carry the same genetic information, and any tissue samples of unknown origin can be compared to candidate individuals. The power of the technique is in exclusion. A single difference between a scene-of-crime sample and a suspect can prove an absence of connection, and exclude the innocent. If a perfect match is found, the probability of finding that same genotype by chance in the population can be calculated, provided the frequencies of the alleles involved are known for that population. Genetic relatedness can also be used to trace family connections and associate children with their natural parents. The same techniques are also used in nonhumans, for example to trace lost or stolen horses and hawks. Using the polymerase chain reaction (PCR), a genetic profile can be obtained from a few cells, and should establish genetic identity beyond reasonable doubt.

Protein comparisons

Proteins may be compared by immunological techniques (e.g. using antibodies to identify blood group or tissue type), or by electrophoresis to screen for charge-to-mass variation caused by amino acid substitutions. These techniques may exclude most of the population.

DNA comparisons

All cells in an individual carry copies of the same DNA sequence, and in sexually reproducing organisms each individual has a unique sequence of DNA in their genome. If DNA sequences from a tissue sample found at the scene of a crime are compared (in sufficient detail) to DNA sequences from a suspect, differences can be found, unless the tissue sample came from the suspect (or his or her identical twin clone)! The variable sequences used for forensic purposes are micro- and minisatellites, typically 10–20 repeats of a 3 or 4 base pair unit. The arrays frequently change their length (repeat number) by mutation, hence their name: variable number tandem repeats (VNTRs). Variation was first detected in restriction fragment length polymorphisms (RFLPs) but this has been superseded by PCR-based techniques. Identical samples give identical fragment length patterns. About 15 million single nucleotide polymorphisms have been identified in the human genome. It is estimated that identifying 70 or 80 of these would identify an individual. Currently the proven method of genetic profiling VNTRs is more practical.

RFLPs: genetic fingerprints

Restriction fragment length polymorphisms occur when the distance between successive restriction enzyme sites varies by mutation. Genetic fingerprinting involves cutting genomic DNA with a restriction enzyme, separating fragments by electrophoresis, then Southern blotting with a multilocus probe. This detects all occurrences of a repeated sequence, which is found at many places in the genome, and so provides many bands in a unique combination. The disadvantages are that it requires a relatively large sample of DNA, and the wealth of data that is obtained is difficult to quantify and communicate in court.

VNTRs: genetic profiles

The standard technique is to use PCR with a unique pair of primers to amplify the DNA around a specific VNTR. This allows individual alleles to be identified and the specific number of repeats in the VNTR in each allele to be counted. The frequency of each allele in the population can be determined, and all the power of traditional population genetic techniques can then be applied to quantify the result. By using sufficient loci with uncommon alleles, the probability of getting a complete match by chance can be made infinitely small. Multiplexed PCR techniques are able to get good results for several loci in DNA from a few cells in a hair follicle or in saliva.

Interpretation of results

A single mismatch between two samples is proof of different origins. If two samples match, the probability of obtaining two identical samples from the population by chance must be calculated. This is difficult for DNA fingerprints because it is difficult to define the position of bands exactly, and the background assumptions can be challenged. (In practice, any two different samples do always look different.) Analysis of VNTRs allows each allele to be classified by the number of repeats in its satellite. The frequency of each allele in a population can be determined, and accurate probabilities calculated for two samples from different people being genetically identical by chance. A certain degree of inbreeding is assumed, and calculations are biased to minimize the chance of making a false association.

Related topics

Unique correlations

The uniqueness of each individual (in a sexually reproducing species) can be used to identify that individual, or any cellular trace of the individual. Similarities and differences can be analyzed directly by examining the DNA or protein molecules, or indirectly by their effects on the phenotype. Describing someone as blue-eyed describes a genetically determined characteristic. It is not very useful in Scandinavia where most people are blue-eyed, but could be very useful in a country where most people are brown-eyed. The usefulness of a characteristic for identification depends upon it being uncommon. Using the polymerase chain reaction (Section E2), a genetic profile can be obtained from a few cells.

Any tissue containing partially intact DNA provides useful samples: semen (in rape cases), blood, hair, skin fragments, mouth epithelial cells (in saliva), or bones in skeletons. Forensic (law-related) science can compare the genetic characteristics of biological fragments found at the scene of a crime (presumed to come from the criminal) to a suspect to exclude or associate them. Any biological sample from a person will match that person in all genetically determined characteristics. The great power of the technique is to exclude the innocent. A single genetic character that does not match can prove that the suspect did not deposit the sample. If enough suitable genetic characteristics do match, that is evidence 'beyond reasonable doubt' that the suspect did deposit the sample. The court can then test the suspect's explanation of this 'contact' with the crime scene.

The genetic similarity between parents and offspring can also be used to confirm or reject their relationship. A child must get one allele of each locus ('gene') from each parent, so half the child's genotype must match each parent. Genetic tests can be used in disputes over paternity to identify or exclude potential fathers. Cases of child-stealing and mix-ups between babies in hospitals can also be resolved. Genetic tests in nonhumans can also be used forensically, for example to detect kangaroo meat in minced beef, and to identify stolen horses and hawks and smuggled parrots. Beef gelatin solution can be injected into chicken meat to improve the weight and texture cheaply. Some suppliers deliberately chemically degrade DNA in the gelatin to avoid identification, but it is also possible to identify a bovine amino acid sequence in the chicken. Murderers have been convicted by genetic fingerprinting of nonhuman material. In one case a seed found in a suspect's vehicle came from the specific individual tree beside the murder victim's body, and in another case cat hairs at a crime scene were shown to be genetically identical to the suspect's cat. Both cases were used to show, with a probability of error of less than one in a million, an association between the suspect and the scene-of-crime that was denied by the suspects. In both cases the suspects were first detected by routine police investigations and the DNA evidence was confirmation.

Protein comparisons

The amino acid sequence of proteins is determined genetically. The first scientific tests were for particular cellular molecules identified immunologically. **Antibodies** are produced by the immune system in response to foreign molecules. It was found that mammals, usually rabbits, sheep, goats, or horses, which were inoculated (injected) with human blood, would produce antiserum containing antibodies which would bind to red blood cells of the type found in the inoculum. These antisera could distinguish the various classes in the A-B-O, rhesus and M-MN-N blood groups, and other characters, according to the type of blood used for the initial inoculum. These antibodies can be used to identify the type of a drop of blood. If blood similar to the victim is found on a suspect, or the other way around, that suggests a physical contact between them. The less common the alleles, the more useful they are in excluding innocent suspects. For example, blood group N is restricted to 1.1% of Navaho Indians, 21% of Caucasians, but is found in 67% of Australian Aborigines. Therefore, it would be useful for excluding 98.9% of Navaho, 80% of Caucasians, but only 33% of Aborigines.

Tissue-type alleles show a great deal of variation, and were used extensively in Argentina to identify the 'Children of the Disappeared.' Children were taken as babies from political dissidents who were murdered. When suspected stolen children were located, their tissue-type alleles were compared with their supposed grandparents (their supposed parents being dead) and where possible with the people claiming to be their parents. This often showed that there was a very high probability that the child was the grandchild of the deceased dissident's parents (Figure 1).

Another technique is **protein electrophoresis**, a process analogous to electrophoresis of nucleic acids (Section E1). A mutation can cause a change in one charged amino acid in a protein, which in turn can change the protein's charge-to-mass ratio, and change the rate at which it migrates in a gel. Variation for such mutations is common. Again the principle is to compare the scene-of-crime sample with the suspect.

DNA comparisons

Every sexually produced individual has a unique combination of sequences in their DNA (a unique genotype) and direct analysis of DNA maximizes detectable variation and minimizes uncertainty. A DNA match of sufficient sequences between crime scene and suspect is virtually certain evidence of contact. The development of techniques using restriction enzymes or PCR made such direct comparison possible (Sections B4 and E1). The techniques used for forensic work are all based on separating DNA fragments by length using electrophoresis. If the DNA sequence was the same in all people, then corresponding segments would always be the same from all individual people. Fortunately length can differ as a result of mutations removing or creating restriction sites, or by insertion or deletion of bases between restriction sites (RFLPs). This commonly occurs by changes in the number of copies present in a block of **variable number tandem repeats** (VNTRs; Section B3). The discovery and mapping of 15 million single nucleotide polymorphisms in humans suggests a new technique for identifying people. It is not currently used for practical reasons. The current genetic profiling system is already working well and proven, and doesn't require as many loci to be tested.

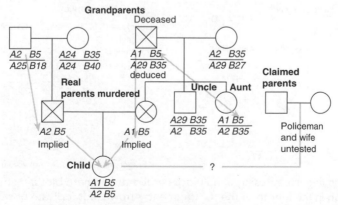

Figure 1. Family tree of *HLA* (tissue type) alleles of presumptive grandparents and abducted grandchild. Gray arrows trace alleles from grandchild to relatives, confirming the presence of the child's alleles in his dead parents' family and showing 99.9% probability of a genuine relationship.

RFLPs: genetic fingerprints

The first forensic technique used on DNA was to examine **restriction fragment length polymorphisms** (RFLPs; Section B1). The restriction fragments can be separated according to length by **gel electrophoresis** then **Southern blotting** the gel using a **multilocus probe** (Section E1). This is a sequence that occurs at multiple sites in the genome, ideally producing as many distinct bands as can be clearly measured on the autoradiograph (Figure 2). If digests of two good samples of DNA from the same person are run side-by-side on the same gel (but separated by an empty lane to avoid the possibility of cross contamination) it is clear that they are identical and unique, with bands of similar intensity clearly aligned. When they are run on separate gels, however, the small unavoidable variations may make this comparison more difficult. The data are difficult to present in court. RFLPs are no longer favored for forensic purposes because: (i) they require relatively large samples of DNA, sufficient to act as a target for the probes in Southern blots, and (ii) the difficulties in calculating the probability of falsely obtaining a match between two samples from different people hinder interpretation. The pattern of bands is a phenotype, not a genotype, so classical population genetics cannot be used to calculate its frequency.

VNTRs: genetic profiles

Forensic investigation currently uses length variation at a set of single loci, each containing one variable number tandem repeat (VNTRs; Section B3). Tetranucleotide repeats (e.g. CAGA) are preferred for forensic work because of the greater difference in length caused by a gain or loss of one repeat, and the greater stability of the sequence when amplified by PCR. PCR primers for an individual locus constitute a **single locus probe**, and the exact number of copies of the repeat in the specific alleles present in the sample is found. Comparing RFLPs, single locus PCR identifies the bands equivalent to B, C, and F in Figure 2 individually and measures the number of repeats in each. Other bands cannot confuse the data. Because each sequence is genetically defined, it can be studied using population genetics (Section D3). A particular VNTR may, for example, have between 12 and 23 repeats. The frequency of each length allele in different racial groups

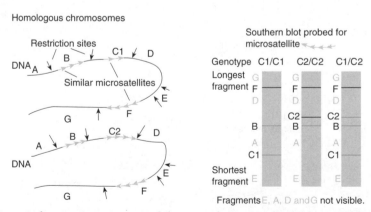

Figure 2. Genetic fingerprint using a multilocus probe to Southern blot a restriction digest, showing variation in the length of the 'C' fragment. Arrowheads indicate position and length of the satellite target sequence in the DNA. Each locus shows one or two bands on the gel, but these bands cannot actually be related to loci in the DNA as illustrated here. Using primers to amplify only the C locus allows its alleles to be specifically identified

can be determined by sampling from the general population. This allows population genetics and Hardy–Weinberg rules to be applied to calculating genotype frequencies (Section D3). This is an important advance and RFLP genetic fingerprint data (multilocus, band lengths) was presented to the courts quite differently from VNTR genetic profiling data (single allele repeat numbers), although scientifically they are two ways of examining similar genetic variation.

The polymerase chain reaction (PCR; Section E2) can amplify DNA from a very low copy number, and a single hair follicle is sufficient to determine the genotype. A heterozygote will produce two bands for one locus. In practice, several pairs of primers may be used together (**multiplexed**) to detect different loci simultaneously. Four primer-pairs produce eight bands from a diploid heterozygous at all four loci. The loci used are chosen from different chromosomes to avoid linkage and ensure independent assortment. They also have a high degree of variability in the population, and have different overall lengths between loci to avoid confusion. Different color labels on different primers allows each locus to be identified and measured automatically, reducing human error (Sections E2 and F2). Current systems use 10 loci (UK: SGM+, Second Generation Multiplex + XX-XY) and 13 loci (USA: the FBI use CODIS, Combined DNA Index System). SGM+ can identify about 1 in 10^{12}; CODIS can detect 1.74×10^{15} different Caucasian genotypes; both can detect more than the number of people that have ever existed, but false matches do still occur because some genotypes are relatively common. Sixteen loci were used to clear one wrongfully arrested man in England. Although a genotype may have a frequency of one in a million, there is a one in ten chance of finding a false match to an innocent person in a database of 100 000 random genotypes.

Interpretation of results

In some countries the adversarial justice system relies on convincing the jury of the rightness of one side of the case and the wrongness of the other. The jury is composed of 12 laypeople (that is people with no qualifications in interpreting the evidence), and the judge is there to uphold the law, not to investigate the case. This can lead to a failure to use evidence effectively, and some of the first cases using genetic fingerprinting evidence were dismissed because the defense raised scientifically unwarranted doubts in the minds of the jury while prosecutors made extravagant claims which they could not support. For these reasons forensic evidence is presented very carefully in a formalized way that is not open to criticism. There have always been cases where scientific proof was ignored. A young actress accused Charlie Chaplin of fathering her child. He lost the case despite the blood-group evidence that he could not possibly be the child's father. Conversely, a suspect was convicted because (after many negative results) a technician spilled the suspect's DNA into the adjacent well used for the scene-of-crime sample, producing identical bands in both lanes. Such an effect of technical incompetence should have been obvious to anyone familiar with the technique. It should be clear from this that presentation of evidence in court is more important than scientific facts.

Genetic fingerprinting works on the principle of exclusion: a single difference between sample and suspect rules out the suspect. The problem is to work out the probability of not excluding someone who is really innocent, but whose sample matches the scene-of-crime sample by chance. For example, suppose that a suspected rapist has an identical genotype to semen from a crime scene. If the frequency of that genotype in the population is 0.01 (1/100) then the probability of the suspect and criminal having the same genotype by chance is 1 in 100. Convicting the suspect on that evidence alone would convict one innocent person in every hundred trials. However, if the frequency of the genotype is 10^{-8} then the probability of a particular suspect matching by chance is 1 in 10^8, but if a database of 100000 innocent people is searched there is a 1 in 1000 chance of wrongly finding a match. This emphasizes the importance of other evidence to implicate the suspect, although current genetic profiling produces much lower odds of false positives.

Traditional fingerprinters did not use statistics; they simply claimed that if enough points matched, then two fingerprints came from the same finger. Recent cases of misidentification of normal fingerprints have been exposed in the UK. Genetics has relied on statistics since Mendel, and attempts are made to calculate the numerical probability that a sample from the scene-of-crime matches the suspect by chance. By merely questioning these calculations, some defense councils managed to win cases by causing doubt. Multilocus data require the measuring of band position to calculate fragment lengths. If 200 mm of a gel can be measured with an accuracy of 1 mm, then only 200 band positions (called bins) can be measured. Bands differing in position by 0.5 mm may be classed together (put in the same bin, hence **binned**), reducing the useful information. The probability that all the bands in two different samples fall into the same bins by chance must be calculated. Realistically it is likely to be very small, but by raising doubts about the validity of the procedures, the method was initially discredited. This was overcome by forensic witnesses being cautious, erring on the side of supporting innocence, and learning what questions they would be asked, and how to answer them.

Using single locus probes to specific VNTRs, individual alleles can be identified, and the Hardy–Weinberg equation (Section D3) can be used to calculate accurately the probability of two unrelated samples being the same genotype by chance. The problem with this is the existence of subpopulations, a form of inbreeding. These have proved to be the biggest objection to genetic profile (VNTR) evidence. Suppose a suspect has a rare allele with a frequency of 0.001, also found at a scene of crime. Only about 1 in 500 people, each with two alleles, will have it. However, one of the suspect's parents must have it, and also half of his brothers, sisters and children. If the family has lived in the community for a long time then there will be cousins and many distant relatives who will also have a much greater than 1/500 chance of having that allele, so the chance of mistaken identity is greater. This can be overcome by giving rare alleles a minimum frequency of 0.05. The court is then given the lower probability of not excluding the innocent suspect if someone quite closely related (e.g. a first or second cousin) was the actual culprit. Races are also subpopulations. Alleles of human locus *WVA* contain between 12 and 21 repeats. Alleles with 12 repeats have only been identified in one racial group. The 15-repeat allele has a frequency around 0.09 in Asians and Caucasians, and 0.2 in Afro-Caribbeans. The highest frequency appropriate to the suspect's racial group or the local population must be used to give the benefit of any uncertainty to the accused. Again, if there is other evidence of racial group, the population genetics calculations can be more accurate.

Probabilities of 1/10000 to 1/1000000 were typically produced using only four loci. Using more loci improves this, and the typical discrimination now exceeds 1 in 10^{12}, more than 1000 times the current population of the world. Nevertheless, spurious matches are still likely to be found in large databases. The same applies if the criminal leaves his name: there are many people in the phone book with the same name. When there is other evidence implicating the accused, genetic profiling evidence becomes overwhelming.

Contamination is becoming more of a problem, because PCR can amplify just a single molecule of DNA, called **low copy number DNA** technique. The skin traces left on a car steering wheel, a sweaty handprint, or a cigarette end are all capable of producing a genetic profile in good conditions. The problem is that someone handling something from a suspect then handling something from the crime scene could accidently transfer DNA from one to the other. This might also arise if evidence collected at a crime and from a suspect was stored together. In one case, a woman's DNA was associated with several serial murders. She was eventually traced as someone working on the production of testing kits, whose DNA had contaminated them.

G2 Biotechnology

Key Notes

Genetic engineering and biotechnology

This field of research uses gene transfer techniques to produce recombinant proteins and genetically modified organisms. The commercial applications of these techniques have led to the development of the biotechnology industry.

Expression of recombinant proteins

Expression vectors direct the transcription and translation of cloned DNA sequences in host cells to produce recombinant proteins. Expression of proteins must be optimized individually. Protein engineering is used to produce variant forms of native proteins with altered properties.

Bacterial expression systems

The bacterium *E. coli* has been used extensively to express foreign proteins. Essential features of bacterial expression vectors include a strong promoter and transcription termination signals, a ribosome binding site, start and stop codons at either end of an open reading frame, and a mechanism for induction of expression. Expressed proteins are protected from degradation by placing extra amino acids at the *N* terminus. Expression levels can be increased by using high-copy-number plasmids but plasmid instability may result. Vectors that integrate into the host chromosome are stable but expression levels are lower.

Eukaryotic protein expression systems

Proteins expressed in eukaryotic cells are post-translationally modified and are more likely to be biologically active. Eukaryotic expression vectors have a similar design to prokaryotic vectors. Many are shuttle vectors that can be propagated in bacterial cells. Yeast, insect cells, and mammalian cells are used as hosts for eukaryotic expression.

Some expression systems allow the expression of the gene to be artificially regulated. Yeasts are easy to manipulate. Cells can be transfected with foreign DNA as protoplasts or by electroporation. Yeast vectors may be episomal or integrated. Baculoviruses infect insect cells. They can be adapted as expression vectors in which transcription of the cloned gene is driven by the strong polyhedrin promoter. Expressed proteins have glycosylation patterns that closely resemble mammalian proteins. Expression in mammalian cells is used to produce recombinant proteins for use as drugs that are an exact match of the native protein.

Safety concerns

There are generalized fears over genetically modified organisms ranging from speculation about unknown possibilities to escape of the transgenes into other 'wild'

Monoclonal antibodies	species. Where there is a definable problem, such as using antibiotic resistance genes as selectable markers, they can be addressed and removed by improved techniques.
	Vertebrates contain millions of clones of lymphocytes. Each clone is capable of producing only a single antibody. In response to infection with a complex antigen such as a bacterium with many epitopes on its surface a large number of lymphocyte clones will be stimulated to produce antibodies. These are polyclonal antibodies. If individual lymphocytes are fused to myeloma tumor cells and grown indefinitely in tissue culture they will produce only a single, monoclonal antibody.

Related topics	(A4) DNA to protein	(F4) Gene therapy
	(B2) Prokaryote genomes	(G3) Transgenics
	(E4) DNA cloning and transfection	

Genetic engineering and biotechnology

These terms refer to a field of research that involves using recombinant DNA techniques to transfer genes from one organism to another. Gene transfer between organisms has two important applications: firstly, to produce large amounts of biologically useful proteins. These include proteins that are used as drugs to treat diseases (Table 1) and proteins or enzymes used in industrial processes (Table 2). Secondly, gene transfer is used to create organisms (plants, animals, and microorganisms) with altered characteristics, for example plants that are resistant to disease or which produce chemicals (e.g. specific starches) for industry (Section G3). These important applications of genetic engineering and biotechnology have enormous commercial potential and have led to establishment of the biotechnology industry. Genetically engineered plants and animals are dealt with in Section G3.

Expression of recombinant proteins

One of the primary aims of genetic engineering and biotechnology is to produce biologically useful proteins. This is achieved by expressing cloned genes in a variety of host cells. Bacteria and yeast as well as animal and plant cells are used. The expression of recombinant proteins requires the use of specialized vectors called **expression vectors** which contain DNA signal sequences that direct the transcription and translation of the cloned sequence by the host cell. Other vector sequences are included which influence the stability of the expressed proteins or direct the host cell to secrete the expressed protein. To ensure high levels of expression, plasmid vectors are used that exist in host cells as multiple copies. The conditions required for efficient expression vary from protein to protein and must be optimized individually. This has led to the development of many diverse expression systems.

The use of recombinant DNA technology to produce proteins allows variant forms of natural proteins to be produced. This is known as **protein engineering**. Cloned gene

Table 1. Recombinant human proteins used as drugs

Protein	Used to treat
α1 Antitrypsin	Emphysema
Calcitonin	Rickets
Chorionic gonadotropin	Infertility
Erythropoietin	Anemia
Factor VIII	Hemophilia
Insulin	Diabetes
Interferons (α, β, γ)	Viral infections, cancer
Interleukins	Cancer
Tissue plasminogen activator	Blood clots
Growth hormone	Growth retardation

sequences can be altered by a procedure called ***in vitro* mutagenesis** and expressed to produce proteins with modified amino acid sequences that have altered characteristics and properties. Protein engineering can be used to produce variants of normal proteins that have properties such as improved heat stability or, for enzymes, altered substrate specificity.

Bacterial expression systems

The first protein expression systems to be developed were based on the use of bacteria as host cells. The bacterium *E. coli* allows many proteins to be produced rapidly and inexpensively and has been used extensively. Bacterial expression systems remain important and are used for the production of most commercially important proteins (Tables 1 and 2).

Vectors for expression of proteins in bacteria have a number of essential features (Figure 1).

* A promoter placed upstream of the cloned sequence to initiate transcription. Selection of a promoter that binds RNA polymerase strongly is necessary to ensure high levels of expression.

Table 2. Recombinant enzymes with industrial uses

Enzyme	Industrial use
Rennin	Cheese making
α-Amylase	Beer making
Bromelain	Meat tenderizer, juice clarification
Catalase	Antioxidant in food
Cellulase	Alcohol and glucose production
Lipase	Cheese making
Protease	Detergents

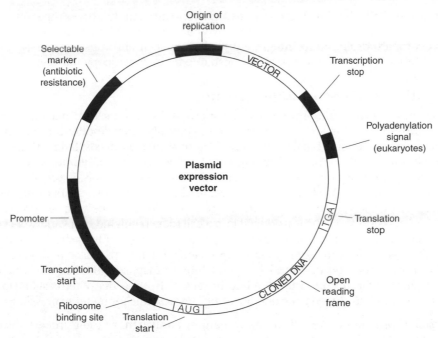

Figure 1. Features of plasmid expression vectors.

- A transcription termination signal placed downstream of the cloned sequence.
- A ribosome binding site (Shine–Dalgarno sequence) placed downstream of the transcription start site. The level of protein synthesis is influenced by how strongly the ribosome binds this sequence.
- The cloned DNA sequence should have an AUG translation initiation codon at the start and a termination codon at the end. The intervening sequence should be an open reading frame. It is very important to establish the correct reading frame to insure the correct amino acid sequence is translated. Vectors usually allow sequences to be cloned in different ways so that each of the three reading frames can be used.
- Vectors should allow expression of the cloned sequence to be induced. Continuous expression creates a drain on the energy resources of the cell which is detrimental to other cell functions. Expression is more efficient if it can be switched on for a limited period then switched off again. Most systems are induced by the addition of a small molecule, often a metabolite or small molecule such as tetracycline, to the bacterial culture. Some systems are induced by altering the culture temperature.

Proteins expressed in bacteria are often unstable due to degradation by protease enzymes in the host cell. Expressed proteins may be protected from degradation by altering the sequence of the vector to place one or more amino acids at the *N* terminus of the expressed protein. The added amino acids may also be used as a tag which allows purification of the protein using antibodies to the tag immobilized on a gel. Vectors are usually designed to allow enzymatic or chemical removal of the tag from the expressed protein following purification.

The level of protein production can often be increased by using plasmid vectors that exist as multiple copies in the host cell. However, the presence of plasmids at high copy number tends to create instability with some cells losing the plasmid. These cells tend to grow faster than the cells that have retained their plasmid and can take over the culture leading to an overall decrease in expression. To overcome this problem of plasmid instability some vectors are designed to integrate into the host cell chromosome where they are stable, although expression levels are reduced due to the lower copy number.

Eukaryotic protein expression systems

Although many eukaryotic proteins can be successfully expressed in bacterial systems, some expressed proteins are unstable or are biologically inactive due to the inability of bacteria to carry out post-translational modifications, such as glycosylation, that occur in eukaryotic cells. In addition, proteins expressed in bacteria are sometimes contaminated with toxins making them unsuitable for use with humans or animals. Expression systems based on eukaryotic host cells have been developed to produce versions of expressed proteins that more closely resemble the native form. Expression systems have been designed to allow the expression of a gene introduced into a cell to be regulated by a variety of external factors. These include chemical or physical agents such as light or heat shock. Examples of chemically regulated expression systems are 'tet-on' and 'tet-off'. These have been developed from the tet operon (Section A7) of a bacterial transposon (Section B2) which conveys resistance to the antibiotic tetracycline. In 'tet-on' the presence of tetracycline induces expression of the introduced gene and in 'tet-off' tetracycline represses expression.

Eukaryotic expression vectors have many features equivalent to their prokaryotic counterparts (Figure 1). These include a promoter and transcription and translation signal sequences. Additional features specific to eukaryotic systems, such as signal sequences for polyadenylation, are also included. Vectors can exist independently of the host cell genome as plasmids or integrated into a host cell chromosome. Some vectors called **shuttle vectors** have prokaryotic sequences that allow them to be manipulated and propagated in bacteria and then transferred to eukaryotic cells for expression.

Three types of eukaryotic host cells are used: yeast, insect cells, and mammalian cells.

Yeast

The yeast *Saccharomyces cerevisiae* is widely used for expression of eukaryotic proteins. A number of features make it especially suitable as a host. Yeasts are easy to culture and their genetics and physiology have been well characterized. They carry out many post-translational modifications and expressed proteins are easily purified following secretion. In addition, yeasts have an endogenous plasmid called the **2μ plasmid** which can be used as a cloning vector. DNA can be introduced into yeast by a variety of methods. Some involve removal of the cell wall to form **protoplasts**. Another method called **electroporation** involves treating cells with pulses of electric current. Yeast expression vectors may exist in cells as plasmids (**episomal vectors**) or integrated into the host cell genome. Episomal vectors give the highest levels of expression but tend to be unstable in large cultures. Examples of recombinant proteins produced by expression in yeast include insulin which is used to treat diabetes and the hepatitis B surface antigen which is used as a vaccine.

Insect cells

Baculoviruses are viruses that infect invertebrates including many insects. They have been adapted for use as eukaryotic expression vectors. Insect cells are used as hosts and are particularly suitable because they produce proteins with glycosylation patterns that are very similar to mammalian proteins. The baculovirus contains a gene for a protein called **polyhedrin** which is transcribed from an exceptionally strong promoter. The gene is not essential for viral replication and can be deleted allowing a foreign gene to be inserted. The baculovirus vector containing the cloned gene is transferred into the host cells where it is expressed at high levels late in infection. Recombinant proteins produced using the baculovirus system include the hormone erythropoietin which is used to treat anemia and the anti-virus agent β-interferon.

Mammalian cells

Recombinant proteins used to treat disease in humans must be an exact match of the native form of the protein. In many cases the recombinant protein is only synthesized and properly modified post-translationally when expressed in mammalian cells. As a consequence, expression vectors that use mammalian cells as hosts have been developed. Mammalian expression vectors are shuttle vectors and can be conveniently propagated in bacteria. Transfer into host cells is achieved by adding the DNA as a precipitate with calcium phosphate or by electroporation. A range of human recombinant protein drugs are produced by expression in mammalian cells. These include growth hormone, the blood clotting protein, Factor VIII, β-interferon, and erythropoietin.

Safety concerns

Safety concerns have limited the overt use of transgenic organisms because of campaigns such as 'frankenfoods' (Section G6) and fears of environmental escapes. Materials produced by transgenic microbes in industrial reactors out of sight have been largely ignored by the public, or approved where they had medical benefits. Cheese making requires the enzyme chymosin (rennin) from rennet which is collected from the stomachs of calves. For many years, vegetarian cheese had to be produced using bovine chymosin from transgenic bacteria, until an equivalent bacterial enzyme was discovered.

The only safety concerns that can be addressed are those with clear practical grounds. One is the use of antibiotic resistance genes inserted with the transgene as a means of selecting cells containing the desired genetic addition. Genetic techniques have been developed to remove the selectable marker after the initial selection process. One technique in potatoes uses the *cre/loxP* system. The transfection vector DNA carries the selectable marker and *cre* gene between two *loxP* sites. The *cre* gene is driven by the heat-shock protein gene *hsp70* promoter. When the cells are heated *cre* is activated and the Cre protein splices out the DNA by recombination (Section C4) between the *loxP* sites, removing itself and the unwanted selectable antibiotic resistance gene.

An interesting development is to synthesize the sequence of DNA needed for insertion in a laboratory then put it directly into the organism, or to engineer it in the organism where it will be expressed to produce a protein. The organism is then not transgenic because the DNA does not come from another species. This removes the unknown element of 'transgenic' and may circumvent the more onerous regulations applied to transgenic organisms. One problem with genetically modified plants is the risk of cross-contamination of food crops by airborne pollen. This is eliminated by using clonal sterile potatoes or animals.

Monoclonal antibodies

Lymphocytes are the cells of the immune system that produce antibodies. In higher animals the population of lymphocytes is made up of millions of different clones, each with the capacity to produce a single, unique antibody against a specific **epitope**. Epitopes are the structural components of antigens that trigger the production of antibodies. When an animal is infected with a bacterium it will respond by producing antibodies against each of the many epitopes present on the surface of the bacterium. The result is a serum containing a variety of antibodies, each specific to a single epitope on the bacterium. Such antibodies are known as **polyclonal**. These can be produced commercially for various clinical, industrial, or research purposes, by inoculating animals and subsequently purifying the antibodies from blood, but there are two major limitations to their use. Firstly, it may be important to differentiate between different epitopes in the material/ organism that is being detected. For example it may be important to distinguish between two closely related strains of virus which share several epitopes, but cause different illnesses. Secondly, as they are produced in animals there may be differences between each batch of antibodies produced. This may have serious problems in quality control. These problems can be overcome by the use of **monoclonal antibodies**. As their name suggests, these antibodies will recognize only unique epitopes.

Monoclonal antibodies are usually prepared using mice, in contrast to the large animals such as sheep used to produce polyclonal antibodies. Monoclonal antibody production starts with inoculation of the mouse with the substance against which antibodies are to be raised. The substance does not need to be highly purified as clones of lymphocytes will be stimulated to produce antibodies against each epitope present in the inoculum. In the mouse, stimulation of lymphocytes is carried out largely in the spleen. The spleen is subsequently removed from the mouse and disaggregated to release individual lymphocytes. These are fused to mouse myeloma tumor cells to form cell hybrids. These hybrids grow indefinitely in tissue culture. Each hybrid cell derives from the fusion of a tumor cell with a single lymphocyte and therefore clones of hybrid cells will each produce only the unique antibody of the parent lymphocyte. In this way cell lines are produced that produce only one monoclonal antibody. The cells can be stored frozen and a sample thawed whenever a new batch of antibody is required.

G3 Transgenics

Key Notes

Genetically modified plants	Gene transfer is being used to create plants with altered characteristics. The Ti plasmid of *Agrobacterium tumefaciens* causes crown gall tumors in plants. The ability of the T DNA region of the Ti plasmid to integrate into plant chromosomes has allowed its use as a cloning vector for the transfer of foreign genes. Alternative gene transfer methods include penetrating plant cells with DNA-coated metal spheres fired from a gene gun. Useful characteristics engineered in plants by gene transfer include resistance to insects, infection by viruses and herbicides.
Transgenic animals	Gene transfer can be used to engineer useful traits in animals. The transgene is introduced into a fertilized ovum or cells of an early stage embryo by microinjection, manipulation of embryonic stem cells or using retroviral vectors. Transgenic offspring result following implantation of the modified embryo in the uterus of a host female. Transgenic animals are used as research tools to study gene function or to produce animal models of human disease.
Pharming	Transgenic animals and plants can be used to produce pharmaceuticals and other valuable products. If the transgene can be engineered to be expressed in lactating females then the product can be purified from milk. Cattle, sheep, pigs, and goats have been used for the production of several pharmaceutical products including anticoagulants. Plants can also be used commercially for production of pharmaceuticals and other products. Trials have used maize, soybean, rice, tobacco, potatoes, and other commercial crop-plants. Human serum albumin and lactoferrin can be commercially harvested from rice.
Transgenic insects	Silkworms that have been genetically engineered to carry copies of the gene for spider silk produce this in the form of silk fibers, making it easy to harvest and use in manufacturing. Denge virus is spread by female mosquitoes. These can be killed if they inherit a terminator gene from males that is only expressed in females, thus wiping out the mosquito population.
Related topics	(G2) Biotechnology (G6) Ethics

Genetically modified plants

Conventional plant breeding has been used for many years with great success to produce crops with increased yield and nutritional value. Genetic engineering, however, provides a more direct method for producing plants with altered characteristics. Plants are especially suited to genetic modification because most plant cells are **totipotent**. This means that an entire plant can be generated from a single genetically modified cell. If the plant is fertile, the modification will be present in the plant seeds allowing further modified plants to be propagated.

A number of systems have been developed for transferring genes into plants. A very successful example is based on the **Ti plasmid** which occurs in the soil bacterium *Agrobacterium tumefaciens*. This bacterium infects dicotyledonous plants (all agricultural crops except cereals) and induces formation of a cancerous growth called a **crown gall** tumor. Transformation of plant cells is due to the effect of the Ti plasmid carried by the bacterium. Part of the Ti plasmid called the **T DNA** integrates into a plant chromosome and expression of T DNA genes causes cell transformation. Vectors for the transfer of genes into plants have been developed based on the use of the Ti plasmid and *A. tumefaciens*. The system is very effective but is restricted by the limited number of plants that are infected by the bacterium.

Alternative methods are now being developed for the transfer of genes into plant cells. One unusual but effective method involves coating gold and tungsten spheres with DNA which are then fired into the plant tissue from a special 'gene gun.' This is known as **biolistics**. The spheres penetrate the cell wall and enter the cells. The DNA is then released from the sphere and integrates into the host cell genome.

A number of useful characteristics are being engineered into plants:

- **Resistance to attack by insects**. Attempts are being made to engineer plants with insecticidal activity. Two strategies have been used. One involves transfer of a gene from the bacterium *Bacillus thuringiensis* which encodes a protein called **protoxin** which is toxic to some insects. The gene was transferred into maize and cotton, and proved to be very effective. Another approach involved engineering expression of **protease inhibitor proteins** which interfere with the insects' ability to digest plant tissue.
- **Resistance to infection by viruses**. Genetic engineering is being used to develop novel mechanisms of virus resistance. Genes encoding antisense copies of viral genes have been transferred to plants but have had only limited success at preventing infection. A more successful strategy involves transfer of a gene encoding a viral coat protein. It is not certain how expression of the viral coat protein inhibits infection.
- **Resistance to herbicides**. Despite the use of herbicides, about 10% of global crop production is lost due to infestation by weeds. In addition, herbicides are expensive, potentially toxic to the environment, and can kill crops as well as weeds. Gene transfer has been used to engineer resistance to herbicides in plants to allow selective killing of weeds. **Glyphosate** is a widely used herbicide that acts by inhibiting an enzyme involved in the synthesis of aromatic amino acids in plants. Resistance to glyphosate has been engineered in plants by transfer of a bacterial form of the enzyme that is unaffected by the herbicide.

Other useful characteristics that have been engineered in plants include: delayed ripening of fruits to increase shelf life (Section A7), tolerance of environmental stresses such as drought, altered pigmentation in flowers, and improved nutritional quality of seeds.

Transgenic animals

Selective breeding has been extensively used to produce domesticated animals with desirable characteristics such as high milk yield or high growth rate. Although selective breeding has been very successful, it becomes difficult to introduce new characteristics without altering existing traits. It is now possible to engineer traits directly in animals by gene transfer. The genetically modified animal is called a **transgenic animal** and the transferred gene is called a **transgene**.

Transgenes can be introduced into animals by three methods. Each involves gene transfer into a fertilized ovum or into cells of an early stage embryo. Modified embryos are then implanted into the uterus of a host animal where they develop into genetically modified offspring. Methods for gene transfer involve:

- **Retroviral vectors.** Retroviruses can be used to infect cells of an early stage embryo prior to implantation. The infected embryo is then transferred into a pseuodpregnant female. The transgene carried by the retroviral vector is efficiently integrated into the host genome but the size of the gene that can be transferred is limited and the use of retroviruses has implications for safety.
- **Microinjection.** This method involves injecting DNA directly into the nucleus of a fertilized ovum viewed under the microscope. Although technically challenging, this is now the most widely used method of producing transgenic animals.
- **Embryonic stem cells.** Cells from the **blastocyst** stage of early mouse embryos can be removed and grown in culture. These are called embryonic stem (ES) cells (Section G4) and have the ability to differentiate into all other cell types. ES cells can be genetically modified in the laboratory and incorporated into blastocysts for implantation (Figure 1). This produces a **chimeric** animal that contains the transgene in only a proportion of its cells. However, as some gametes are likely to contain the transgene it can be passed to animals which contain it in all their cells.

Transgenic animals are used as tools in research and for the production of recombinant proteins. There are two main research applications:

- **Studying gene function.** Transgenic technology was perfected using mice in the early 1980s. Since then, hundreds of genes have been introduced into mice. By examining the characteristics of the genetically modified animal, it is possible to obtain information about the function of the transferred gene. This has contributed greatly to our understanding of gene regulation and function. A related technique involves using gene transfer to disrupt genes. **Knockout mice** which lack functional forms of specific genes are produced which provide information about the function of the knocked-out gene.
- **Model systems for human disease.** Transgenic animals can be created which simulate human diseases in which defective genes play a role. Whole animal models can be used to follow the onset and progression of diseases and provide a system for testing new drugs that may be useful for treatment. This approach has been used to develop models for diseases such as Alzheimer's and arthritis, and transgenic mice, **oncomouse**, engineered to carry oncogenes and other genetic defects that make them prone to develop cancer are important tools in cancer research.

Pharming

Conventional production of drugs important in medicine or agriculture is expensive. Several pharmaceutical companies have seen the advantage of using genetically engineered animals or plants as **'bio-reactors'** in the manufacture of such products.

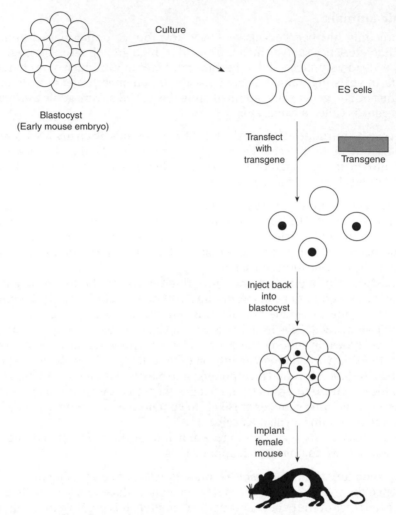

Figure 1. Gene transfer using embryonic stem (ES) cells.

In animals it is very advantageous if the transgene can be engineered with appropriate regulatory sequences (Section A7) so that it can be expressed by lactating females. This method of producing recombinant proteins has a number of advantages. Milk is renewable, is produced in large quantities, and can be collected without harm to the animal. The expressed proteins closely resemble their human versions and are easily purified since milk contains only a small number of proteins. In 1997 a transgenic cow was reported to yield significant levels of human alpha lactalbumin. A number of animal platforms including sheep, goats, pigs, and rabbits are now used to produce a range of human products. Anti-coagulants produced in goats have been approved for human use by the US Food and Drug Administration and the European Medicines Agency. Other proteins produced by this method include the blood clotting protein Factor IX, and the plasma protein α1-antitrypsin.

Plants have been recognized as a potentially significant alternative as platforms for pharming. They are easily harvested and although the target proteins may be more difficult to purify than those produced in milk, they can be produced more cheaply. Most trials have been carried out using engineered maize plants. However, other species have specific advantages. Tobacco is not used as a food crop so there is no risk of genetically engineered plants reaching the human food chain, and bananas are sterile and cannot spread pollen carrying the transgene. Over 400 studies have been carried out although there is now a trend to use glasshouse environments rather than open field sites for these experiments. Human products such as lactoferrin, transferrin, and lyzozyme and a number of antibodies have been successfully produced in a variety of plant species and are commercially available. There are hopes that transgenic bananas could be used to produce 'edible' vaccines.

Rice has been genetically engineered to produce human serum albumin (HSA), the main constituent of plasma used to treat serious burns and fluid loss. The worldwide use of this requires about 500 tons per year which nearly all comes from human plasma donors and is itself in high medical demand, The rice produces 2.75 g of HSA per kilogram of seed, so 182 000 tons of rice would supply the total requirement, and the remaining 99.7% of the rice could be used industrially as starch or as animal feed. There is a risk of antigenic effects from contaminating rice proteins, but none were found. Use of plants on a large scale allows cheaper production with far less toxic contamination than bacterial fermentation.

Pharmaceutical products are not the only commodities to have been produced in transgenic plants. The potato variety Amflora has been modified specifically for industrial purposes and approved for commercial use in Europe. Antisense RNA (Section A7) has been used to block productions of an enzyme that makes amylose starch. The potatoes then produce pure amylopectin starch which is used to produce biodegradable plastic, yarn, glossy paper, glue, and other industrial products. The current European production of potato starch exceeds 2 million tonnes per year. Cyanophycin is a polymer of aspartic acid with industrial uses. The gene for its production, *cyanophycin synthetase*, can be introduced into the same potatoes as amylopectin and cyanophycin purified alongside the starch. This combination makes the crop more valuable without using more resources.

Transgenic insects

Two interesting applications of transgenic technologies in insects are described, one of which has industrial applications and the other is in the field of environmental health. Spider silk has tensile strength greater than steel for its weight. It has a significant range of industrial and medical applications if it could be produced in large enough quantities. Initially the gene was cloned in goats and produced in milk, but recently the system has been improved and the gene has been introduced into silkworms in such a way that they spin the protein into the silk they produce. This saves having to purify it from milk and then spin it into fibers.

The second example uses transgenes to eradicate populations of mosquitoes that spread Denge virus. This technique involves inserting a **terminator gene** that, when expressed, will destroy flight-muscle cells in mosquitoes. The gene is regulated by a 'tet-off' expression system (Section G2) and engineered to only be expressed in females. The female mosquitoes are responsible for spreading the virus. Large numbers of these engineered mosquitoes can be bred in cages as long as tetracycline is present in drinking water. The terminator gene remains unexpressed. Removal of tetracycline expresses the gene in female mosquitoes and they cannot fly. The males are then released in an area with a mosquito problem. They are healthy and breed with native females, but the female progeny of these pairings cannot fly and die before they can mate. As the frequency of females continues to drop, the population soon collapses.

G4 Cloning animals

Key Notes

Background to cloning	Clones of animals were first produced in amphibians. Nuclei from adult cells were transferred to unfertilized eggs that had been enucleated. A proportion of these survived to adulthood. In mammals early embryos may be split and introduced into surrogate mothers.
Cloning mammals by nuclear transplantation	Recently the nuclear transfer process has been successfully carried out with several mammalian species. Only a very small proportion of cloned eggs survive to birth. Cloned individuals often show side effects including obesity, shortened lifespan, and cancers. The cloned individuals may inherit changes that have occurred in the lifetime of the donor cell. Cloning of primates is problematical, and attempts to clone humans are illegal.
Stem cells	Embryonic stem cells can be obtained from the inner cell mass of embryos. These cells are pluripotent and can develop into most tissues of the body. They could be produced for therapeutic cloning, and used to treat patients, but this raises ethical issues. Induced pluripotent stem cells can be derived from adult cells by transfection of growth factor genes and may provide an alternative without associated ethical issues.
Related topics	(G2) Biotechnology (G3) Transgenics (G6) Ethics

Background to cloning

Since the 1960s it has been known that **clones** (genetically identical replicate organisms) can be made using amphibians. Nuclei were removed from eggs and replaced by diploid nuclei taken from adult tissues. The eggs were then allowed to develop. A proportion of these eggs were able to progress to fully mature individuals, although often this was only after several rounds of nuclear transfer from successive early embryos. In addition an established practice in animal breeding is to split cattle embryos at an early stage and return the fragments to surrogate mothers. For example an 8-cell embryo can be divided into four 2-cell embryos. In this way a number of identical calves are produced. This technique, which is analogous to the development of identical twins, has recently been applied to monkeys. Embryo splitting allows the number of individuals with a particularly favorable genotype to be increased more rapidly than would be the case naturally. This is particularly important in breeding programs for large farm animals.

Cloning mammals by nuclear transplantation

It has become possible to extend the **nuclear transplantation** procedure, originally carried out in amphibians, to mammals. There were a number of technical problems to be overcome, including manipulating the much smaller mammalian eggs and the fact that the fertilized mammalian egg has to implant in the wall of the uterus rather than develop in an extra-cellular environment. The first mammal to be cloned in this way was 'Dolly the sheep.' As with the amphibian procedure, the first step is to remove the nucleus from an unfertilized egg cell. The donor cell may be taken directly from a mature animal or even from cells grown in culture. The nucleus from this cell is then transferred to the recipient egg. It is necessary to ensure that the donor and the recipient cells are both in the resting 'G zero' stage of the cell cycle (Section B5). The resulting diploid egg can be allowed to divide in culture to produce a blastocyst or directly introduced into a female that has been treated with hormones to simulate pregnancy. If successful, the embryo will implant in the uterine wall and eventually come to term. It will have a set of chromosomes with the same genetic makeup as the donor animal.

Different techniques are used to transfer the nucleus to the enucleated egg cell. In the case of 'Dolly' the donor cell and the egg were fused together. This results in an egg containing the donated nucleus and two distinct cytoplasms, one from the donor and one from the egg cell. Remember that the mitochondria contain DNA (Section C7). Thus, mitochondrial genes both from the egg and the adult cell may be present in the cloned individual. Alternatively the donor nucleus can be removed from its cell and microinjected into the egg.

Although nuclear transfer has been used successfully in cloning several mammalian species including sheep, mice, cattle, cats, monkeys, and certain endangered species, the success rate remains low. Cloned eggs may cease development at several stages before birth. Increased success has been reported where, after nuclear transfer, the egg is prevented from dividing for a short period. This is called the **Honolulu method**. This presumably allows for chromatin remodeling to be carried out (Section A8). The choice of tissue from which the donor nucleus is obtained may also be important in determining the frequency of successful cloning.

Cloned animals have proved fertile, but some health problems have been encountered. They often display obese phenotypes, although these are not passed on to their offspring. In addition they show shortening of the telomeres at the ends of chromosomes (Section B4). This is because the adult cell that was 'cloned' will have had a reduction in telomere length due to its age. Thus, the chromosomes of the cloned individual are effectively older than the animal's chronological age. This may lead to problems in premature aging and the development of cancers. Examples of both have been reported. The somatic cell that is used as a donor may have undergone mutations or persistent epigenetic alteration (Section A8). These will be reflected in the cloned individual. Plants cloned from somatic tissue show considerable variation. This is known as **somaclonal variation**, and is probably also true, at least to some extent, for animals.

Cloning in humans remains very controversial. On the basis of the results already observed in cloned domestic animals it is unlikely that cloned humans could expect a normal lifespan and health. Results from research on primates such as monkeys have been problematical. Despite this there have been several claims to have successfully cloned humans. None have been substantiated. Attempts to clone humans are illegal in the USA and the UK. However, cloned human embryos may have a role to play in stem cell therapy.

Stem cells

Stem cells provide a potential therapeutic route to replacement of damaged tissues such as cardiac muscle, or nerves. All tissues of the adult body contain stem cells, but these adult stem cells are few in number and capable of developing into only a limited number of tissues. As described in Section G2, embryonic stem cells can be isolated from the inner cell mass (ICM) of the blastocyst. These cells are **pluripotent**, capable of developing into almost all tissues of the body. They can be grown in culture and genetically modified. Unfortunately, despite their potential, they cannot be used to treat a patient as, even if it was ethically acceptable to create the embryos necessary to produce them, they would be rejected as 'foreign' by the patient's immune system.

A potential way exists, however, to produce embryonic stem cells that would not be rejected. This requires the cloning of a cell taken from the patient. The adult cell would be fused with an enucleate human egg cell exactly as is done in the cloning of animals. However, the development of the embryo would be terminated at the blastocyst stage and the ICM used to initiate a line of embryonic stem cells. This procedure is known a **therapeutic cloning**. The stem cells could be used to treat the adult who donated the original cell. They could proliferate and replace cells in any damaged tissue. As the egg contained no nucleus the only genes in the stem cells would be those inherited from the adult donor and hence there could be no problem of rejection.

Obviously, such approaches would raise major ethical problems, and are not yet in use, although results from several early experiments in mouse models have proved successful. Recently attention has turned to an alternative approach which involves reprogramming (Section A8) of adult cells after removal from the body and growing in tissue culture, so that they regain the ability to develop into a wide range of cell types. Reprogrammed cells are referred to as **induced pluripotent stem** cells (iPS). Their production requires transfection of genes for several transcription factors into the adult cells. One of these is *c-myc*, a gene known to be involved in cancer (Section F3). At present there is a risk with all iPS cell lines that they will develop into tumors. However, currently methods to produce iPS cells that do not involve transfection with genes appear to be successful.

At present most successful clinical treatments involving stem cells use adult stem cells only.

G5 Pharmacogenetics

Key Notes	
Pharmacogenetics	Pharmacogenetics is the practice of examining the response of individual genotypes to specific drugs and adjusting the treatment accordingly. These differences arise from differences in the expression levels or structure of proteins affecting drug uptake, metabolism, or the drug target. In the USA, over 100 000 people die and over two million are adversely affected by reactions to drugs. Many promising drugs cannot be used because a few patients would suffer a severe reaction to them. Some drugs only work on a part of the population, so are wasted when prescribed to many patients. A similar study involves cancer treatment, where mutation considerably alters the specific genomes of the cancer cells and affects their response to treatment. The deciding factor in pharmacogenetics is the cost-benefit trade-off of how beneficial the drug, or how serious its potential harm, and how expensive the drug is (if it is likely to be ineffective) compared to the cost of doing genetic tests before using the drug.
Examples	Enzymes which normally degrade drugs are missing in most of the cases where drugs fail, so that normal doses become effectively massive overdoses. Examples are butyrylcholinesterase which breaks down a muscle relaxant; glucose-6-phosphate deficiency and sensitivity to certain antimalarial drugs; cytochrome P450 oxidase family (CYP) of liver enzymes that detoxify (or activate) a range of drugs; and thiopurine S-methyltransferase (TPMT) which metabolizes thiopurines used in various chemotherapies.
Ethics	The application of screening to patients before drug treatment may not be given for reasons related to ethics. The cost of testing may be prohibitive; it may be cheaper to start the drug carefully and see what happens, stopping if there are adverse signs; or the drug may be deemed too risky to use at all. There are also issues of civil rights. Testing for genetic variants that affect drug metabolism would produce a genetic profile which could be used to identify the subject. Racial differences in reaction to drugs are common, but to withhold a drug from an ethnic group who would not benefit from it has been construed as racist.
Related topics	(B1) Concepts of genomics (D1) Population genetics (B3) Eukaryote genomes and evolution (C8) Quantitative inheritance (E5) Bioinformatics

Pharmacogenetics

Pharmacogenetics is about matching drug treatments to patients to achieve the best therapeutic response with the least harmful side effects. Individuals vary in the levels of particular enzymes which metabolize the drugs, and this can affect their uptake, metabolism to an active form, or degradation and excretion. The drug's target molecule may vary in quantity or may carry a mutation that is phenotypically neutral but that stops the drug affecting it. Psychoactive drugs are usually aimed at modifying the levels of activity in specific areas of the brain by modifying either the levels of neurotransmitters or the response of downstream neurons. The activities of all the components in neural pathways are very variable, which is one reason why humans have different personalities. Often the balance of components, for example dopamine and serotonin, is more important than the absolute level of either. The dose of the common anticoagulant warfarin needed to reduce blood clotting varies between about 2 mg to over 20 mg per day, and an error of 25% can have life-threatening consequences. By one calculation, over 100 000 people die in the USA each year from a reaction to medication, and over two million are harmed by a reaction to drugs. There are also wasted drugs. Some are not used because a small fraction of patients show adverse reactions. Others are used ineffectively because some patients do not respond to them. Cancer cells are very heavily modified by mutation (Section F3), including loss of individual genes, inappropriate expression, and massive overexpression. Analysis of expression levels of key proteins in specific cancers allows the selection of chemotherapies that will be most effective. There is also a high risk of some patients being sensitive to chemotherapeutic agents.

Examples

The first case of drug sensitivity was discovered in the 1950s with suxamethonium chloride. This is a muscle relaxant used to cause paralysis to patients during surgery. It is metabolized by butyrylcholinesterase (alias pseudocholinesterase) so quickly that less than 10% of an intravenous injection reaches the muscle junction, and muscle activity returns to normal about 5 min after a single injection. One in 3500 Caucasians has a low activity, however, and can remain paralyzed for eight hours, during which they require artificial respiration. The absence of the enzyme butyrylcholinesterase in homozygotes does not produce any know impairment in normal conditions and it seems to be redundant.

Another classical example of genetic drug predisposition concerns commonly used anti-malaria drug 6-methoxy-8-aminoquinoline. In a significant minority of patients it caused varying levels of hemolysis. The sensitivity correlated with deficiency in glucose-6-phosphate dehydrogenase in patients' red cells. Mutations in this X-linked gene are common. It is estimated that over 300 000 000 people worldwide carry a mutation in this allele. Ironically, the high frequency of the allele is due to the fact that it delivers increased resistance to malaria. This is an example of **balanced selection** (Section D4) which explains why the mutation is common in certain geographical regions.

Liver contains many detoxifying enzymes in the cytochrome P450 oxidase family (CYPs) which metabolize many drugs. They are very variable because their substrate ranges overlap, and they are not under strong individual selection. CYP2D6 is an example with over 75 allelic variants, some ineffective. It occurs in multiple copies in about 30% of people in some East African groups, rendering the painkiller codeine ineffective. It is also essential for metabolism of the anticancer drug tamoxifen. The non-steroidal anti-inflammatory drug ibuprofen is said to be ineffective as a pain killer in about two-thirds of patients, although they receive placebo benefits. It is likely that the CYP enzymes are responsible, but as there are no extreme cost or health issues involved, no one is going to examine it, and genetic screening (Section F2) would cost more than the drug.

Chemotherapy agents are by nature toxic, killing dividing cells, and, like radiation, the difference between effective treatment with bearable side effects and excessive harm to the patient is narrow. The drugs 6-mercaptopurine and azathioprine are thiopurine drugs used to inhibit purine nucleotide synthesis, and are used for cancer chemotherapy, and as immune disease suppressors. They are most toxic to rapidly dividing cells, and can inhibit growth of cancer cells and proliferating immune cells. They are also toxic to normal cell division, and the dose is critical. They are metabolized by the enzyme thiopurine S-methyltransferase (TPMT) which has very variable activity. About 5% of thiopurine treatments fail because of toxicity, typically toxicity to bone marrow, in patients with low levels of activity.

Ethics

Any attempt to apply genetic knowledge to humans runs a risk of ethical objections (Section G6). It might seem sensible to look at the genotype of a person before prescribing a potentially useless or potentially toxic drug, but costs matter. If the drug is cheap and does not work, little is lost. If the toxic effects are not too serious it is usually cheaper to start with a low dose and see what happens, halting treatment if it goes wrong. In many cases there is insufficient knowledge on which to base an effective test, and either the drug is not used at all, or it is used carefully on new patients, so the majority benefit quickly, and the few with adverse reactions do not suffer unduly. Another problem is the thought of invasion of privacy or racism. Screening an individual for a range of mutations that would affect drug metabolism would incidentally provide a genetic profile which could be used to identify a tissue sample. This would allow the data to be used forensically (Section G1) as part of a criminal investigation. The enzyme polymorphisms involved often produce different frequencies in different ethnic groups, for instance a drug that works on Europeans may not work on Africans, or *vice versa*. There would be no point in treating someone from the wrong ethnic group with that particular drug, but this can lead to claims of racial discrimination.

G6 Ethics

Key Notes

Overview

The new genetic knowledge and techniques now available make it possible to do things which were impossible before. These range from procedures that are clearly beneficial to science-fiction horror, but with a continuous spectrum in between. Moral decisions require individuals and societies to draw lines between good that is permitted (or the lesser of two evils) and bad, which is not allowed. These decisions are arbitrary, based on comparing the benefits with the disadvantages, and affected by cultural and religious attitudes (e.g. the death penalty for murder). We are not drawing those lines, but are posing the questions and relating some counter arguments to assist you to decide what is allowed, what is not, and, most important, why you make that decision.

Privacy and profit

The increased availability of information about individuals' genotypes raises questions of privacy (who should be allowed to know) and who should profit from the knowledge. Should insurance companies have a right to information about a client's genetic risks? People who know they have a high risk of death or disease might take out larger insurance policies, putting up premiums. People who are at low risk would like lower premiums. Insurance companies would like to balance premiums to risk, but people at high risk might get no insurance at all. Employers might not employ people with genetic 'weaknesses,' and genetic 'inferiors' might be stigmatized by society. Pharmaceutical firms might exploit DNA sequence data from individuals, families, or tribes to gain insight into a genetic condition, develop drugs, and then make a large profit. Should the provider of the sequence have a share? Should the state keep genetic profiles of convicted criminals, or the whole population, to identify perpetrators of new crimes, or identify missing persons or anonymous corpses, or as a permanent identity tag, impossible to lose or forge?

Moral choices

In all the moral decisions relating to genetics, the problem is where to draw the line on intervention. Screening for genetic diseases identifies individuals with inherited disorders. This allows selective abortion of fetuses that are at high risk of being born with genetic defects. Similarly, where a family history is known, *in vitro* fertilization and preimplantation diagnosis can be used to ensure only 'healthy' embryos are implanted. Some people would use religious or ideological grounds to ban all intervention; however most people would accept intervention if the baby was to be born severely handicapped, certain to die after a few days or months of distress. Most disputes arise with conditions where the child can expect some years of normal or happy life. For example, should intervention be used to prevent the birth of babies who would be born healthy but with no prospect of reaching adulthood (e.g. Tay–Sachs disease) or who will have only slight physical or mental disabilities (e.g. Down's syndrome), or conditions that are fully treatable (e.g. phenylketonuria). At the trivial extreme, should preimplantation intervention be used to select a child of a particular sex, eye color, or skills? Such activity is already prevalent in some societies where brides provide dowries, and sons are preferred. Many million female babies (per generation) are already aborted, killed, or allowed to die from neglect, without the use of genetic technology.

Potential harm

Genetic modification of individuals or populations carries hypothetical risks. In humans, there are objections against modifying the gene pool by producing designer babies. This activity would only be affordable by rich people, and make little global impact, but could create a genetic elite. Another possibility is the accidental or deliberate creation of a lethal pathogen. Genetic manipulation and cloning of animals and crop plants is of more immediate relevance. There is a specific objection to creating transgenic species by transferring genes from one species to another. This fundamentalist objection is to 'playing God' and changing creation. Objectively, the two important considerations are the cost/benefit ratio, and the comparison with the alternative, typically current practice. There are many perceived risks. The mere act of adding or changing a single gene may have unforeseen consequences. The gene and product are thoroughly known when modification is done by genetic manipulation. This makes genetic manipulation safer than conventional breeding that involves the introduction of hundreds of unknown genes, or mass mutation, then selection for the desired phenotype with no investigation of hidden changes. Suffering may occur in genetically modified animals with human genetic diseases (e.g. cystic fibrosis in mice) or with accelerated muscle growth or milk production in cattle. The costs and benefits must be assessed. There are already regulations regarding animal experiments, but none on conventional breeding which can produce gross abnormalities (e.g. small dogs whose eyes tend to fall out, cows with gross udders). Crop plants producing internal pesticides may also be toxic to harmless species, but the conventional alternative is chemical sprays that are generally worse. Plants resistant to herbicides allow chemical removal of weeds, in turn reducing support for nonpest wildlife; however the whole purpose of agriculture is to replace natural ecosystems with human crops. Fire and the plough have a much greater impact, and have been used for thousands of years. Greater efficiency of production on land already cultivated reduces the need to extend cultivation (e.g. into rain forests) and can allow poorer soils to be returned to a natural state.

Related topics	(B3) Eukaryote genomes	(F4) Gene therapy
	(F1) Genetic diseases	(G1) Genetics in forensic science
	(F2) Genetic screening	(G3) Transgenics

Overview

The new genetic knowledge and techniques now available make it possible to do things which were impossible before. Beneficial examples are disease treatment and prevention, healthcare individually tailored to a patient's genotype, and improved food production. Some examples are more ambiguous: embryonic diagnosis of genetic defects, extensively prolonging human lifespan, cloning genetically identical individuals, and some potential examples are of dubious morality. A science-fiction example would be cloning an army of aggressive, uncaring soldiers. Ethical values indicate what is right and what is wrong, and are matters of choice, derived from comparing the benefits to the disadvantages, making value judgments based on religious and social attitudes (is the death penalty acceptable for murder?). Thus they vary between individuals and cultures. The uses of genetics fall on a spectrum from 'obviously' good to 'obviously' bad, and the critical ethical decisions are about where to draw the line between the two. In this section we are not going to draw those lines. Rather we present the questions, and some of the arguments from both sides, and make comparisons with similar or alternative procedures current in society. This should guide you in considering where the lines should be, and more importantly why you rate a particular procedure to be allowable or immoral.

Privacy and profit

There is a debate about the ownership of the rights to genetic information, both concerning the details of individual's genetic makeup, and the right to profit globally from knowledge of DNA sequence data.

Individuals can gain benefit from knowing that they have a genetic disease if it allows them to adjust their life to reduce the risks. Someone genetically predisposed to cancer can be screened regularly because early diagnosis increases the chance of successful treatment; a couple who both carry a genetic disease can take steps to avoid having affected homozygous children. When there is no cure, however, knowledge may cause stress and depression but no benefit. Huntington's chorea is caused by an autosomal dominant mutation, so children with an affected parent have a one in two chance of being affected. Many refuse testing. Would you want to find out in your twenties or thirties that you will lose your mental faculties in your forties and die of dementia?

Someone who knows they are at high risk could take out excessive health or life insurance knowing they will probably gain a large benefit. This exploits the other policyholders, who must pay larger premiums to meet the costs. If insurance companies have a right to disclosure so that they know as much as the policyholder, they can increase their premiums, or refuse insurance. This would create a genetic underclass, ultimately forced to rely on help from the state. This already happens to the extent that companies can ask about the fate of family members and gather some indication of risk. At the other extreme, people whose genotype suggests a low risk might expect to pay lower premiums, and might be offered them in exchange for test results.

Tested individuals might be stigmatized and unable to marry, just as relatives of mentally ill people may be now. Employers might reject workers likely to be absent or to underperform through illness. Psychological tests are already used to screen workers. As we find more genetic markers indicative of mental tendencies, these might also be used by employers. Results of such tests could be used in education or work to identify those occupations to which individuals were supposedly most suited. Done by choice, it may be beneficial, but by coercion or force it is seen as an unacceptable infringement of freedom, akin to creating a worker class in ants.

Who owns the rights to a DNA sequence? As an example, a gene for a genetic disease is identified in a family (Sections F1 and F2). The researchers then patent the sequence so that any hospital using those sequence data for diagnosing patients must pay royalties. The sequence was there all the time; there is no new invention. In any case, does the sequence belong to the original family? The risk of losing patents has led to many randomly sequenced DNA fragments being patented when no one knows what they do, in fact many are so inaccurate as to be useless. To be patented, an invention must be new, nonobvious, thoroughly described, and useful. The use could be spurious and imaginary (e.g. a use for a protein described as an additive in shampoo). One view is that revealed sequence data should be unpatentable, but novel drugs developed by using the discovered structure of the coded protein should be patented. In general, patenting is easier in the USA than in Europe.

What use should the community or the state make of genetic profiles? A genetic profile is much more reliable than a photograph, cannot be forged, and can be written exactly, taking less than one line. The US Department of Defense has stored tissue samples of all personnel so that the remains of soldiers can always be identified, even bones. In the UK, samples can be taken from felons at arrest, being destroyed upon acquittal. In the USA, laws vary between states: Virginia takes samples from all felons, but some states only take samples for sex crimes and murder. Under-resourcing in some states led to a backlog of half a million samples (in late 1999) awaiting typing, including thousands of rape samples urgently needed to check suspects. From the other side, many prisoners convicted of rape years ago, but claiming innocence, can now ask for genetic profiling of the original samples to prove their innocence and wrongful imprisonment. Spain is storing genetic data from relatives of missing persons to identify corpses of unknown people.

The current protocols only collect data from convicted criminals, and may over-represent some classes and ethnic groups. Is this fair? How long should the data be kept? There have been suggestions that everyone should be profiled, although cost would require a gradual introduction, probably by routine testing of newborns. Strict regulation of access and use of the data would be required. A proposed benefit would be the early identification of young minor criminals. Traces left at their first crimes could identify them from the database. Early apprehension and corrective measures might stop some from progressing into habitual or serious crime, and dramatically reduce the total number of crimes.

Pharmacogenetics examines the role of genetics in determining how patients respond to drug treatment. There is benefit in giving drugs to patients where they will be beneficial but not where they will either not work or worse, produce dangerous side effects. Collecting the necessary genetic data inevitably produces an identifiable genetic profile, arousing concerns about privacy. When a particular group or race would not benefit from a particular drug, there is no point in giving it to them, but withholding it may be construed as racism (Section G5).

Moral choices

Moral decisions, in genetics as elsewhere, depend upon the balance between the perceived good and bad aspects of those decisions. In the case of genetics, it is the uses to which genetic techniques and knowledge may be put. As with all moral choices, the problem is in drawing the correct line in difficult decisions, which are ultimately a matter of personal opinion.

The availability of screening for genetic disease allows the identification of susceptible individuals before any symptoms are visible. This includes preimplantation embryos; the balls of cells formed soon after fertilization. The moral question is: in what circumstances should fetuses be aborted, or embryos produced *in vitro* be selected or rejected for implantation? At one extreme, some groups oppose all intervention while some only oppose abortion. Preimplantation diagnosis and selection are most acceptable, because there is usually a surplus of embryos, and it is sensible to discard the least healthy ones. (Is it less acceptable to choose the 'best' ones?) The more difficult question is: which genetic conditions are severe enough to warrant intervention, and which are not? The considerations are severity, age of onset, and the effectiveness of treatment. If a child will be born totally mentally and physically disabled and soon die, there is little objection to abortion. If the child will be born normal, but will suffer debilitating brain degeneration, blindness, and death by the age of three (e.g. Tay–Sachs disease, hexosaminidase A deficiency), only those with strong ideological grounds would oppose abortion, and many couples would have no qualms about preimplantation selection. Less clear cases are those involving some degree of disability, at a level at which care will be needed and an independent life is impossible. A severe mental or physical handicap that requires almost constant care puts an enormous strain on parents and siblings, and often leads to abandonment, separation, or suicide. In these cases, should the parents have discretion and choice? More problematic cases are conditions such as Down's syndrome, where individuals are of low intelligence but are usually cheerful, often have treatable physical disorders of the heart and kidneys, and can live relatively normal, high quality lives into middle age without excessive family support. How should physically disabling conditions be approached? Muscle wasting diseases can confine sufferers to wheelchairs, but many disabled people have very active lives (witness the Para Olympic Games, or the physicist, Stephen Hawking) and are appalled at the thought that a human life would be ended to avoid their condition. Other accident victims do commit suicide to avoid life in their wheelchair. Phenylketonuria is fully treatable by controlled diet, so does it warrant any intervention? There are cases where embryos have been selected, both to be healthy and to be potential donors of bone marrow or stem cells for genetically afflicted elder offspring. Does this matter as long as they are valued in their own right as individuals?

Should intervention be allowed to choose the sex of a child? Or eye color? In societies where brides provide dowries and sons are preferred, some twenty-five million females are missing, aborted after ultrasound scan, or killed as babies, either deliberately or by neglect. Eventually there is the prospect of cosmetic genetics and designer babies produced by genetic manipulation to introduce favored genetic traits (e.g. eye color, musical ability). This is deprecated or illegal in most developed countries, and currently has too low a success rate to be usable.

Potential harm

The possible effects of applying genetic techniques (apart from moral considerations) act at all levels: the individual, the population or species, and the entire ecosystem.

For **individual humans**, there are risks of unforeseen side effects of genetic manipulation (but this is in any case illegal in many countries). Another proposed risk is altering the human gene pool by genetic manipulation to produce favored characteristics in babies, or selection against embryos carrying deleterious alleles, even heterozygotes. This would reduce variation, and could remove alleles beneficial to carriers (e.g. sickle cell anemia carriers are relatively tolerant of malaria, cystic fibrosis carriers are thought to be tolerant of some enteric infections). Given the expense of these techniques, they are unlikely to affect significantly the global human population, although there are fears of producing a genetic 'superclass' from rich elites. Gene therapy (Section F4) may have unforeseen side effects, but so do conventional drug therapies. A unique aspect of genetic manipulation is that genetic changes to the germline will affect future generations, whereas conventional treatments expire with the patient. Objections have been raised to cloning multiple copies of a single human individual by nuclear transplant into enucleated eggs, as in 'Dolly the sheep.' Cloning is one possible cure for infertility that does not use unrelated gametes. Genetically it is equivalent to the natural production of identical twins (but of different ages). Is cloning more objectionable than using an extramarital egg or sperm donor for *in vitro* fertilization, which could be described as technical adultery?

There is a real risk of **producing a virulent pathogen**, accidentally or deliberately, either against humans or wildlife. The ease of manipulation of microbes increases this risk. A harmless mouse virus was engineered for use as a humane infectious contraceptive to control mouse populations but this unexpectedly made the virus lethal to mice. Should it be destroyed, or used to control mice, analogous to the use of myxomatosis virus to control rabbits?

Genetic manipulation and cloning of animals and plants is more widespread but has less sensational effects. The fundamentalist objection against 'playing God' and 'interfering with nature/creation' is essentially religious ideology. Humans have been hunting species to extinction and clearing natural ecosystems to grow genetically improved crops and animals since the Stone Age. The correct assessment for modern genetic techniques is by comparison with current conventional practice.

Ever since the Stone Age, humans have selected for increased quality and quantity in domesticated crops and animals. In the Bronze Age, around 4000 BC, Jacob knew how to breed for color in sheep, whatever his explanation (Genesis 30, v 37–43). Improvement in quality and yield of wheat, maize, cotton, wool, beef, milk, and other products were essential for the growth of civilization. Many of the world's population only exist because of the huge improvements in yield in the 'Green Revolution' of the 1950s and 1960s. These used X-rays to cause massive mutation and chromosomal rearrangement, and genetic tricks to introduce new genes (e.g. from weed grasses into wheat). All improvements involved selecting mutated (i.e. genetically modified) organisms, and no consideration was given to any hidden changes that might harm consumers. The hypothetical risks of genetically modified crop plants must be compared to the huge ecological damage that has been caused by introducing alien crops, weeds, and pests between continents (e.g. dogs, cats, rats, rabbits, cactus, and bullfrogs in Australia).

Transgenic organisms (incorrectly termed genetically modified organisms, GMOs) incorporate one or more genes from another species. These genes are individually identified so their effects can be predicted and monitored. Stringent tests and controls are imposed, so genetically engineered crops are as safe as or safer than those produced traditionally. Nevertheless, there is extensive opposition to the technique. This is especially strong among Western Europeans, who are relatively wealthy and have an excess of cheap food, and therefore see no benefit in GMOs. Left-wing newspapers have deliberately encouraged opposition as part of a campaign against multinational corporations and global capitalism. Supermarkets in the UK stopped selling most products containing transgenic organisms, although those products had been popular, successful, and harmless for several years.

There are three primary modifications made to plants:

(i) Plants are given a gene to produce a specific toxin internally. An example is *Bacillus thurengiensis* toxin (bt) that specifically kills lepidopteran (moth, butterfly) caterpillars. This is especially effective against caterpillars which live inside plants, protected from sprays (e.g. cotton bollworm, maize stem borer). Nontarget species that eat the pollen may be affected, but the alternative is repeated spraying with powerful conventional insecticides, which kill all invertebrates and can injure and kill birds, animals and humans. Organic farmers use bt-toxin as a spray, which has greater effects on wildlife, especially those eating weeds rather than the crop. Thus, transgenic pesticides are much safer than conventional sprays, and reduce pesticide use.

(ii) Plants are made resistant to a specific herbicide. The crop can then be sprayed with that herbicide to kill the weeds, increasing yield. This increases herbicide use, and, by reducing weeds, reduces the habitat for some wildlife. This is the purpose of agriculture, to replace weeds with crops, and wildlife may be best served by using less land more efficiently. Extra use of herbicides may increase human exposure.

(iii) Genes may be introduced to improve productivity or quality. Specific examples include: (a) Engineering maize to produce citrate from its roots. This chelates aluminum which otherwise poisons the roots. Phosphate in the soil is simultaneously released as a fertilizer. About 40% of the world's arable land suffers from this problem. It occurs in tropical arid lands and it includes most regions prone to chronic famine. (b) Introducing genes for nutrients. For example, people living on polished rice suffer vitamin A deficiency. About a million children a year go blind as a result and a large fraction then dies. 'Golden' rice containing genes to make vitamin A could alleviate this scourge. It is also true that the introduction of a small amount of fresh vegetables and fruit into the diet would have an equally beneficial effect, but that would require intervention at the level of the individual peasant family.

Perceived risks

(i) Transfer of transgenic genes to weeds causes concern. This requires wild plants closely related to the crop, capable of hybridization. Note that (so far) all transferred genes already exist in natural organisms present in the environment and could have transferred to new species already. Wild plants making new internal pesticides could become more successful, escaping natural population control by insect herbivores, and causing change to ecosystems. This would probably be less catastrophic than some escaped garden plants (rhododendron and Himalayan Balsam in the UK). Herbicide resistant weeds would only be at an advantage in farmland where that specific herbicide would then become less effective. Transfer of genes can be eliminated by making the crop sterile, but the objection to this is that poor farmers could not then save their own seed.

(ii) The introduced protein may be allergenic. Stringent tests are applied, the proteins must be digested quickly when eaten, and chemically unlike any known allergen. For comparison, some cereals (including wheat and barley) and peanuts are so allergenic they would never progress to field trials if they were produced today by genetic modification.

(iii) The selectable antibiotic resistance marker genes used in the process may transfer to pathogens. Such markers will tend not to be used in new products, and in any case, they are usually antibiotics that are too toxic to use to treat humans, and are already present in common soil bacteria. Techniques have now been developed to remove the selectable antibiotic resistance marker genes from plants once they have served their purpose, and before release.

Other specific modifications may have specific risks (Section G3).

Suffering of animals engineered as models of human disease (e.g. diabetes, cystic fibrosis, cancer) could be comparable to suffering in affected humans, and must be balanced against medical benefits. These benefits are large, because both experimental investigations of the disease and treatment trials are faster and more informative on animals. A counterpoint is normal breeding that has produced deformed dogs with flattened faces and breathing problems, and short small faces without adequate space for eyes, and with congenital hip dysplasia. There are epileptic tumbler pigeons, cattle and horses with muscles enhanced by overactive nerves and therefore prone to seizures, cattle with oversize udders prone to physical damage and mastitis, and chickens laying eggs a size too big. These problems all originate from traditional breeding techniques. More stringent controls are in place for laboratory animals.

Further reading and useful websites

There are many comprehensive textbooks of molecular biology and biochemistry and no one book that can satisfy all needs. Different readers subjectively prefer different textbooks and hence we do not feel that it would be particularly helpful to recommend one book over another. Rather we have listed some of the leading books which we know from experience have served their student readers well. A great deal of information is now available on-line and we have recommended a range of web sites that will be useful to individuals studying genetics. In particular, we have included links to recent Nobel Prizes awarded for discoveries in genetics, and a free educational resource provided by Nature Publishing Group containing up to date information on key topics in genetics and cell biology.

General reading

Brown, T.A. (2011) Introduction to Genetics: A Molecular Approach, 1st edn. Garland Science, New York.

Klug, W.S. and Cummings, M.R., Spencer, C.A. and Palladino, M.A. (2011) Concepts of Genetics with Mastering Genetics, Pearson Education, New Jersey.

Krebs, J.E. (2009) Lewin's Genes X. Oxford University Press, Oxford.

Ridley, M. (2003) Evolution, 3rd edn. John Wiley & Sons, New York.

Snustad, D.P. and Simmons M.J. (2008) Principles of genetics, 5th edn. Wiley and Sons, New York.

Strachan, T. and Read, A.P. (2011) Human Molecular Genetics 4th edn. Garland Science, New York.

More advanced reading

The following selected articles are recommended to readers who wish to know more about specific subjects. In many cases they are somewhat advanced for first year students but are very useful sources of information for essays, seminars or dissertations, and for subjects that may be studied in later years.

Section A

Avner, P. and Heard, E. (2001) X-chromosome inactivation: counting, choice and initiation. *Nature Reviews Genetics* **2**, 59–67.

Barrick, J.E. and Breaker, R.R.(2007) The power of riboswitches. *Sci. Am.* **296,** 36–43.

Bayne, E.H. and Allshire, R.C. (2005) RNA-directed transcriptional gene silencing in mammals. *Trends Genet.* **21**, 370–373.

Birchler, J.A., Riddle, N.C., Auger, D.L. and Veitia, R.A. (2005) Dosage balance in gene regulation: biological implications. *Trends Genet.* **21**, 219–226.

Brodersen, P. and Voinnet, O. (2006) The diversity of RNA silencing pathways in plants. *Trends Genet.* **22**, 268–280.

Browler, V. (2011) Unravelling the cancer code. *Nature Outlook,* **471**, S12–S13

Caceres, J.F. and Kornblihtt, A.R. (2002) Alternative splicing: multiple control mechanisms and involvement in human disease. *Trends Genet.* **18**, 186–93.

Cedergren, R. and Miramontes, P. (1996) The puzzling origin of the genetic code. *Trends Biochem. Sci.* **21**, 199–200.

Chamary, J.V. & Hurst, L. The price of silent mutations. (2009) *Sci. Am.* **300**, 34–42.

Gerber, A.P., Keller, W. (2001) RNA editing by base deamination: more enzymes, more targets, new mysteries. *Trends Biochem. Sci.* **26**, 376–384.

Gibbs, W. W. (2003) The unseen genome beyond DNA. *Sci. Am.* **289**, 78–85.

Gutierrez-Preciado, A., Jensen, R.A., Yanofsky C. and Merino, E. (2005) New insights into regulation of the tryptophan biosynthetic operon in Gram-positive bacteria. *Trends Genet.* **21**, 432–436.

Henikoff, S., Furuyama, T. and Ahmad, K. (2004) Histone variants, nucleosome assembly and epigenetic inheritance. *Trends Genet.* **20**, 320–326.

Landry, J., Mager, D.L. and Wilhelm, B.T. (2003) Complex controls: the role of alternative promoters in mammalian genomes. *Trends Genet.* **19**, 640–648.

Latchman, D.S. (2001) Transcription factors: bound to activate or repress. *Trends Biochem. Sci.* **26**, 211–213.

Latham, K.E. (2005) X chromosome imprinting and inactivation in preimplantation mammalian embryos. *Trends Genet.* **21**, 120–127.

Lau, C.N. and Bartel, D.P. (2003) Censors of the Genome. *Sci. Am.* **289**, 26–33.

Li, O., Barkess, G. and Qian, H. (2006) Chromatin looping and the probability of transcription. *Trends Genet.* **22**, 197–202.

Martin, D.I.K. (2001) Transcriptional enhancers – on/off gene regulation as an adaption to silencing in higher eukaryotes. *Trends Genet.* **17**, 444–448.

Morison, I.M., Ramsay, J.P. and Spencer, H.G. (2005) A census of mammalian imprinting. *Trends Genet.* **21**, 457–465.

Nelson, C.L. and Bartell, D.P. (2003) Censors of the genome. *Sci. Am.* **289**, 26–33.

Ptashne, M. (2005) Regulation of transcription: from lambda to eukaryotes. *Trends Biochem. Sci.* **30**, 275–279.

Sharp, P.A. (2005) The discovery of split genes and RNA Splicing. *Trends Biochem. Sci.* **30**, 279–281.

Tjian, R. (1995) Molecular machines that control genes. *Sci. Am.* **272**, 38–45.

Villarreal, L.P. (2004) Are viruses alive? *Sci. Am.* **291**, 76–81.

Vitreschak, A.G., Rodionov, D.A., Mironov, A.A. and Gelfand, M.S. (2004) Riboswitches: the oldest mechanism for the regulation of gene expression. *Trends Genet.* **20**, 44–50.

Wilkins, J.F. (2005) Genomic imprinting and methylation: epigenetic canalization and conflict, *Trends Genet.* **21**, 356–365.

Yanofsky, C. (2004) The different roles of tryptophan transfer RNA in regulating *trp* operon expression in *E. coli* versus *B. subtilis? Trends Genet.* **20**, 367–374.

Section B

Bevian, M. (2011) Endless variation most beautiful. *Nature,* **447**, 415–416.

Bhattacharyya, M.K. and Lustig, A.J. (2006) Telomere dynamics in genome stability. *Trends Biochem. Sci.* **31**, 114–122.

Cann, A.J. (2011). *Principles of Molecular Virology,* 5th edn. Academic Press, London.

Castillo-Davis, C.I. (2005) The evolution of noncoding DNA: how much junk, how much func? *Trends Genet.* **21**, 533–536.

Cordaux, R., Hedges, D.J. and Batzer, M.A. (2004) Retrotransposition of *Alu* elements: how many sources? *Trends Genet.* **20**, 464–467.

Dale, J.W. (2004). *Molecular Genetics of Bacteria,* 4th edn. John Wiley, Lewes.

Dunleavy, E., Pidoux, A. and Allshire, R. (2005) Centromeric chromatin makes its mark. *Trends Biochem. Sci.* **30**, 172–175.

Gee, H.(2008) The amphioxus unleashed. *Nature,* **453,** 999–1000.

Gerstein, M. and Zheng, D. (2006) The real life of pseudogenes. *Sci. Am.* **295**,30–37.

Grewal, S.I.S. and Elgin, S.C. (2002) Heterochromatin: new possibilities for the inheritance of structure. *Current Opinion in Genetics and Development,* **12**, 178–187.

Hancock, J.M. (2005) Gene factories, microfunctionalization and the evolution of gene families. *Trends Genet.* **21**, 591–595.
Kiene, R.P. (2008) Genes in the glass house. *Nature,* **456,** 179–180.
Latham, K.E. (2005) X chromosome imprinting and inactivation in preimplantation mammalian embryos. *Trends Genet.* **21**, 120–127.

Section C

Birky, G.W. (2001) The inheritance of genes in mitochondria and chloroplasts; laws, mechanisms and models. *Annual Review of Genetics,* **35**, 125–148.
Carroll, S.B., Prud'homme, B. and Gompel, N. Regulating evolution. *Sci. Am.* **298**, 38–45.
Kleckner, N., Storlazzi, A. and Zickler, D. (2003) Coordinate variation in meiotic pachytene SC length and total crossover/chiasma frequency under conditions of constant DNA length. *Trends Genet.* **19**, 623–628.
Koopman, P. (2005) Sex determination: a tale of two *Sox* genes. *Trends Genet.* **21**, 367–370.
Lichten, M (2008) Thoroughly modern meiosis. *Nature* **454**, 421–422.
Lightowlers, R.N., Chinnery, D.F., Turnbull, D.M. and Howell, N. (1997) Mammalian mitochondrial genetics: heredity, heteroplasmy and disease. *Trends Genet.* **13**, 450–454.
Meyer, B.J. (2000) Sex and the worm: counting and compensating X-chromosome dose. *Trends Genet.* **16**, 247–253.
Mittwoch, U. (2006) Sex is a threshold dichotomy mimicking a single gene effect. *Trends Genet.* **22**, 96–100.
Siracusa, L.D. (1994) The agouti gene: turned on to yellow. *Trends Genet.* **10**, 423–427.
Sykes, B. (2003). *Adam's Curse; a future without men*, Bantam Press, London. A good explanation of the evolutionary biology of the Y chromosome.
Vaimin, D. and Paithoux, E. (2000) Mammalian sex reversal and intersexuality: deciphering the sex-determination cascade. *Trends Genet.* **16**, 488–494.
Zeh, J.A. and Zeh, D.W. (2005) Maternal inheritance, sexual conflict and the maladapted male. *Trends Genet.* **21**, 281–286.

Section D

Bamshad, M.J. and Olson, S. E. (2001) Does race exist? *Sci. Am.* **289**, 50–57.
Burke, J.M. and Arnold, M.L. (2001) Genetics and the fitness of hybrids. *Annual Review of Genetics,* **35**, 31–52.
Cavalli-Sforza, L.L. (2000). *Genes, Peoples and Languages.* Allen Lane, Penguin Press, London.
Ho, S.Y.W. and Larson, G. (2006) Molecular clocks: when times are a-changin. *Trends Genet.* **22**, 79–83.
Jegalian, K. and Lahn, B.T. (2001) Why the Y is so Weird. *Sci. Am.* **284**, 42–47.
Kreitman, M. and Di Rienzo, A. (2004) Balancing claims for balancing selection. *Trends Genet.* **20**, 300–304.
Louis, E.J. (2009) Origins of reproductive isolation genes. *Nature* **457,** 549–550.
Mallet, J. (1995) A species definition for the modern synthesis. *Trends Ecol. Evol.* **10**, 294–299.
Nature Insight: Evolution (2009) Series of articles on aspects of evolution. *Nature,* **457**, 807–848.
O'Brein, S.J. & Johnston,W.E.(2007) The evolution of cats. *Sci. Am.* **297**, 68–75.
Openheimer, S. (2003). *Out of Eden; the peopling of the world.* Constable, London.
Pollard, K.S. (2009) What makes us human? *Sci. Am.* **300**, 24–37.
Scientific American Special Issue on Evolution (2009) **300**, Number 1.

Stix, G. (2008) Traces of a distant past. *Sci. Am.* **299**, 38–45.

Sykes, B. (2002). *The Seven Daughters of Eve*, W.W. Norton, London. (Traces the origins of modern European populations through their mitochondrial DNA.)

Whitton, J. (2000) Polyploid incidence and evolution, *Annual Review of Genetics*, **34**, 401–438.

Wong, K. (2005) The Littlest Human. *Sci. Am.* **292**, 40–49.

Wong, K. (2009) Rethinking the hobbits of Indonesia. *Sci. Am.* **301**, 46–53.

Section E

Altschul, S.F., Gish, W., Miller, W., Myers, E.W. and Lipman, D.J. (1990) Basic local alignment search tool. *Journal of Molecular Biology*, **215**, 403–410.

Brown, T.A. (2010). *Gene Cloning and DNA. Analysis: An Introduction*, 6th ed. Wiley-Blackwell, London.

Lesk, A.M. (2008). *Bioinformatics*, 3rd. Edn. Oxford University Press, Oxford.

Lipman, D.J. and Pearson, W.R. (1985) Rapid and sensitive protein similarity searches. *Science* **227**, 1435–1441.

Mardis, E.R.(2008) The impact of next-generation sequencing technology on genetics. *Trends Genet.* **24**, 133–141.

Mullis, K.B. (1990) The unusual origin of the polymerase chain reaction. *Sci. Am.* **262**, 36–43.

Primrose, S.B., Twyman, R.M. and Old, R.W. (2006). *Principles of Gene Manipulation*, 7th Edn., Blackwell-Wiley, London.

The Chipping Forecast, Nature Genetics **21** supplement, 1–60 (1999) an issue devoted to DNA chips [see also December 2002, Volume 32 and June 2005, Volume 37 for updates].

The nobel prize winning lectures for DNA sequencing are available at these links.

http://www.nobelprize.org/nobel_prizes/chemistry/laureates/1980/sanger-lecture.html

http://www.nobelprize.org/nobel_prizes/chemistry/laureates/1980/gilbert-lecture.pdf

Section F

Albertson, D.G. (2006) Gene amplification in cancer, *Trends Genet.* **22**, 447–455.

Aplan, P.D. (2006) Causes of oncogenic chromosomal translocation, *Trends Genet.* **22**, 46–55.

Bjornsson, H.T., Fallin, M.D. and Feinberg, A.P. (2004) An integrated epigenetic and genetic approach to common human disease. *Trends Genet.* **20**, 350–358.

Brown, Y.L. and Brown, A.S. (2004) Alanine tracts: the expanding story of human illness and trinucleotide repeats. *Trends Genet.* **20**, 51–58.

Chanock, S.J.& Hunter,D.J. (2008) When the smoke clears. *Nature* **452,** 537–538.

Clarke, F.M. & Becker, M.W. (2006) Stem cells: the real culprits in cancer. *Sci. Am.* **295**, 35–41.

Collins, F.S. & Barker, A.D. Mapping the cancer genome. *Sci. Am.* **296**, 32–39

Daujat, S., Neel, H. and Piette, J. (2001) MDM: life without p53. *Trends Genet.* **17**, 459–464.

Davidson, D.J., Rolfe, M. (2001) Mouse models of cystic fibrosis. *Trends Genet.* **17**, S29–37.

Devilee, P., Clenton-Jansen and Cornelisse, C. (2001) Ever since Knudson. *Trends Genet.* **17**, 569–573.

Estevs, F.J. and Hortobagyi, G.N. (2008) Gaining ground on breast cancer. *Sci. Am.* **298**, 34–41.

Haken, R. and Mak, T.W. (2001) Animal models of tumor-suppressor genes. *Annual Review of Genetics*, **35**, 209–241.

Jiricny, J. and Nyström-Lahti, M. (2000) Mismatch repair defects in cancer. *Current Opinion in Genetics & Development*, **10**, 157–161.

Nissim-Rafinia, M. and Kerem, B. (2005) The splicing machinery is a genetic modifier of disease severity. *Trends Genet*, **21**, 480–483.

Nurenburger, J.I. & Bierut, L.J. (2007) Alcoholism and our genes. *Sci. Am.* **296**, 30–37.

Palfi, S. & Bechir, J. Genetics lends a hand. *Nature* **453**, 863–864.

Piggins, H.D. (2011) Zooming in on a gene. *Nature*, **471**, 455–456.

Reich, D.E. and Lander, E.S. (2001) On the allelic spectrum of human disease. *Trends Genet.* **17**, 502–510.

Rubinsztein, D.C. (2002) Lessons from animal models of Huntington's disease. *Trends Genet.* **18**, 202–209.

Section G

Capecchi, M.R. (1994) Targeted gene replacement. *Sci. Am.* **270**, 34–41.

Cyranoski, D. (2008) Stem Cells: Five things to know before jumping on the iPS bandwagon. *Nature* **452**, 406–407.

Hochedlinger, K.(2010) Your inner healers *Sci. Am.* **302**, 29–35.

Horgan, J. (1994) High profile – The Simpson case raises the issue of DNA reliability. *Sci. Am.* **241**, 33–36.

Lanza, R.P., Dresser, B.L. and Damiani, P. (2000) Cloning Noah's Ark. *Sci. Am.* **283**, 66–71.

Nature (2002) volume 418, Issue 6998 devoted to 'Food and the future'.

Maxman, A (2011) Playing the odds. *Nature Outlook,* **474**, S9–S10

McGowan, E., Eriksen, J. and Hutton, M. (2006) A decade of modelling Alzheimer's disease in transgenic mice. *Trends Genet.* **22**, 281–289.

Nassar, N. and Ortiz, R. (2010) Breeding cassava to feed the poor. *Sci. Am.* **302,** 60–65.

Reilly P. (2001) Legal and public policy issues in DNA forensics. *Nat. Rev. Genet.* **2**, 313–317.

Ronald, P.C. (1997) Making rice disease resistant. *Sci. Am.* **277**, 68–73.

Rothstein, M.A. (2008) Keeping your genes private. *Sci. Am.* **299**, 40–45.

Soto, C., Estrada, L. and Castilla, J. (2006) Amyloids, prions and the inherent infectious nature of misfolded protein aggregates. *Trends Biochem. Sci.* **31**, 150–155.

Trivedi, B. (2011)The wipeout gene. *Sci Am.* **305**, 51–57.

Tzfira, T., Li, J., Lacroix, B. and Citovsky, V. (2004). *Agrobacterium* T.-DNA integration: molecules and models. *Trends Genet.* **20**, 375–383.

Velander, W.H., Lubon, H. and Drohan, W.N. (1997) Transgenic livestock as drug factories. *Sci. Am.* **276**, 54–59.

Wilmut, I. (1998) Cloning for Medicine. *Sci. Am.* **279**, 30–35.

Useful Web Sites

A great number of web sites relating to genetics are available. Sites relevant to students of genetics fall into two major groups: databases and general information sites. Below a small number of sites are reviewed under these headings.

Database Sites

There are a number of databases holding information on gene maps, DNA sequences and protein structures. Many of these can be accessed from the **National Center for Biotechnology Information (NCBI)** http://www.ncbi.nlm.nih.gov/. **Online Mendelian Inheritance in Man (OMIM)** now has its own improved website: http://www.omim.org/ This is a catalog of human genes and genetic disorders. A European site giving access to genome databases is **Ensembl** (http://www.ensembl.org/). **The Wellcome Trust Sanger Institute** also hosts a range of interesting material including access to a range of genome and genetic databases under the scientific resources tab at http://www.sanger.ac.uk/ .

Nobel Prizewinners:

Genetics often contributes to the Nobel Prize awards. The lectures given by the prize-winners to their general audiences can be obtained from http://www.nobelprize.org/nobel_prizes/medicine/laureates/

Mario R. Capecchi, Sir Martin J. Evans and Oliver Smithies "for their discoveries of principles for introducing specific gene modifications in mice by the use of embryonic stem cells". http://www.nobelprize.org/nobel_prizes/medicine/laureates/2007/

Andrew Z. Fire and Craig C. Mello "for their discovery of RNA interference - gene silencing by double-stranded RNA" /2006/

Sydney Brenner, H. Robert Horvitz and John E. Sulston "for their discoveries concerning genetic regulation of organ development and programmed cell death" /2002/

Leland H. Hartwell, Tim Hunt and Sir Paul M. Nurse "for their discoveries of key regulators of the cell cycle". /2001/

Edward B. Lewis, Christiane Nüsslein-Volhard and Eric F. Wieschaus "for their discoveries concerning the genetic control of early embryonic development". /1995/

Richard J. Roberts and Phillip A. Sharp "for their discoveries of split genes" /1993/

General Information Sites

These may prove useful for general information, keeping abreast of new developments, in writing essays or making presentations. Several contain opportunities to test your knowledge in self-assessment quizzes.

Scitable: http://www.nature.com/scitable This provides an excellent set of resources in the form of short articles on a wide range of topics in genetics, cell biology and related subjects.

The Human Genome Organisation: http://www.hugo-international.org/ hugo is a generally useful site containing both human chromosome maps and information on bioethics.

Genetic Science Learning Center: http://learn.genetics.utah.edu/ This site has a broad range of useful information aimed at students, families and teachers.

Dolan DNA Learning Center: http://www.dnalc.org/ covers some basic areas of genetics information, with considerable information on inherited diseases and social aspects of genetics. Site has good animations.

Genetics Education Center University of Kansas: http://www.kumc.edu/gec/ provides links to a wide range of genetic information.

Howard Hughes Medical Institute (HMMI); Biointeractive: http://www.hhmi.org/biointeractive/ contains some very useful material including Transgenic Fly Lab and DNA interactive which contains interviews with some of the pioneers of molecular genetics.

The Biology Project, University of Arizona: http://www.biology.arizona.edu/default.html has very useful information on karyotyping and tests on this topic and other areas of human genetics.

International Society for Stem Cell Research: http://www.isscr.org/public/index.htm.

Nature – free and subscription articles on the Y chromosome: and links to other chromosomes: http://www.nature.com/nature/focus/ychromosome/index.html.

National Geographic *The Genographic Project*: https://genographic.nationalgeographic.com/genographic/index.html provides an index to the site and https://genographic.nationalgeographic.com/genographic/lan/en/atlas.html follows human migrations by the distribution of DNA markers. Each track is clickable to give information about its history and frequency

Virtual Library on Genetics: https://public.ornl.gov/hgmis/genetics/default.cfm contains many excellent links to a variety of sources of information on genetics.

BBC Science: http://www.bbc.co.uk/sn/ this site contains useful genetic information some of which is linked to television programs.

15 Evolutionary Gems http://www.nature.com/nature/newspdf/evolutiongems.pdf A resource for those wishing to spread awareness of evidence for evolution by natural selection

Stem cell information: http://stemcells.nih.gov/ Comprehensive information about stem cells provided by The National Institutes of Health, U.S.A.

The NHS National Genetics Education and Development Centre: http://www.geneticseducation.nhs.uk

Abbreviations

3D	three-dimensional	dGTP	2'deoxyguanosine 5'-triphosphate
5BU	5-bromouracil	DIG	digoxigenin
AAV	adeno-associated virus	DMD	Duchenne muscular dystrophy
ADA	adenine deaminase		
ADP	adenosine diphosphate	DNA	deoxyribonucleic acid
AIDS	acquired immune deficiency syndrome	DNMT(1,2,3)	DNA methyltransferase (with variant number)
ATP	adenosine triphosphate	dNTP	deoxynucleotide triphosphate
BAC	bacterial artificial chromosome	ds	double-stranded
BLAST	Basic Local Alignment Search Tool	dTTP	2'-deoxythymidine 5'-triphosphate
bp	base pairs	ES	embryonic stem
BMD	Becker muscular dystrophy	ESC	embryonic stem cell
		EST	expressed sequence tags
BrdU	bromodeoxyuridine	ETS	external transcribed spacers
bZLP	basic leucine zipper		
cAMP	cyclic adenosine monophosphate	F	fertility (plasmid)
CAP	catabolite activator protein	F'	fertility plasmid carrying host DNA
cdk	cyclin dependent kinase		
cDNA	complementary or copy DNA	FISH	fluorescent *in situ* hybridization
CF	cystic fibrosis	FMR1	Fragile X mental retardation 1
CFTR	cystic fibrosis transmembrane conductance regulator	FSHD	fascioscapulohumeral muscular dystrophy
cM	centiMorgan	GDP	guanosine diphosphate
CNS	central nervous system	GMO	genetically modified organism
CODIS	Combined DNA Index System	GTP	guanosine triphosphate
Col	colicin plasmid	GWAS	genome wide association studies
Contig	contiguous sequence of overlapping DNA fragments	HD	Huntington's disease
		HFr	high frequency recombination strain
CYP	cytochrome P450		
DALI	Distance Matrix ALIgnment	HIV	human immunodeficiency virus
dATP	2'-deoxyadenosine 5'-triphosphate	HLH	helix-loop-helix
		hnRNA	heterogeneous nuclear RNA
dCTP	2'-deoxycytosine 5'-triphosphate	Hox	homeobox (genes)
dd	dideoxy (nucleotide)	HSA	human serum albumin
ddNTP	dideoxynucleotide triphosphate	HUGO	human genome organization

ICM	inner cell mass	RNA	ribonucleic acid
ICR	internal control region	RNAi	RNA interference
IGF2	insulin-like growth factor type 2	ROS	reactive oxygen species
IGFR	insulin-like growth factor receptor	RPE65	Retinal pigment epithelial protein, 65 kiloDaltons
IL2RG	Interleukin 2 receptor, gamma	rRNA	ribosomal RNA
iPS	induced pluripotent stem	RTPCR	reverse transcription polymerase chain reaction
ITS	internal transcribed spacers	SAR	scaffold attachment regions
kb	kilo base	SCE	sister chromatid exchanges
kbp	kilo base pair	SCID	severe combined immunodeficiency
kDa	kiloDaltons	SGM+	Second Generation Multiplex
LCR	locus control region.		
LINE	long interpersed nuclear elements	SINE	short interspersed nuclear elements
LTR	long terminal repeat	siRNA	small interfering RNA
MCS	multiple cloning site	SNP	single nucleotide polymorphism
mDNA	mitochondrial DNA		
MECP	methy-CPG binding protein	SnRNP	small nuclear ribonucleoproteins
miRNA	micro RNA	ss	single-stranded
MOF	males absent on the first	SSB	single-strand binding
mRNA	messenger RNA	SSR	simple sequence repeat
MSL	male-specific lethal	STS	sequence tagged sites
NOR	nuclear organizer region	TF	transcription factor
OMIM	Online Mendelian Inheritance in Man	TIC	transcription initiation complex
ORF	open reading frame	Tm	melting temperature (of DNA etc.)
PAC	P1 artificial chromosome		
PCR	polymerase chain reaction	TPMT	thiopurine S-methyltransferase.
PKU	phenylketonuria		
pms	postmeiotic segregation	tRNA	transfer RNA
pp	pyrophosphate	URL	uniform resource location
PSI-BLAST	Position Sensitive Iterated BLAST	UV	ultraviolet
QTL	quantitative trait locus	VNTR	variable number tandem repeats
R	resistance (plasmid)		
RF	replicative form	X-gal	5-bromo-4-chloro-3-indolyl-β-D-galactopyranoside
RFLP	restriction fragment length polymorphism		
RISC	RNA-induced silencing complex	YAC	yeast artificial chromosome

Index